D1382908

EVOLUTIONARY BIOINFORMATICS

EVOLUTIONARY BIOINFORMATICS

By

Donald R. Forsdyke

Queen's University, Kingston, Ontario, Canada

Library of Congress Control Number: 2006923099

ISBN-10: 0-387-33418-1 e-ISBN-10: 0-387-33419-X
ISBN-13: 978-0-387-33418-9

Printed on acid-free paper.

Printed in the United States of America.

9 8 7 6 5 4 3 2 1

springer.com

In memory of Erwin Chargaff whose studies of the base composition "accent" of DNA underlie everything in this book, and of Susumo Ohno who first understood that DNA "speaks" in palindromic verses; with many thanks to Vinay Prabhu whose neat little 1993 paper showed the way.

Erwin Chargaff (1905-2002)

Contents

Prologue

To Select is Not To Preserve

> "So long as uncertainty exists, there is nothing to be so much avoided as that sort of clearness which consists in concealing difficulties and overlooking ambiguities."
>
> The original "JBS" (John Burdon Sanderson 1875) [1]

It is a truth universally acknowledged, that a single tract of text in possession of charms attractive to potential readers, must be in want of good binding and secure lodging. Universally acknowledged perhaps, but is charm alone sufficient to guarantee the latter? The loss of many works of antiquity with the burning of the great library of Alexandria suggests not. Texts that had been selected by generations of priests as most fitting to survive were not preserved. Selection did not guarantee preservation.

In the decade following James Watson and Francis Cricks' discovery of the structure of the genetic text in 1953, it was shown that certain DNA segments could be copied into RNA segments and that the information there encoded could then be translated into proteins. These were units of function that bestowed on an organism the discrete characters that Gregor Mendel had shown to be passed from generation to generation. Genes were defined as DNA segments encoding specific proteins, and were named accordingly. A protein in one organism might function better than the same protein in another organism, and on this basis the former organism might preferentially survive, as would the genes it contained. Thus, the natural selection of Charles Darwin would favour the survival of genetic texts because of the functions they encoded.

However, in 1966 George Williams proposed a more fundamental definition of the gene, not as a unit of function, but as a unit of preservation. A gene passed intact from generation to generation because it was preferentially preserved. A decade later Richard Dawkins popularized this idea as the "selfish gene." Yet, although there were already significant clues that preservation might require more than just successful function, neither Williams nor Dawkins would acknowledge agencies other than natural selection. Difficulties and ambiguities were shrugged off. Surely natural selection, by optimizing function, should keep organisms, and hence the genes they contained, away from fires (enemies without). What threat could Darwinian natural selection not evade? That to select *was* to preserve had been made abundantly clear in the full title of Darwin's great work – *On the Origin of Species by Means of Natural Selection, or the Preservation of Favoured Races in the Struggle for Life*? Williams thought that even the threat of dismemberment by the internal cutting-and-pasting of DNA segments known as recombination (enemies within) would be evaded by natural selection.

The genic paradigm, with its faith in the supreme power of natural selection, reigned until the 1990s when, as part of various genome projects, computers began to spew forth DNA sequences and evolutionary bioinformatics emerged as a non-gene-orientated bioinformatics with roots dating back to the nineteenth century. Alas, when it came to evading recombination, natural selection did not guarantee preservation. The prevention of dismemberment was too important to be left to natural selection. Yet the paradigm shift implicit in this "new bioinformatics" was not generally acknowledged. The genic juggernaut sped effortlessly from the twentieth century into the twenty-first.

That Sort of Clearness

Genome sequencing projects are expensive. Several were completed with great expedition, not because of the scientific interest, not because a better understanding of genomes would make future gene manipulations safer, but because of the perception of those able to influence funding priorities that biology is about genes, and that the ability to identify and manipulate genes would be furthered by sequencing constellations of genes in the form of entire genomes. *Everyone* knows that genomes are about genes! So bioinformatics must be about genes and their products. Marketing the prospect of manipulating genes was easy when cures for cancer, schizophrenia and much more seemed within reach.

In marketing, simple messages work. "Concealing difficulties and overlooking ambiguities" becomes a way of life. The fanfare announcing that two research teams, one privately sponsored and one publicly sponsored, had sequenced the human genome, led Sydney Brenner, one of the great pioneers in molecular biology, to declare in 2002 [2]:

> "Sequencing large genomes has nothing to do with intellectual endeavour. The creative work was done earlier by Fred Sanger and by others who improved the technology. The rest is about two things: money and management. As the various projects developed, their demand for money increased. The … sequencing

> project quickly consumed most of the funds available for [other] genome work at a time when money was short."

On a book written by a leader of one of the sequencing teams, Brenner commented:

> "[He]...doesn't tell us anything about the genomes he sequenced. What did he find there that excited him? What did he learn about genes, about life, about evolution, about worlds to come? It is the play *Hamlet* without Hamlet."

While intending to disparage neither the great organizational and technical skills involved, nor the ultimate rewards that may accrue, my following comment in a leading medical journal reinforced Brenner's point [3]:

> "The overtaking of publicly funded research teams by Celera Genomics in its completion of the sequencing of the human genome smacks somewhat of the Sputnik episode. The USSR scaled up the payload of their rocket and put a man into space before the USA, but failed to make parallel progress in other technologies to capitalize on this advance. In a recent paper ... the CEO of Celera correctly states that "*understanding* the genome may help resolve previously unanswerable questions". But Celera has merely made available the full text of the genome a little sooner than we might otherwise have had it. Those of us working to *understand* the genome no more need the full text of the genome than you would need the full text of a dictionary if you were trying to understand how a dictionary works."

A Buck or Two

So yes, let us applaud the progress in sequencing the human genome as a historic step. But let this not distract from the real issues. There is disquietude. At the beginning of the 21st century the AIDS pandemic spreads to all sectors of society. A strange element, named "prion," lingers in latent form and slowly strangles its hosts, as it did the mad cows from which it seems to have arisen. Genes from genetically modified seed spread to normal crops. In Philadelphia, a genetically engineered virus (adenovirus) is used to treat a relatively minor genetic disease and the patient is dead three days later. In Paris, treatment of a life-threatening genetic disease with a genetically engineered virus (retrovirus) ends up with the "good" gene (that was meant to replace a "bad" gene) disrupting control of another gene that controls cell proliferation ("insertional mutagenesis"), so initiating cancer [4].

These are not acts of bioterrorism. They are acts of Nature and/or of well-intentioned people who think risk-benefit ratios are low enough to justify "concealing difficulties and overlooking ambiguities" – and perhaps making a buck or two as well.

The Dawn

Yet in 1991, at the dawn of the new era, Sydney Brenner set out the agenda quite clearly [5]:

> "Searching for an objective reconstruction of the vanished past must surely be the most challenging task in biology. I need to say this because, today, given the powerful tools of molecular biology, we can answer many questions simply by looking up the answer in Nature – and I do not mean the journal of the same name. ... In one sense, everything in biology has already been 'published' in the form of DNA sequences of genomes; but, of course, this is written in a language we do not yet understand. Indeed, I would assert that the prime task of biology is to learn and understand this language so that we could then compute organisms from their DNA sequences. ... We are at the dawn of proper theoretical biology."

In similar vein, biochemist Peter Little pointed out [6]: "It would be a great mistake to see the approach of enumeration of coding sequences as a substitute for the goal of the Human Genome Project. The goal is explicit ... to understand the information content of our DNA." And, under the heading "Languages of the Chromosomes and Theoretical Biology", molecular biologist Leroy Hood and his associates pointed to the numerous functions of DNA [7]:

> "Each of these functions places informational constraints upon our chromosomes. In a sense, these constraints represent distinct and sometimes overlapping languages. Deciphering some of these languages, and others that we do not have the imagination to envision, will require increasingly powerful computational and mathematical tools."

Homo bioinformaticus

Thus, from many quarters there were calls to cast aside genocentric blinkers and return to the fundamentals. This book explores the fundamental, sensed in England more than a century ago by George Romanes and William Bateson, that genomes are more than genes. Genomes are information channels, and genic information is but one of the forms of information which compete for genome space. Identifying these forms, revealing conflicts, and showing how the genomes of different species balance competing pressures on genome space, are the tasks of evolutionary bioinformatics. These pages point the way and hint at rich rewards to follow.

You will find here that computation and mathematics are more than mere tools. In many respects, biological systems *are* computational systems [8], as indeed may be the universe itself [9]. Understanding computational systems (e.g. computers) may help us understand biological systems (e.g. you and me), and perhaps much more. In many respects, biology is more a branch – yes, a humble branch – of the discipline known as "information science," than a component of a new hybrid discipline – bioinformatics.

Sadly, in the twentieth century many "wet" laboratory-based biomedical scientists looked on "in silico" laboratory-based computer scientists as capable only of tamely writing data-handling programs, rather than as capable of providing profound insights. Sometime in the twenty-first century a new breed of biomedical scientist, *Homo bioinformaticus*, will emerge. But current progress depends largely on partnership between the bio-people and the info-people. To facilitate this, biomedical scientists must eschew jargon and describe their perceived problems in simple, clearly defined, terms, and informatists must strive to make their art intelligible.

"Informatics" implies computing. To many, anything to do with computers sounds mathematical, and is therefore arcane and incomprehensible. Calls for "increasingly powerful computational and mathematical tools" tend to frighten people away. But there is very little actual mathematics in this book. Happily, I can retrieve sequences from on-line databases, and I can add, and substract, and multiply, and divide. That is all *Homo bioinformaticus* will need to do to engage in the new "informatics." And a first year college biology background should suffice for the "bio."

Further Veils?

Will evolutionary bioinformatics still be "new" a decade (or even a century) hence? While so often one learns that to predict is to be wrong [10], there are substantial reasons for believing there will never be a *newer* bioinformatics with respect to the fundamentals. Unlike physics, where the conquest of each fundamental appears as the mere lifting of a veil to further fundamentals beneath, it is likely there is nothing more fundamental in biology than the sequence of bases in a nucleic acid. To understand fully that sequence of four letters – **A**, **C**, **G** and **T** – is the goal of evolutionary bioinformatics. In this context, there can be nothing newer.

Of course, bioinformatics itself is bigger than this. It is literally the understanding of all biology in informational terms. When we focus on the bases of nucleic acids we call it genomics. When we focus on the proteins encoded by genes in genomes, we call it proteomics. Much of this book is about genomics, broadly interpreted to cover nucleic acids (DNA and RNA) as they operate both in cell nuclei (the usual site of DNA), and in the cytoplasm (the usual site of RNA)

Scope

"The historical mode of studying any question is the only one which will enable us to comprehend it effectually." These words of the Victorian sheep-breeder and polymath, Samuel Butler [11], remind us that the ideas presented in this book once did not exist. In the minds of humans vaguely groping through a mist of incoherency, they emerged first, just as ideas. This was followed by an often much slower elaboration and definition of terms, which were eventually gathered into text. This was the creative process that happened in the past, and is ongoing. By understanding the past we prepare ourselves for participating in the ongoing present. To understand the past we look to biohistorians. However, for the all-too-human reason that workers at interfaces between disciplines need more than 24 hours in their days, they have not generally served us well. Palaeontologist Stephan Jay Gould pointed out [12]:

> "Many historians possess deeper knowledge and understanding of their immediate subject than I could ever hope to acquire, but none enjoy enough intimacy with the world of science … to link this expertise to contemporary debates about the causes of evolution. Many more scientists hold superb credentials as participants in current debates, but do not know the historical background. … A small and particular – but I think quite important – intellectual space exists, almost entirely unoccupied, for people who can use historical knowledge to enlighten … current scientific debates, and who can then apply a professional's 'feel' for the doing of science to grasp the technical complexities of past debates in a useful manner inaccessible to historians (who have therefore misinterpreted, in significant ways, some important incidents and trends in their subject)."

Gould might have added that the biohistorian's dilemma is that many historical stones must be left unturned. Ideally, for complete objectivity, the turning should be random. This not being possible, then knowledge of contemporary debates can act as a powerful guide (heuristic). Of necessity, such debates contain speculative components and the lines between speculation and fact are never absolute. Cherished ideas that may have withstood centuries of debate (e.g. Newtonian physics) are sometimes later seen to have been merely provisional.

Since its name was coined relatively recently, it might be thought that bioinformatics has a short history. It is shown here that the history of bioinformatics extends back to Darwin and beyond. This is made explicit in Chapter 1, *Memory – A Phenomenon of Arrangement*, which draws attention to the characteristics shared by different forms of stored information, be they mental, textual, or hereditary, and notes that our understanding of forms we are familiar with (e.g. textual), may help our understanding of forms we are less familiar with (e.g. hereditary).

The quantitative relationships between **A**, **C**, **G** and **T** were first determined in the USA in the middle decades of the twentieth century by the biochemist Erwin Chargaff [13]. The regularities he discovered underlie much of the new bioinformatics. *Chargaff's First Parity Rule* (Chapter 2) considers features of the Watson-Crick structure of DNA that facilitate accurate transmission of hereditary information with the appropriate detection and correction of errors. This requires various levels of redundancy, the most obvious of which is that DNA molecules generally come in duplex form. At least two copies of the DNA message travel from one generation to the next. Yet, Crick questioned the potency of natural selection as an evolutionary force, and pointed to mechanisms involving recombination that might accelerate evolution.

Chapter 3, *Information Levels and Barriers*, focuses on the informational aspect of heredity, noting that information often comes in quantal packets of restricted scope, like a book. The present book is about bioinformatics, not deep-sea diving, and it would be difficult to unite the two themes in a common text. Expanding the scope of information may require formation of new packets. This means that packets must be distinguished and demarcated by appropriate informational barriers. Thus, the "primary" information that a packet contains can be distinguished from the "secondary" information that demarcates and protects. There are different *levels* of information.

Chapter 4, *Chargaff's Second Parity Rule*, introduces conflict between different types and levels of information, as exemplified by poetry, prose and palindromes. Each of these imposes constraints on information transfer that enhance error-detecting power (e.g. if a poem does not rhyme you soon notice it). In Chapter 5, *Stems and Loops*, this theme is further explored in terms of the general ability of duplex DNA molecules, by virtue of the palindromes they contain, to extrude single-stranded segments as duplex-containing stem-loop structures with the potential to "kiss" similar duplex structures. Paradoxically, the fittest organisms are those that can optimize the potential to form stem-loops, sometimes at the expense of other functions. The reader is left wondering why Nature has taken such pains to install stem-loops in all life forms.

Chapter 6, *Chargaff's Cluster Rule*, points to another informational constraint manifest as the preferential loading of loops in stem-loop structures with one type of base ("purine-loading"). This should decrease the loop-loop "kissing" interactions between single-stranded nucleic acids that may precede their pairing to form double-stranded nucleic acids (duplexes). Again, the reader is left wondering. Why should it be necessary to prevent the formation of such duplexes?

Chapter 7, *Species Survival and Arrival*, compares changes in DNA *primary* information that affect an organism's structure or physiology (the *conventional phenotype*) and provide for linear evolution within a species, with changes in DNA *secondary* information that affect the *genome phenotype* and have the potential to provide an initial demarcation that preserves, and hence permits the creation of, new species (branching evolution). The flexibility of the genetic code allows some independence of *within*-species and *between*-species evolution, but that flexibility has its limits.

The frequency of each base "letter" in a sample of DNA can be counted. Chapter 8, *Chargaff's GC-Rule*, considers the frequencies of the bases **G** and **C** as the "accent" of DNA. The accent of human DNA is different from the accent of worm DNA or of daisy DNA. Differences in DNA accents work to impede the "kissing" homology search between stem-loops extruded from potentially recombining DNA molecules during formation of gametes (e.g. spermatozoa and ova) in the gonads (e.g. testis and ovary). This can result in hybrid sterility, an interruption of the reproductive cycle within a species. Such reproductive isolation of certain species members can originate a new species. "Kissing" turns out to be a powerful metaphor, since it implies an exploratory interaction that may have reproductive consequences.

Remarkably, the very same forces that serve to isolate genomes within a group of organisms (i.e. to originate a new species) also serve to isolate genes within a genome (i.e. to originate a new gene). Different genes have different "accents" that preserve them as inherited units. While at the time not knowing that they could differ from each other in this way, Williams defined genes as units that could resist dismemberment by recombination, rather than as units of function. This distinguished gene *preservation* from gene *function* and led to the question whether an agency other than natural selection was involved in the preservation (isolation) of genes and genomes. In the past it has been remarked that Variation offers and Natural Selection *then* selects. Thus, to select is not to vary. Now we can say that, through its role in preserving variations, DNA's **GC** "accent" offers and Natural Selection *then* selects. Thus, to select is not to preserve.

The needs of "selfish genes" and "selfish genomes" can sometimes conflict. Chapter 9, *Conflict Resolution*, shows that the finite nature of genome space makes it likely that all demands on that space cannot be freely accommodated. It is unlikely that any DNA can be dismissed as "junk" or that any mutation is truly "neutral." The multiple pressures acting at various levels on genomes require that there be trade-offs between different forms and levels of information. This can explain introns and low complexity elements (Chapter 10). The latter are prominent in the genomes of herpesviruses and malaria parasites, which are considered in Chapter 11. Here it is shown that some features of the amino acid sequence of proteins do not serve protein function; instead they reflect the needs of the encoding nucleic acid. The amino acids are mere "place-holders."

Organisms exist in an environment of predators, both external (enemies without, who remain without) and internal (enemies without, whose strategy is to get within). These coevolve with their prey or host as they track them through evolutionary time. Thus, the full understanding of the genomes of any biological species requires an understanding of the genomes of the species with which it has coevolved. Chapter 12, *Self/Not-Self?*, turns from intracellular conflict *within* genomes to intracellular conflict *between* genomes. It is shown that, through purine-loading, "self" RNA molecules could avoid "kissing" and forming double-stranded RNA with other "self" RNAs. This would allow the RNA repertoire, derived from conventional and hidden "transcriptomes," to provide a selection of RNA immunoreceptors that would "kiss" viral (not-self) RNAs to generate double-stranded RNAs, so activating intracellular alarms. Likewise, intracellular protein molecules are shown in Chapter 13, *The Crowded Cytosol*, to constitute a protein "immunoreceptor" repertoire. Although reactions between immunoreceptor proteins and viral proteins are likely to be weak, the crowded cytosol provides a conducive environment for this.

Chapter 14, *Rebooting the Genome*, reinforces the point made throughout the book that the need to detect and correct errors has been a major evolutionary force, driving DNA to "speak in palindromes" and splitting members of species into two sexes. This, in turn, required sex chromosome differentiation, as a consequence of which the phenomena known as Haldane's rule and dosage compensation emerged. The role of *The Fifth Letter* (Chapter 15) then assumed a new importance.

Finally, the *Epilogue* turns to science politics and the great opposition to the people whose work is the subject matter of the book. There is all-too-often an all-too-human lack of objectivity that the reader should recognize. This has been with us in the past, is with us at present, and will doubtless persist. The flame of the torch of (what I believe is) enlightenment, has barely escaped extinguishment as it has passed from William Bateson to his son Gregory, and to Richard Goldschmidt, and to Stephen Jay Gould. The latter's disagreements with Dawkins have received wide public attention [14]. Perhaps the greatest contribution of the present book is that it presents both evidence and a theoretical basis for a resolution of these disagreements. Thus, after a century of discontent, evolutionists may at last become able to speak with one voice.

Goals

Who will read this book? Hopefully all who visit my bioinformatics website daily from many parts of the world [15]. The site has been up-and-running since 1998 as a primary resource in its own right, and to supplement my books. Those unfamiliar with the interpretation of graphical data and the laws of chance and probability may find Appendix 1 of help. Each chapter has a summary of the main points so that readers who wish to skip should not miss much (e.g. Chapter 1, which summarizes a history they may have partly read in my previous book – *The Origin of Species, Revisited* [16].

In my opinion much of the confusion in the evolution literature is semantic, so plain language is employed wherever possible, and usages are carefully defined (check the Index to locate definitions). Anticipating readers with a variety of backgrounds, for many of whom English may not be a first language, common usages (vernacular usages) have been preferred, while the corresponding more technical usages follow in round brackets (parentheses). There are abundant quotations, which sometimes contain my explanatory comments in square brackets. Unless otherwise stated, all italicization is mine.

To make the text accessible to a wide audience I have adopted a "swimming pool" approach so that moving between chapters should be somewhat like moving from the shallow to the deep end. More advanced readers who skip the early chapters should note that they are not shallow in substance. Rather they provide important images and metaphors that later chapters will flesh out in chemical terms. For advanced readers who happen to be still wedded to the classical Darwinian paradigm, the early chapters are mandatory! Less advanced readers eager to read the latest on junk DNA, introns, sex and speciation, are urged to be patient. Nevertheless, having splashed around in the early chapters, they should not hesitate to move on. Even though they may find some of the details daunting, they will not drown. Rather they will gain a general overview that I hope will satisfy. Those who cannot abide quantitation, however much simplified, should be able to skip Parts 2 and 4 without losing the essence of the argument. Thus, there is a relatively easy route through the book – Parts 1, 3, 5 and 6.

My ambitious hope is that, apart from being of interest to the general reader, the book will become a definitive resource on genome bioinformatics and evolution that will complement the numerous gene-orientated texts both currently available and to be written. As such it should serve the needs of all engaged in the biomedical sciences at college level and higher, from among whom *Homo bioinformaticus* may eventually emerge. It is not a "hands on" text. The other texts do that very well. Rather, it is a "heads on" text, designed to inspire ways of thinking that might make your hands more productive, if you are so inclined. In short, you are invited to escape the twentieth century box of gene reductionism and become part of the revolution that we may, for the time-being at least, refer to as "the new bioinformatics."

So – I am at the end of my Prologue. Hopefully, you now have some feeling for the subject of the book, and are tempted to read on. But, is that all? What about "being," "soul," "ultimate meaning," – concerns touching the essence of our existence? Are such matters beyond our present scope? It is difficult to improve on the words of Leslie Dunn in the introduction to his *A Short History of Genetics* [17] when anticipating that some might consider his text lacking in philosophical penetration:

> "It is not that I, or another who may review the development of the science of heredity, may not be interested in what it [all] means, but rather that before we ask the ultimate question we should know what has caused it to be asked."

This book continues the quest to "know what has caused it to be asked" [18].

Part 1 Information and DNA

Chapter 1

Memory – A Phenomenon of Arrangement

> "A little learning is a dang'rous thing; drink deep, or taste not the Pierian spring."
>
> Alexander Pope [1]

Because genes could explain so much in the biomedical sciences, in the twentieth century it was tempting to assume they could explain everything. So, for many, the bioinformatic analysis of genomes came to mean the bioinformatic analysis of genes and their products. However, bioinformatics slowly became less gene-centred. Evolutionary bioinformatics – the "new bioinformatics" – views genomes as channels conveying multiple forms of information through the generations from the distant past to the present. As with information channels in general, genomes can become overloaded so that conflicts between different forms of information can arise, and "noise" can delay information transmission and imperil its accuracy. The tasks of the new bioinformatics are to identify the forms of information that genomes transmit, and to show how conflict and noise are dealt with so that reliable transmission is sustained. The results should complete Charles Darwin's work of understanding how, from simple organic molecules, "endless forms most beautiful and most wonderful have been, and are being, evolved"[2].

Apart from the satisfaction of this, there is an urgent practical purpose. Heeding not Pope's admonition to "drink deep", we have forged from fragile foundations new gene-based technologies that, while of immense potential benefit, are being applied to agricultural and medical problems with but a cosmetic nod towards possible hazards. We can only hope that the goals of the new bioinformatics will be attained before biotechnologists, with massive industrial support, open Pandora's box.

Priorities

Evolutionary bioinformatics is only now new because the seductive simplicity of the genic paradigm (Fig. 1-1) and all that it promised, distorted perceptions, and hence, priorities. Thus a *gene-centred* bioinformatics was the first to emerge. The new, *non-gene-centred* bioinformatics emerged later. It is to the twenty-first century, somewhat what genetics was to the twentieth. The words of William Bateson – the biologist who, struggling for recognition of genetics, brought Gregor Mendel to the attention of the English-speaking world – still seem apt [3]:

> "If the work which is now being put into these occupations were devoted to the careful carrying out and recording of experiments of the kind we are contemplating, the result ... would in some five-and-twenty years make a revolution in our ideas of species, inheritance, variation, and other phenomena."

Likewise Charles Darwin had declared in 1872 when praising a book on natural history [4]: "How incomparably more valuable are such researches than the mere description of a thousand species." This might well be paraphrased today with respect to "a thousand genes" and their products.

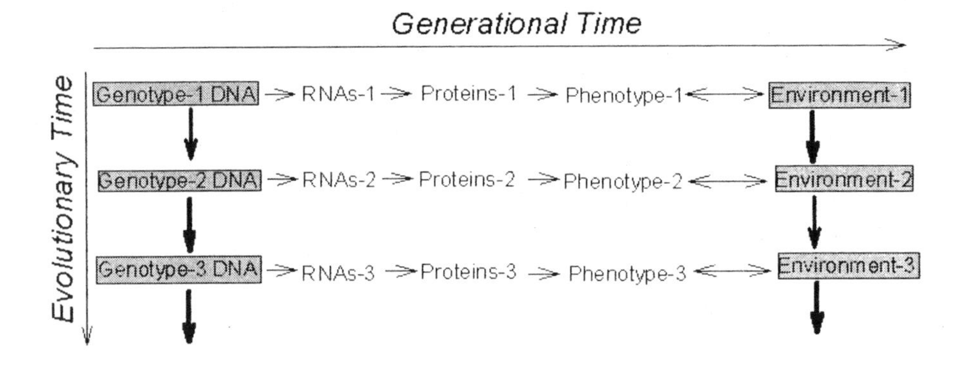

Fig. 1-1. The genic paradigm. Genomes and environments persist but change over evolutionary time (heavy vertical arrows). Within a given generation the genetic program (contained in the DNA "archive") is *transcribed* to provide instructions (messenger RNA templates) that are *translated* into proteins. These are largely responsible for the anatomical and physiological characters (phenotype) with which organisms interact with the prevailing environment. Genotypes (i.e. constellations of genes) that most successfully track environmental changes (i.e. produce successful phenotypes) tend to persist

Of course, descriptions of species, and descriptions of genes, are important, and should be well supported. We have many complete genomic sequences and we need many more. The point is that humankind is likely to progress optimally when *all* aspects of science are supported – not just those that, being perceived as important by vested interests, provide a safe academic haven not only for serious researchers, but also for the intellectually non-curious and the politically ambitious. While some were ecstatically comparing the sequencing of the human genome to the discovery of the Rosetta Stone, molecular geneticist William Gelbart soberly suggested [5]:

> "It might be more appropriate to liken the human genome to the Phaestos Disk: an as yet undeciphered set of glyphs from a Minoan palace on the island of Crete. With regard to understanding the **A**'s, **T**'s, **G**'s and **C**'s of genomic sequence, by and large, we are functional illiterates."

Without apology we will begin our tasting of the "Pierian spring" by examining the historical roots of our subject. So often the understanding of a subject requires the understanding of its history. Yet, to really understand its history, one must understand the subject. This paradox implies that the study of a subject and of its history must go hand in hand. Furthermore, by studying history we learn about *process*. By understanding the process of past discovery we prepare ourselves for understanding, and perhaps participating in, future discoveries.

History is about people, ideas and technologies. Our sources on these are usually written sources. However, often ideas arrive before there are suitable words to convey them. So the examination of the history of ideas can become an examination of the words people coined and used, and of what they meant by those words in particular circumstances. This is no light task but, as I hope to show, is well worth the effort.

Textual and Hereditary Information

At the heart of the new bioinformatics are the ideas of "information" and "memory" (i.e. stored information). To build a house you need bricks and mortar and something else – the "know-how", or "information," as to how to go about your business. You need a plan (i.e. again, stored information). Victorians like Charles Darwin understood this. In the context of the "building" of animals and plants, the *idea* of "information," but seldom the word, recurs in their writings.

Classical scholars tell us that Aristotle wrote that the "*eidos*", the form-giving essence that shapes the embryo, "contributes nothing to the material body of the embryo but only communicates its program of development" [6]. Darwin's mentor, the geologist Charles Lyell, may have had some intuition of this when in 1863, in his *Geological Evidences on the Antiquity of Man* [7], he wrote in a chapter entitled "Origin and Development of Languages and Species Compared:"

> "We may compare the persistence of languages ... to the force of inheritance in the organic world... . The inventive power which

coins new words or modifies old ones ... answers to the variety making power in the animate creation."

Gregor Mendel may have thought similarly when wondering in 1865 why one pea plant looked so much like its parents, but another did not [8]. He considered that "development must necessarily proceed in accord with a law," and that "distinguishing traits ... can only be caused by differences in the *composition* and *grouping* of ... elements." One of the few familiar with Mendel's work, the botanist Carl von Nägeli in Munich [9], agreed:

> "To understand heredity, we do not need a special independent symbol for every difference conditioned by space, time or quality, but a substance which can represent every possible combination of differences by the fitting together of a limited number of elements, and which can be transformed by permutations into other combinations."

Darwin in 1868 considered the relative contributions of information in the male gamete (pollen), and in the female gamete (ovule), to the seed from which a new plant would emerge [10]. Thus, he wrote of "the peculiar *formative* matter" that was contained in male gametes and was needed for "the full development of the seed" and for "the vigour of the plant produced from such seed." He noted that "the ovules and the male element have *equal* power of transmitting every single character possessed by either parent to their offspring." Furthermore, phenomena such as the regeneration of an entire plant from leaf cuttings (vegetative propagation) suggested that the "formative elements" were "not confined to the reproductive organs, but are present in the buds and cellular tissue of plants." So "the child, strictly speaking, does not grow into the man, but includes *germs* which slowly and successively become developed and *form* the man."

He equated "germs" with the "formative matter," which "consists of minute particles or gemmules." These were smaller than cells, were capable of independent multiplication by "self-division," and were responsible for the generation of new organisms. Thus: "An organism is a microcosm – a little universe, formed of a host of self-propagating organisms, inconceivably minute and numerous as the stars in heaven."

The major Victorian advocate of Darwin's views, biologist Thomas Huxley, anticipated modern ideas on "selfish genes" [11] when writing in similar vein in 1869 [12, 13]:

> "It is a probable hypothesis that, what the world is to organisms in general, each organism is to the molecules of which it is composed. Multitudes of these, having diverse tendencies, are competing with each other for opportunities to exist and multiply;

and the organism as a whole, is as much a product of the molecules which are victorious as the Fauna or Flora of a country is the product of the victorious beings in it."

Mental Information

Darwin's ideas were based on three fundamental principles – variation, heredity and natural selection. *Variation* caused an individual to differ in certain characters from others. *Heredity* caused characters to be transmitted from parents to children. *Natural selection* caused some individuals to survive and reproduce better than others. Of these three – variation, heredity and natural selection – only the basis of the latter was understood. For the Victorians, heredity and variation were unknowns to be entrusted to the researchers of future generations. The words "heredity" and "variation" were handles with which to manipulate these principles, irrespective of their underlying mechanisms. One Victorian, however, thought otherwise (Fig. 1-2).

Fig. 1-2. A self-portrait (1878) of Samuel Butler (1835-1902)

Initially heredity was seen as the process by which characters (which we would now call the phenotype) were passed *forward* from parent to child. However, around 1870, this perspective was shifted by the physiologist Ewald Hering in Prague, and by Samuel Butler in London. They put the onus on the embryo, which *remembers* its parents. The newly formed embryo is a passive recipient of parental information (which we now call the genotype), and this information is used (recalled) by the embryo to construct itself (i.e. to regenerate phenotype from genotype). Heredity and memory are, in principle, the same. *Heredity is the transfer of stored information.* This powerful conceptual leap led to new territory. Evolutionary processes could thenceforth be thought of in the same way as mental processes. Hering [14] considered that:

> "We must ascribe both to the brain and body of the new-born infant a far-reaching power of remembering or reproducing things which have already come to their development thousands of times in the persons of their ancestors."

More simply, Butler wrote [15]:

> "There is the reproduction of an idea which has been produced once already, and there is the reproduction of a living form which has been produced once already. The first reproduction is certainly an effort of memory. It should not therefore surprise us if the second reproduction should turn out to be an effort of memory also."

Perhaps sitting across from Karl Marx in the Reading Room of the British Museum in London, Butler devoured the evolution literature of his time and challenged the men of science. Bateson, who himself opposed a too simplistic interpretation of Darwin's theory, came to regard Butler as "the most brilliant, and by far the most interesting of Darwin's opponents" [16]. Butler thought "that a hen is only an egg's way of making another egg" [17]. In the twentieth century, zoologist Richard Dawkins thought similarly that a body is only a gene's way of making another gene [11]. An effort at reading Butler's seemingly convoluted writings about plant seeds in *Erewhon* (the title is close to "nowhere" backwards) is well rewarded [18]:

> "The rose-seed did what it now does in the persons of its ancestors – to whom it has been so linked as to be able to remember what those ancestors did when they were placed as the rose-seed now is. Each stage of development brings back the recollection of the course taken in the preceding stage, and the development has been so often repeated, that all doubt – and with all doubt, all consciousness of action – is suspended. ... The action which

> each generation takes – [is] an action which repeats all the phenomena that we commonly associate with memory – which is explicable on the supposition that it has been guided by memory – and which has neither been explained, nor seems ever likely to be explained on any other theory than the supposition that there is an abiding memory between successive generations."

Hering wanted his strictly materialist position be understood [14]: "Both man and the lower animals are to the physiologist neither more nor less than the matter of which they consist." He held that ideas that disappeared from consciousness would be stored in some material form in the unconscious memory, from which they could be recalled to consciousness. Thus, "the physiology of the unconscious is no 'philosophy of the unconscious'."

Hering considered that actions that were repeated were more likely to be stored in memory, and hence more accurately recalled, than actions that were performed only once or twice. So practice-makes-perfect with respect both to psychomotor functions, such as reproducing music on the piano, and to hereditary functions, such as reproducing a cell from a parent cell, and reproducing an organism from a parent organism (asexual reproduction) or parent organisms (sexual reproduction)[14]:

> "But if the substance of the germ can reproduce characteristics acquired by the parent during its single life, how much more will it not be able to reproduce those that were congenital to the parent, and which have happened through countless generations to the organized matter of which the germ of today is a fragment? We cannot wonder that action already taken on innumerable past occasions by organized matter is more deeply impressed upon the recollection of the germ to which it gives rise than action taken once only during a single lifetime."

Among the recollections of the germ would be instinctual (innate) actions. A pianist, having practiced a piece so well that it can be played virtually unconsciously, does not, as far as we know, pass knowledge of that piece through the germ line to children. Yet, some psychomotor activities of comparable complexity are passed on [14]:

> "Not only is there reproduction of form, outward and inner conformation of body, organs, and cells, but the habitual actions of the parent are also reproduced. The chicken on emerging from the eggshell runs off as its mother ran off before it; yet what an extraordinary complication of emotions and sensations is necessary in order to preserve equilibrium in running. Surely the supposition of an inborn capacity for the reproduction of these intri-

> cate actions can alone explain the facts. As habitual practice becomes a second nature to the individual during his single lifetime, so the often-repeated action of each generation becomes a second nature to the race."

In 1884 in his *Mental Evolution in Animals*, Darwin's ex-research associate, the physiologist George Romanes, was happy to consider instinctual activities, such as bird migration and nest-making, as requiring inherited ("ready-formed") information [19]:

> "Many animals come into the world with their powers of perception already largely developed. This is shown ... by all the host of instincts displayed by newly-born or newly-hatched animals. ...The wealth of ready-formed *information*, and therefore of ready-made powers of perception, with which many newly-born or newly-hatched animals are provided, is so great and so precise, that it scarcely requires to be supplemented by the subsequent experience of the individual."

But, much to Butler's frustration, he would not move beyond this [20]:

> "Mr. Romanes ... speaks of 'heredity as playing an important part *in forming memory* of ancestral experiences;' so that whereas I want him to say that the phenomena of heredity are due to memory, he will have it that the memory is due to heredity .... Over and over again Mr. Romanes insists that it is heredity which does this or that. Thus, it is '*heredity with natural selection which adapt* the anatomical plan of the ganglia;' but he nowhere tells us what heredity is any more than Messrs. Herbert Spencer, Darwin and Lewes have done. This, however, is exactly what Professor Hering, whom I have unwittingly followed, does. He resolves all phenomena of heredity, whether in respect of body or mind, into phenomena of memory. He says in effect, 'A man grows his body as he does, and a bird makes her nest as she does, because both man and bird remember having grown body and made nest as they now do ... on innumerable past occasions. He thus reduces life from an equation of say 100 unknown quantities to one of 99 only, by showing that heredity and memory, two of the original 100 unknown quantities, are in reality part of one and the same thing." [Butler's italics]

Periodical Rhythms

In 1880 in his book *Unconscious Memory* Butler employed the word "information" in the context of memory [21]:

"Does the offspring act as if it remembered? The answer to this question is not only that it does so act, but that it is not possible to account for either its development or its early instinctive actions upon any other hypothesis than that of its remembering, and remembering exceedingly well. The only alternative is to declare ... that a living being may display a vast and varied *information* concerning all manner of details, and be able to perform most intricate operations, independently of experience and practice."

In 1887 in *Luck or Cunning* [22], Butler showed a good understanding of the concept of information, and of its symbolic representation in a form conditioned by factors both internal and external:

"This idea [of a thing] is not like the thing itself, neither is it like the motions in our brain on which it is attendant. It is no more like these than, say, a stone is like the individual characters, written or spoken, that form the word 'stone.' ... The shifting nature ... of our ideas and conceptions is enough to show that they must be symbolic and conditioned by changes going on within ourselves as much as those outside us."

Later, he was more poetic [23]:

"Some ideas crawl, some run, some fly; and in this case words are the wings they fly with, but they are only wings of thought or of ideas, they are not the thought or ideas themselves."

Again emphasizing "the substantial identity between heredity and memory," Butler noted that "Variations [mutations to the modern reader] ... can only be perpetuated and accumulated because they can be inherited." Echoing Huxley, in 1880 he considered the unicellular amoeba [21]:

"Let us suppose that this structureless morsel of protoplasm is, for all its structurelessness, composed of an infinite number of living molecules, each one of them with hopes and fears of its own, and all dwelling together like Tekke Turcomans, of whom we read that they live for plunder only, and that each man of them is entirely independent, acknowledging no constituted authority, but that some among them [DNA to the modern reader] exercise a tacit and undefined influence over the others.

Let us suppose these molecules capable of memory [of storing information], both in their capacity as individuals [individual molecules], and as societies [groups of molecules], and able to transmit their memories to their descendents, from the traditions

> of the dimmest past to the experiences of their own lifetime. Some of these societies will remain simple, as having no history, but to the greater number unfamiliar, and therefore striking, incidents [mutations to the modern reader] will from time to time occur, which, when they do not disturb memory so greatly as to kill [lethal mutations], will leave their impression upon it [accepted mutations]. The body or society will remember these incidents, and be modified by them in its conduct, and therefore … in its internal arrangements. ... This memory of the most striking events of varied lifetimes [inherited information] I maintain … to be the differentiating cause, which, accumulated in countless generations, has led up from the amoeba to man … .
>
> We cannot believe that a system of self-reproducing associations should develop from the simplicity of the amoeba to the complexity of the human body without the presence of that memory which can alone account at once for the resemblances and differences between successive generations, for the arising and the accumulation of divergences – for the tendency to differ and the tendency not to differ."

Like most others, Butler discarded the "Russian doll" ("germs within germs" or "preformation") model of inheritance [15]:

> "When we say that the germ within the hen's egg remembers having made itself into a chicken on past occasions … do we intend that each single one of these germs was a witness of, and a concurring agent in, the development of the parent forms from their respective germs, and that each of them therefore, was shut up with the parent germ, like a small box inside a big one? If so, then the parent germ with its millions of brothers and sisters was in like manner enclosed within a grandparental germ, and so on till we are driven to admit, after even a very few generations, that each ancestor has contained more germs than could be expressed by a number written in small numerals, beginning at St. Paul's and ending at Charing Cross. … Therefore it will save trouble … to say that the germs that unite to form any given sexually produced individual were not present in the germs, or with the germs, from which the parents sprang, but that they *came into* the parents' bodies at some later period."

If the germs that united to form the parents were no longer present, then this, of course, raises the question as to how *new* germs "came into" the parents. Butler continued [15]:

> "We may perhaps find it convenient to account for their intimate acquaintance with the past history of the body into which they have been introduced by supposing that by virtue or assimilation [metabolism] they have acquired certain *periodical rhythms* already pre-existing in the parental bodies, and that the communication of the characteristics of these rhythms determines at once the physical and psychical development of the individual in a course as nearly like that of the parents [inheritance] as changed surroundings [environment] will allow."

So the material forms taken by the parental information are no longer present in their child, but during its life "periodical rhythms" are communicated to new material forms that will allow the generation of future children (the grandchildren of the original parents). Were the original parental material forms – the "papers" upon which the "plan" for their child was written – completely destroyed in the child so that fresh germs had to be reconstructed from scratch, like an architect taking measurements from an existing building in order to prepare a plan for a new, identical, building? Probably not. The term "periodical rhythms" suggests that Butler had some sort of copying in mind [15]:

> "For body is such as it is by reason of the characteristics of the vibrations that are going on in it, and memory is only due to the fact that the vibrations are of such characteristics as to catch on to, and be caught onto by, other vibrations that flow into them from without – no catch, no memory."

So, although the original parental material form of the plan (paper) would no longer be present, its information would have been copied to create identical plans on new papers. Thus, there would be a "recurrence" of the "rhythm" [15]:

> "I see no way of getting out of this difficulty so convenient as to say that a memory is the reproduction and *recurrence* of a rhythm communicated directly or indirectly from one substance to another, and that where a certain rhythm exists there is a certain stock of memories [stored information], whether the actual matter in which the rhythm now subsists was present with the matter in which it arose, or not."

Butler considered that "matter … introduced into the parents' bodies during their life histories [i.e. food]… goes to form the germs that afterwards become their offspring," but he was troubled as to whether this matter should be considered as living or non-living, for "if living, it has its own *memories and life-histories* which must be *cancelled and undone* before the assimilation and the becoming *imbued with new rhythms* can be complete. That is to say it must become as near to non-living as anything can become" [15].

Recall that he considered cells to consist of "living molecules, each one of them with hopes and fears of its own, and all dwelling together like Tekke Turcomans" [21]. As we shall see in the next chapter, if we think of polymeric DNA as "living", and the subunits (letters) from which DNA is constructed (assimilated) as "non-living," then Butler's abstraction is far from wild. Food contains specific DNAs of other organisms, which encode their specific characteristics ("memories and life histories"). These polymeric DNAs are degraded ("cancelled and undone") in digestive systems to subunits ("non-living") that are absorbed into bodies where they are repolymerized in a new order ("imbued with new rhythms") to produce copies of the bodies' own DNAs ("living"). He may have been thinking of the set of letters that a nineteenth century printer would assemble on a printer's block when preparing a newspaper. After a print run the letters would be removed from the block and then reutilized (i.e. assembled together in a different order) for the next day's print run. Thus, when considering verbal communication Butler wrote [23]:

> "The spoken symbol ... perishes instantly without material trace, and if it lives at all does so only in the minds of those who heard it. The range of its action is no wider than that within which a voice can be heard; and every time a fresh impression is wanted the type must be set up anew."

Informational Macromolecules

That the "composition and grouping of ... elements" might require some material basis was recognized by many nineteenth century writers, including the English philosopher Herbert Spencer, Nägeli, and Darwin's cousin, Francis Galton [24, 25]. It was noted that if the basis were material, since "the ovules and male element have equal power of transmitting every single character" (i.e. maternal and paternal contributions to offspring were equal), then the quantity of that material could be no more than could be contained in the smallest gamete, which was usually the male gamete (pollen or spermatozoon; Fig. 1-3).

In 1866 the English chemist William Crookes commented on the probable nature of the poisonous "particles" of the cattle plague that was ravaging Europe at that time. The particles could not be observed with the microscopes then available. He deduced that they were not bland, but "organized" to a degree much greater than that of the chemical poisons with which people were then familiar [26]:

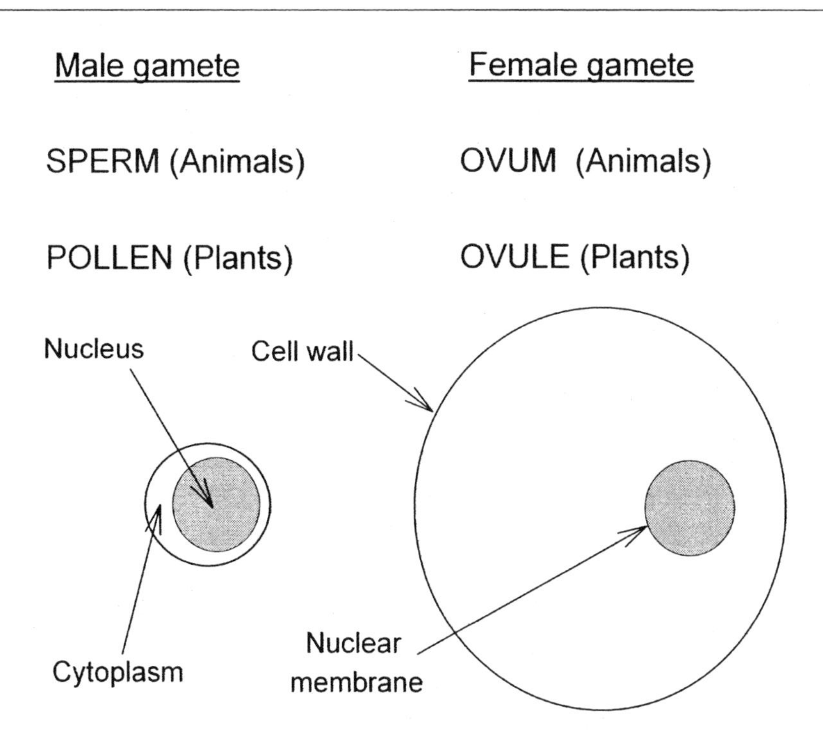

Fig. 1-3. If heredity has a non-compressible, non-stretchable, material basis, then, since male and female contributions to our genetic heritage are approximately equal, the quantity of hereditary material cannot be more than is contained in the smallest of the two gametes, which is usually the male gamete (spermatozoon or pollen grain). Thus much of the material in the female gamete (ovum or ovule) cannot be directly concerned with the transmission of hereditary characters

> "The specific disease-producing particles must moreover be organized, and possess vitality; they must partake of the nature of *virus* rather than of *poison*. No poison yet known to chemists can approach, even in a faint degree, the tremendous energy of the active agent of infectious diseases. A poison may be *organic*, but it is not *organized*. It may kill with far greater rapidity than the virus of infection, but, unlike this virus, it cannot multiply itself in the animal economy to such an extent as to endow within a few hours every portion of its juices with the power of producing similar results. A virus, on the contrary, renders the liquids of an infected animal as virulent as the original germ."

Viruses, as we now know them, were first identified and named decades later. Crookes used the word in the Latin sense understood by his contemporaries, meaning, "poison" or "toxin." But his personal understanding appears modern in that he implied that a virus would use the resources available in "the animal economy" to multiply itself. The word "germ" meant a biological entity with the potential to sprout or "germinate" (like a plant seedling), and as such the word was used in the context of organisms we now refer to as viruses and bacteria. Darwin's notion of the organism as "a little universe, formed of a host of self-propagating organisms" seems to have derived from this emerging understanding of the minuteness and great proliferative powers of the microorganisms responsible for the cattle plague. In his 1866 testimony to a government enquiry, John Simon, the Medical Officer of the Privy Council suggested [27]:

> "The several zymotic diseases are aetiologically quite distinct from one another. ... How their respective first contagia arose is, ... quite unknown. This, in pathology, is just such a question as in physiology is 'the origin of species.' Indeed ... it is hardly to be assumed as certain that these apparently two questions may not be only two phases of one. Hourly observation tells us that the contagium of smallpox will breed smallpox, that the contagium of typhus will breed typhus, that the contagium of syphilis will breed syphilis, and so forth, – that the process is as regular as that by which dog breeds dog, and cat cat, as exclusive as that by which dog never breeds cat, nor cat dog."

In 1868 Darwin advanced his "provisional theory of pangenesis," and named a minute fundamental unit of heredity a "gemmule" [10]. By this time, on the continent of Europe microscopic investigations had distinguished cell nuclei and the phenomena of cell division (28). Galton in 1872 [29] and August Weismann in 1885 noted that for many organisms there was a potential separation of the hereditary material in germ-line cells (that produced gametes) from that contained in other body cells (the cells of the mortal "soma")[30]:

> "Splitting up of the substance of the ovum into a somatic half, which directs the development of the individual, and a propagative half, which reaches the germ-cells and there remains inactive, and later gives rise to the succeeding generation, constitutes *the theory of the continuity of the germ plasm.*"

Bateson considered that Butler had arrived at "this underlying truth" in 1878 "by natural penetration" [16, 17]. The Dutch botanist Hugo de Vries in his *Intracellular Pangenesis* [31] boldly took the issue to the molecular level in 1889, noting that an *observed* character of an organism could be "*determined by* a single *hereditary character* or a small group of them." These determining "hereditary characters" are "units, each of which is built up of numerous chemical molecules," and are "arranged in rows on the chromatin threads of the nucleus." The units which, rather than "gemmules," he named "pangens," could, for example, confer the "*power of assuming* a red colour" in a plant. This "*power of assuming*" was held to be different from the actual "*production* of ... the red colouring matter," which was "more or less in the same manner as the action of enzymes or ferments." Thus, de Vries distinguished three hierarchically arranged elements: hereditary character units (later known as the genotype), executive functions catalyzing chemical changes (associated with enzyme-like elements), and observed characters themselves (later known as the phenotype; Greek: *phainomai* = I appear).

The Swiss physiologist Johann Friedrich Miescher, discoverer of what we now call DNA, suggested in 1892 that, by virtue of structural differences, one class of "huge molecules" should suffice to convey the hereditary character units [32]:

> "In ... huge molecules ... the many asymmetric carbon atoms provide a colossal amount of stereo-isomerism. In them all the wealth and variety of heredity transmissions can find expression just as all the words and concepts of all languages can find expression in twenty-four to thirty alphabetic letters. It is therefore quite superfluous to make the egg and sperm cell a storehouse of countless chemical substances each of which carries a particular heredity quality.... My own research has convinced me that the protoplasm and the nucleus, far from consisting of countless chemical substances, contain quite a small number of chemical individuals which are likely to be of a most complicated chemical structure."

In 1893 in *An Examination of Weismannism* [33] Romanes, summarized what, for many at that time, were obscure abstractions:

> "But the theory of Pangenesis does not suppose the future organism to exist in the egg-cell *as a miniature* (Romanes' italics): it supposes merely that every part of the future organism *is represented* in the egg-cell by corresponding material particles. And this, as far as I can understand, is exactly what the theory of germ-plasm [of Weismann] supposes; only it calls the particles 'molecules,' and seemingly attaches more importance to the matter of variations in their arrangement or 'constitution,' whatever these vague expressions may be intended to signify."

Along similar lines, in 1894 in *Materials for the Study of Variation* [34], Bateson stated:

> "The body of one individual has never *been* the body of its parent,... but the new body is made again new from the beginning, just as if the wax model had gone back to the melting pot before the new model was begun."[Bateson's italics]

Later, agreeing with de Vries, Bateson considered enzymes affecting colour [35]:

> "We must not suppose for a moment that it is the ferment [enzyme], or the objective substance [colour], which is transmitted [from generation to generation]. The thing transmitted can only be the *power or faculty to produce* the ferment or objective substance."

Rather than "hereditary characters," Bateson spoke of "character units," "genetic factors," or "elements." Grasping for what we now know as DNA, in 1913 in his *Problems in Genetics* [36] he elaborated on the terms "genotype" and "phenotype" that had been introduced by the Danish botanist Wilhelm Johannsen:

> "Of the way in which variations in the ... composition of organisms are caused we have ... little real evidence, but we are beginning to know in what such variations must consist. These changes must occur either by the addition or loss of [genetic] *factors* [to/from the gametes and hence to/from the offspring formed by those gametes]. We must not lose sight of the fact that, though the [genetic] factors *operate by* the *production* of: (i) enzymes, (ii) bodies on which these enzymes can act [substrates], and (iii) intermediary substances necessary to complete the enzyme action [enzyme cofactors], yet these bodies themselves [all three of the above] *can scarcely themselves* be genetic factors, but *consequences* of their existence."

The genetic factors could be newly acquired and had a continuity from generation to generation, being apportioned symmetrically during cell division. Bateson continued:

> "What then are the [genetic] factors themselves? Whence do they come? How do they become integral parts of the organism? Whence, for example, came the *power* which is present in the White Leghorn of destroying ... the pigment in its feathers? That power is now a definite possession of the breed, present in all its germ-cells, male and female, taking part in their symmetrical divisions, and passed on equally to all, as much as [is passed on] the protoplasm or any other attribute of the breed."

It seems that Bateson never stated that a genetic factor (i.e. the genotype) conveyed "*information for*", but only conveyed the "*power to cause*" enzymes, substrates and enzyme-cofactors to appear in the offspring (and thus generate the phenotype). However, in addresses in 1914 and later he showed that he sensed that genetic factors (now interpreted as "genes") might be some "phenomenon of arrangement" (now interpreted as "sequence"), and that variation (now interpreted as mutation; Latin, *muto* = I change) could occur when the arrangement changed. Indeed, he almost latched on to the idea of a genetic code linking the observed characters of an organism (its phenotype) to "elements upon which they depend" (now interpreted as genes)[37]:

> "The allotment of characteristics among offspring is ... accomplished ... by a process of cell-division, in which numbers of these characters, or rather the elements upon which they depend, are sorted out among the resulting germ-cells in an orderly fashion. What these elements, or factors as we call them, are we do not know. That they are in some way directly transmitted by the material of the ovum and of the spermatozoon is obvious, but it seems to me unlikely that they are in any simple or literal sense material particles. I suspect rather that *their properties depend on some phenomenon of arrangement*.... That which is conferred in variation must rather itself be a change, not of material, but of arrangement. ... By the re-arrangement of a very moderate number of things we soon reach a number of possibilities practically infinite."

Hering had spoken similarly, albeit more abstractly, in 1870 [14]:

> "It is an error, therefore, to suppose that such fine distinctions as physiology must assume, lie beyond the limits of what is conceivable by the human mind. An infinitely small change in position on the part of a point, or in the *relations* of the parts of a segment of a curve to one another, suffices to alter the law of its whole path, and so in like manner an infinitely small influence exercised by the parent organism on the molecular disposition of the germ may suffice to produce a determining effect upon its whole further development... An organized being, therefore, stands before us a product of the unconscious memory of organized matter, which, ever increasing and ever dividing itself, ever assimilating new matter and returning it in changed shape to the inorganic world, ever receiving some new thing into its memory, and transmitting its acquisitions by the way of reproduction, grows continually richer and richer the longer it lives."

So, we can interpret "*power to cause*" as showing that Bateson (like Darwin, Galton, Hering, and Butler) understood the information concept, although he never used the word in this context (but he used the word in other contexts). He understood that different arrangements can convey different pieces of information. He was prepared to believe his factors were molecules of the type we would today call "macromolecules," but he was far from calling them "*informational* macromolecules". There was not so much a conceptual gap as a semantic gap. The story of how the semantic gap was later bridged by Erwin Schrödinger and others has been told many times [38–41], and will not be retold here.

Translation

Regarding the "difficulty of conceiving the transition of germinal substance into somatic substance," Romanes considered "a transition of A into B ... where B is a substance which differs from A almost as much as a woven texture differs from the hands that weave it"[33]. But he could not see A and B as languages. He could not see a transition of this magnitude as a *translation* from one form of information to another. The elements of language are, according to Butler [23]:

> "A sayer, a sayee, and a convention, no matter what, agreed upon between them as inseparably attached to the idea which it is intended to convey – these comprise all the essentials of language. Where these are present there is language; where any of them are wanting there is no language."

The need for a code (an agreed convention) to relate the information carried from generation to generation in the gametes (derived from germ cells in the gonads) to the actual characters displayed by an organism (e.g. parts of the body) was hinted at by Weismann in 1904 [42]:

> "The germ substance never arises anew, but is always derived from the preceding generation – that is ... the continuity of the germ plasm. ...We ... assume that the conditions of all the parts of the body [are] ... reflected in the corresponding primary constituents of the germ plasm and thus in the germ cells. But, as these primary constituents are *quite different* from the parts themselves, they would require to vary in quite a different way from that in which the finished parts had varied; which is very like supposing that an English telegram to China is there received in the Chinese language."

This implied that one form of information, namely the "primary constituents," must be *translated* into another form, namely, "the parts." We would now relate "primary constituents" to DNA (the genotype), and "the parts" (the phenotype) to certain classes of macromolecules, especially proteins.

The four main classes of macromolecules found in living systems, – nucleic acids, proteins, complex fats (lipids), complex sugars (carbohydrates) – all convey information. However, the former two are special in that the latter two are dependent upon them for their construction. Nucleic acids and proteins construct both themselves and lipids and carbohydrates (Fig. 1-4).

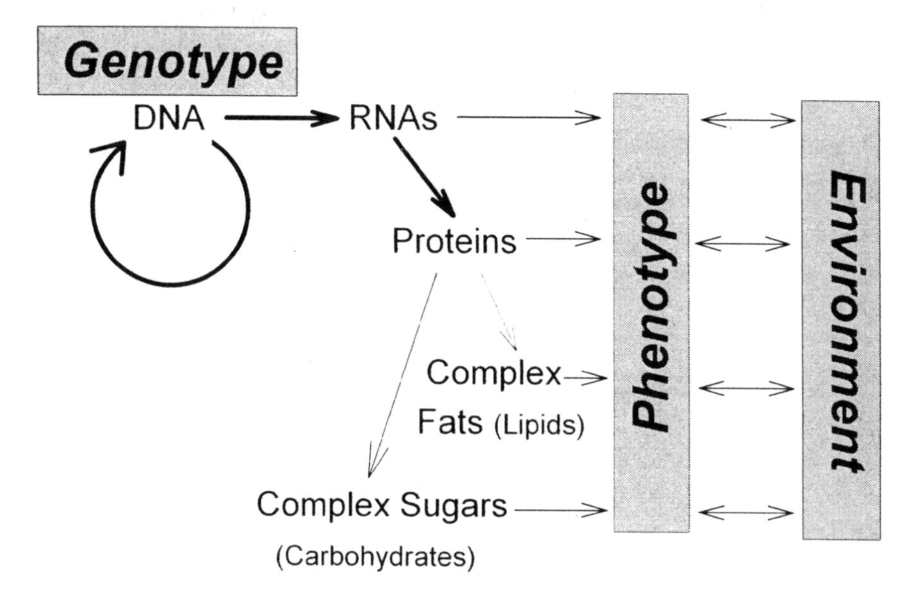

Fig. 1-4. Template-dependent reactions (heavy arrows), which require nucleic acids, result in the production from subunits (base-containing nucleotides, or amino acids) of "informational macromolecules" (DNA, RNA and proteins). Proteins (enzymes) catalyse both these template-dependent reactions and the mechanical assembly from subunits (e.g. simple sugars, fatty acids) of other macromolecules, namely complex fats (lipids) and complex sugars (carbohydrates). When a nucleic acid acts as a linear template for the formation of another nucleic acid, information is copied (*replicated* or *transcribed*). When a nucleic acid acts as a linear template for the formation of a protein, one form of information is decoded (*translated*) to give rise to another form of information. RNAs, proteins, lipids and carbohydrates collectively compose the conventional Darwinian phenotype through which members of a species interact with (track) their changing environment through the generations. The curved arrow indicates the self-reproducing nature of the genotype

Nucleic acids and proteins have sequences that are so varied that the "machines" that construction them (enzymes which are themselves mainly proteins) have to be programmed with linear information templates (nucleic acids). On the other hand, lipids and carbohydrates have regularities in their structures that allow their mechanical assembly by enzymes in the absence of linear information templates. Thus, nucleic acids and proteins can be considered as "informational macromolecules," to indicate that they contain a higher order of information than lipids and carbohydrates.

Butler supposed "a memory to 'run' each gemmule" [17], but, of course, he knew the material basis of the storage of inherited information no more than he knew the material basis of the storage of mental information. We now know that DNA is the former, but, at the time of this writing, still have little understanding of the latter. If challenged, Butler would probably have speculated that both forms of storage were the same (Occam's principle; see Chapter 3). Current evidence does not support this. Just as DNA contains the information for the construction, by some chain of events, of, say, a hand, so DNA should contain the information for a particular pattern of cerebral "wiring" (an "anatomical plan of the ganglia") that would, by some chain of events, be manifest as a particular instinct. How such germ-line memories relate to memories acquired during individual lifetimes ("cerebral alterations") is still unknown. Romanes pointed out in 1881 [20]:

> "We can understand, in some measure, how an alteration in brain structure when once made should be permanent, ... but we cannot understand how this alteration is transmitted to progeny through structures so unlike the brain as are the products of the generative glands [gametes]. And we merely stultify ourselves if we suppose that the problem is brought any nearer to solution by asserting that a future individual while still in the germ has already participated, say, in the cerebral alterations of its parents."

Butler seems to have been the first to extrapolate the principle that "higher" forms of life evolve from "lower" forms, to suggest that eventually, beyond man, machines made by man would take over the planet [43]. Today we would equate this with computers. Butler's genius was in seeing that inheritance and mental function, like today's computers, are all "modes of memory" [17]:

> "Life is that property of matter whereby it can remember. Matter which can remember is living; matter which cannot remember is dead. *Life, then, is memory*. The life of a creature is the memory of a creature. We are all the same stuff to start with, but we remember different things, and if we did not remember different things we should be absolutely like each other. As for the stuff itself, we know nothing save only that it is 'such as dreams are made of'." [Butler's italics]

Variation

"Memory" refers either to the physiological process by which stored information is retrieved, or to the actual stored information itself. In the former sense, the storage of information is implicit. In the latter sense, memory *is* stored information. Whatever the sense, the concepts of memory and of stored information are very close. At some point in time information is selected and assigned to a store, and at a later point in time it may be retrieved. But storage of information is not enough. Information must be *safely* stored. To select and to store is not to preserve. On a hot summer's day, a package of ice-cream stored in your shopping bag will not be preserved. Stored information must also be *preserved* information. When preservation is not perfect, there is *variation.*

Butler declared that "the 'Origin of Variation,' whatever it is, is the only true 'Origin of Species'," [17]. He knew the mechanisms by which Nature's expressions vary from generation to generation, no more than he knew the mechanisms by which mental expressions vary from day to day. Nor did he know the scope of variation [34]. Variation can only occur in that which already exists, and the nature of that which already exists, limits the scope of its variation. Just as the scope of a dice is limited by its structure, so the scope of variation of a living form is limited. From time to time a human is born with six fingers, but, as far as we know, no human has ever been born with a hundred fingers, or with feathers (see Chapter 3).

The Victorians delighted in games of chance. They appreciated that variation in landing position of a balanced dice is random. The variation in landing position of an unbalanced dice is partly non-random. So, from first principles, it was supposed that variation in living forms might be either a chance event, or biased. Whether unbiased or biased, a variation might be beneficial, deleterious, or somewhere in between. Within its scope, the dice of life might be multifaceted allowing fine gradations in variation. Alternatively variations might be quite discrete (e.g. the usual six-sided dice). And would the dice of life be thrown? If so, by what agency? If not thrown, then any variation that occurred would be "spontaneous" and perhaps unbiased. If the dice of life were thrown, then there would be a thrower. Would this agency be internal or external to the organism? In either case, could the agency direct the bias to the adaptive advantage of the organism? In other words, could the agency "design" the organism?

Among these alternatives, current evidence shows (see Chapter 7) that variation is a spontaneous property of matter, but that it may be affected, both quantitatively and qualitatively, by external agents, such as X-radiation. Variation is usually unbiased. It is not directed. There is no design by means of variation, either internal or external. If an organism is closely adapted to its environment then a variation is unlikely to be beneficial. If the environment has recently changed dramatically, then a variation is more likely to be beneficial. But variations, *per se*, do not occur "for the good of the organism" (teleology).

However, the parallels he had drawn between memory and heredity suggested to Butler a set of alternatives that had some plausibility at the time. He opted for an agency *internal* to the organism that would, in small steps, bring about variations that would accumulate to the advantage of the organism. This set of alternatives had gained some scientific respectability in both England (e.g. Spencer), and continental Europe (e.g. Nägeli) and some came to call it "orthogenesis," implying an innate tendency to progressive development [44]. Butler's case rested on grounds that were both aesthetic and logical.

Darwin's argument that natural selection would cruelly send the weaker to the wall and select the fittest to survive sufficed "to arouse instinctive loathing; ... such a nightmare of waste and death is as baseless as it is repulsive" [23]. To buttress this feeling, Butler turned to an argument that will be made here, in a different context, in Chapter 12. Microscopic studies of unicellular organisms, such as amoebae, showed them to possess organelles analogous to the organs of multicellular organisms. Thus, an amoeba, far from being an amorphous mass of protoplasm, was seen to extrude "arms" (pseudopodia), fashion a "mouth", and digest its prey in a prototypic stomach (digestive vacuole). If it could achieve this degree of sophistication, then perhaps it could, in an elementary way, also think? Butler was quite open with his premises [45]:

> "*Given* a small speck of jelly with some power of slightly varying its actions in accordance with slightly varying circumstances and desires – *given* such a jelly speck with a power of assimilating other matter, and thus, of reproducing itself, *given* also that it should be possessed of a memory and a reproductive system ...".

When something was counter-intuitive Butler spurned common-sense, but when it served his purpose he often appealed to "people in ordinary life," or "plain people" [23]:

> "The difference between Professor Weismann and, we will say, Heringians consists in the fact that the first maintains the new germ-plasm, when on the point of repeating its developmental process, to take practically no cognisance of anything that has happened to it since the last occasion on which it developed itself; while the latter maintain that offspring takes much the same kind of account of what has happened to it in the persons of its parents since the last occasion on which it developed itself, as people in ordinary life take things that happen to them. In daily

> life people let fairly normal circumstances come and go without much heed as matters of course. If they have been lucky they make a note of it and try to repeat their success. If they have been unfortunate but have recovered rapidly they soon forget it; if they have suffered long and deeply they grizzle over it and are scared and scarred by it for a long time. The question is one of cognisance or non-cognisance on the part of the new germs, of the more profound impressions made on them while they were one with their parents, between the occasion of their last preceding development and the new course on which they are about to enter."

Thus, Butler downplayed chance ("luck") and championed an *intrinsic* capacity for bias ("cunning") as the means by which advantageous characters acquired by parents would be transmitted to their children. In short, he appealed to the doctrine of the inheritance of acquired characters that had been advocated in France in 1809 by Jean Baptiste Pierre de Lamarck. Butler even went so far as to assert, in Romanes' words [20], "that a future individual while still in the germ has already participated ... in the cerebral alterations of its parents." However, like Hering, Butler maintained a strictly materialist position. He used the words "intelligent" and "design," often separately, and sometimes together, but never in a way as to suggest the involvement of an agency *external* to the organism [23] (see Appendix 3):

> "The two facts, evolution and design, are equally patent to plain people. There is no escaping from either. According to Messrs. Darwin and Wallace, we may have evolution, but are in no account to have it mainly due to intelligent effort, guided by ever higher and higher range of sensations, perceptions and ideas. We are to set it down to the shuffling of cards, or the throwing of dice without the play, and this will never stand. According to the older men [e.g. Lamarck], the cards did count for much, but play was much more. They denied the teleology of the time – that is to say, the teleology that saw all adaptation to surroundings as part of a plan devised long ages since by a quasi-anthropomorphic being who schemed everything out much as a man would do, but on an infinitely vaster scale. This conception they found repugnant alike to intelligence and conscience, but, though they do not seem to have perceived it, they left the door open for a design more true and more demonstrable than that which they excluded."

Butler soared beyond the comprehension of most of his contemporaries on the wings of his conceptual insight that heredity was the transfer of stored information. Although his dalliance with Lamarckism has not won support, the robustness of the underlying idea led him close to solutions to fundamental biological problems such as the origin of sex and the sterility of hybrids (see Chapter 3), and aging (see Chapter 14). In his times, Butler's *correct* equation of heredity with memory was beyond the pale, but his *incorrect* Lamarckism seemed plausible. Spencer was, at heart, a Lamarckian and, at Darwin's behest, Romanes spent several years fruitlessly seeking to show that gemmules containing information for acquired characters could be transferred from normal parental tissues (soma) to the gonads (germ-line). And as late as 1909 the German evolutionist, Ernst Haeckel, was proclaiming that "natural selection does not of itself give the solution of all our evolutionary problems. It has to be taken in conjunction with the transformism of Lamarck, with which it is in complete harmony" [46].

Latency

One good reason why a character is not perceived in an organism is that the information for that character is not present. Yet, information can be present but remain undetected because it is not expressed. Such information is referred to as "latent." Francis Galton suggested in 1872 that individuals consist of "two parts, one of which is *latent*, and only known to us by its effects on posterity, while the other is *patent*, and constitutes the person manifest to our senses" [29]. Darwin [10] believed that "every character which occasionally reappears is present in a latent form in each generation," and in 1877 Butler, tongue in cheek, quoted him with approval in this respect [17]:

> "We should expect that reversion should be frequently capricious – that is to say, give us more trouble to account for than we are either able or willing to take. And assuredly we find it so in fact. Mr. Darwin – from whom it is impossible to quote too much or too fully, inasmuch as no one else can furnish such a store of facts, so well arranged, and so above all suspicion of either carelessness or want of candour – so that, however we may differ from him, it is he himself who shows us how to do so, and whose pupils we all are – Mr. Darwin writes: 'In every living being we may rest assured that a host of long-lost characters lie ready to be evolved under proper conditions' (does not one almost long to substitute the word 'memories' for the word 'characters?) 'How can we make intelligible, and connect with other facts, this wonderful and common capacity of reversion – this power of calling back to life long-lost characters?' Surely the answer may be hazarded, that we shall be able to do so when we can make intelligible the power of calling back to life long-lost memories. But I

> grant that this answer holds out no immediate prospect of a clear understanding."

Sometimes the information for an entire organism can be latent, as in the case of the AIDS virus genome that, under certain circumstances, becomes merged with human genomes but is not expressed. Darwin was thinking of characters, distributed among members of a species, some of which might remain latent for generations until suddenly they re-emerge among the members of a new generation. There are also characters that are usually present in all members of a species, but are only observed at certain times in the life cycle, or under particular environmental conditions. Thus, information for a butterfly is latent in its caterpillar. Human secondary sexual characters can be considered latent until adolescence.

Increased pigmentation of the skin occurs in white people exposed to excess sunlight; this inducible "tanning" is a latent ability. When white people in the northern hemisphere move south their skins darken. On the other hand, for black people, pigment formation is a permanent, "constitutive," ability. When black people move north their skins do not whiten. Black pigment (melanin) protects skin cells against radiation and in equatorial climes there is still sufficient penetration of light to assist formation of the essential nutrient, vitamin D. When black people move north they risk vitamin D deficiency. Thus, there is a trade-off between the need to protect against radiation, implying more melanin, and the need to form vitamin D, implying less melanin. It is possible that hairless hominoids first evolved in hot equatorial climes, and were constitutively black. Through natural selection (see Chapter 7) those that migrated to and fro, from south to north and back again, acquired the ability to regulate melanin production [47].

All this implies that the expression of a piece of information can be regulated. It may, or may not, be expressed throughout a lifetime (constitutively), or its expression may be confined to certain periods of life or to certain environments. Information may be expressed fully or partially, sometimes in a precisely calibrated fashion, and sometimes more randomly. Just as a light from a forest dwelling may be perceived only when its path is not blocked by trees, so the penetrance of one piece of information (from within the organism to the phenotype perceived from outside the organism) can sometimes depend on the expression of other pieces of information that accompany it within the organism.

Summary

Heredity requires stored information (memory). Early ideas that this information storage might be "a phenomenon of arrangement" can now be interpreted in terms of the linear arrangement of bases in nucleic acid molecules. Typically, genes contain information for the construction of executive units (proteins), which facilitate chemical reactions whose products give rise to the observed features of an organism – its phenotype. Thus, the twentieth century's obsession with genes is understandable. However, the gene-centred viewpoint spawned a corresponding bioinformatics very different from the emergent, non-gene-centred, bioinformatics. The tasks of evolutionary bioinformatics – the new bioinformatics – are to identify *all* forms of hereditary information, both genic and non-genic, and show how conflicts between different forms are accommodated to permit the proper functioning of biological entities (e.g. genes, cells, individuals, species, genera). Because different forms of information, be they mental, textual or hereditary, have common characteristics, our understanding of the familiar forms (e.g. textual information) may assist our understanding of less familiar forms (e.g. hereditary information). For example, information, once selected and assigned to a store, must be *preserved.* If preservation is imperfect, stored information will change, and anything dependent on that information may also change. When hereditary information changes, parental characters may vary in offspring, and biological evolution can occur. Samuel Butler arrived at many of these truths "by natural penetration."

Chapter 2

Chargaff's First Parity Rule

> "The formation of different languages and of distinct [biological] species, and the proofs that both have been developed through a gradual process, are curiously the same. ... Languages, like organic beings, can be classed in groups under groups. ... A language, like a species, when once extinct, never ... reappears. The same language never has two birthplaces... . The survival or preservation of certain favoured words in the struggle for existence is natural selection."
>
> Charles Darwin (1871)[1]

Consider a creature, whose attributes you are probably familiar with, perhaps pausing in a posture similar to that which you are now adopting:

The cat sat on the mat.
The cat sat on the mat.

You might guess that, since the information is repeated and rhythms, it forms part of some artistic endeavour that we might refer to as poetry. Alternatively, the author may have had little faith in the typesetters and, to make assurance doubly sure, sent the message twice. It is with the latter explanation that we are most concerned at this point.

Error-Detection

Since type-setting errors are usually random and rare, it is likely that, if an error were to occur, it would affect only one of the above sentences. Instead of "mat" on the top line you might have seen "hat." Coming across the two parallel sentences for the first time, and knowing that the repetition was deliberate, how would you decide which was the correct sentence? You might read the top sentence first as the "sense" text, and then read the bottom sentence to check that each of the letters of the alphabet is faithfully matched by its "sense" equivalent (i.e. "a" is always matched by "a;" "t" is always matched by "t," etc.). Thus, you would check that there is *parity* between the two parallel lines.

Finding that "h" in "hat" in the top line is mismatched with "m" in "mat" in the bottom line, you would know that an error had occurred. But, in the absence of other information you would not be able to decide whether "h" in the top line was correct, or that it should be corrected to "m" based on the information in the bottom line. All you would know was that there had been an error. If for some reason you were forced to decide, you might toss a coin. But, in the absence of other information, if you accepted the coin's guidance there would only be a 50:50 chance of your being correct.

In-Parallel Redundancy

To increase the chance of accurate error-*detection*, and hence of accurate error-*correction*, the author might have repeated the sentence, in-parallel, three times. If two lines contained the word "mat" and only one line the word "hat," then you would prudently choose the former. Your choice would be even more likely to be correct if the author repeated the sentence, in-parallel, four times, and only one line had the word "hat."

All this requires much redundant information, which both takes up space in the medium conveying the message (in this case, the printed page), and imposes extra labour on the reader. For some purposes it might suffice merely to detect that an error has occurred. Having been alerted, you might then be able to consult other sources of information should the need to distinguish between "hat" and "mat" be critical. For example, there are 25 possible alternatives to "h" as the first letter in "hat." Of the resulting words – aat, bat, cat, dat, eat, fat, gat, etc. – several can be excluded because there is no English dictionary equivalent, others can be excluded syntactically (e.g. "eat" is a verb not an object), and others can be excluded contextually (e.g. neighbouring text might refer to mat, not to hat).

Thus, there is much to be gained by duplication, but with increasing levels of redundancy (triplication, quadruplication, etc.) the gains are less evident. At face value, this appears to be a strategy adopted by biological systems for accurately transferring information from generation to generation. Genetic messages are sent as duplexes, but with a "twist" in more than one sense.

DNA Structure

Contrasting with the 26 letter English alphabet, the DNA alphabet has four letters – the bases **A** (adenine), **C** (cytosine), **G** (guanine), and **T** (thymine). Thus, a message in DNA might read:

TACGACGCCGATAGCGTCGTA (2.1)

With duplex in-parallel redundancy, the message could be sent from generation to generation as:

TACGACGCCGATAGCGTCGTA (2.2)
TACGACGCCGATAGCGTCGTA

We might refer to this as "sense-sense" pairing since, like the cat sentences above, both sentences read the same (i.e. **A** is matched with an **A**, and **T** is matched with a **T**, etc.). However, when arriving at their model for the duplex structure of the DNA in 1953, James Watson and Francis Crick [2] took into account a "rule" enunciated by biochemist Erwin Chargaff. He and his co-workers had found that bases did not match themselves. They matched other bases. In DNA, base **A** is quantitatively equivalent to base **T**, and base **G** is quantitatively equivalent to base **C**. Chargaff speculated in 1951 that this regularity might be important for DNA structure, noting [3]:

> "It is almost impossible to decide at present whether these regularities are entirely fortuitous or whether they reflect the existence in all DNA preparations of certain common structural principles, irrespective of far-reaching differences in their individual composition and the absence of an easily recognizable periodicity."

In 1952 Canadian biochemist Gerard Wyatt went further, suggesting a spiral structure [4]:

> "If you have a spiral structure" ... [it is quite possible to have the bases] "sticking out free so that they don't interfere with each other. Then you could have a regular spacing down the backbone of the chain, in spite of the differences in sequence."

Later he added [5]:

> "One is tempted to speculate that regular structural association of nucleotides of adenine with those of thymine and those of guanine with those of cytosine ... in the DNA molecule requires that they be equal in number."

If the top message were "sense," the bottom message could be considered as "antisense." The above "sense" message could then be sent in duplex form as:

TACGACGCCGATAGCGTCGTA "sense" (2.3)
ATGCTGCGGCTATCGCAGCAT "antisense"

Error-detection would still be possible. In this "sense-antisense" error-detection system, errors would be detected when an **A** was matched with **G**, **C** or another **A**, rather than with **T**. Similarly, if **G** was matched with **A**, **T** or another **G**, rather than with **C**, another error would have been detected.

That a base would not match itself was also right for chemical reasons. Just as the letters of the standard alphabet come as either vowels or consonants, so the bases of DNA are either purines (**A** and **G**) or pyrimidines (**C** and **T**; Table 2-1).

	R (Purines)	**Y** (Pyrimidines)
W (Weak)	**A** (Adenine)	**T** (Thymine)
S (Strong)	**G** (Guanine)	**C** (Cytosine)

Table 2-1. Symbols for groups of the four main bases in DNA. When picking symbols for collectivities of bases some form of logic is attempted. Thus, since pu<u>r</u>ines and p<u>y</u>rimidines both begin with the same letter, the second consonants **R** and **Y** are employed. Watson-Crick base-pairing involves interactions (hydrogen-bonded, non-covalent) that are either "weak" (**W**) in the case of **A**-**T** base-pairs, or "strong" (**S**) in the case of **G**-**C** base-pairs

Vowels and consonants often match or "complement" to the extent that vowels separate consonants giving words a structure, which facilitates their pronunciation. Purines are bigger than pyrimidines, and the chemical models that Watson and Crick constructed required that a purine always match or "complement" a pyrimidine. A molecular complex of two purines would be too big. A molecular complex of two pyrimidines would be too small. The solution is that the purine **A** pairs with the pyrimidine **T** and the purine **G** pairs with the pyrimidine **C**. By match we mean an actual structural (i.e. chemical) pairing. Although your eyes can detect that **A** on one line matches **T** on the other, inside our cells it is dark and there are no eyes to see. Matching is something molecules do for themselves by recognizing complementary shapes on their pairing partners, just as a key recognizes the lock with which it "pairs."

The key-lock analogy will serve us well in the present work; however, pairing may require other molecular subtleties such as similar molecular vibrations, or resonances [6]. To try to visualize this, in 1941 the geneticist Herman Muller likened molecular mixtures to imaginary mixtures of floating electromagnets each charged with an alternating current of a particular frequency [7]. Since magnet polarity would depend on the direction of current flow, the polarity of each magnet would be constantly changing at a frequency determined by the frequency of the alternating current.

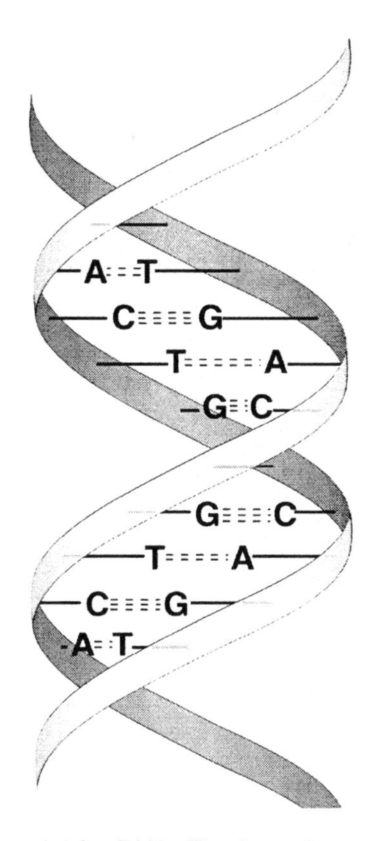

Fig. 2-1. Double helix model for DNA. The base "message" is written on two helical strands, which are shown here as twisted ribbons – the "medium." Bases are arranged internally so that an **A** on one strand pairs with a **T** on the other strand (and vice versa), and a **G** on one strand pairs with a **C** on the other strand (and vice versa). The bases are attached to the strands by very strong (covalent) bonds, whereas the base-pairing interactions involve weaker (non-covalent) bonds (shown as dashed lines). Chemically, the bases are like flat discs that are "stacked" on top of each other within the helical strands, like a pile of coins (rouleau). These stacking interactions stabilize the double-helical structure, and being largely entropy-driven (see Chapter 12), become greater as temperature increases. However, in solution at high temperatures (e.g. 80°C) this can be overcome, and the two strands separate (i.e. the duplex "melts") to generate free single strands. This figure was kindly adapted by Richard Sinden from his book *DNA Structure and Function* [8]

> "If we had a heterogenous mixture of artificial electromagnets, floating freely about and having different frequencies of reversal of sign, those of the same frequency would be found eventually to orient towards and attract one another, specifically seeking each other out to the exclusion of others."

Of course, the final twist of Watson and Crick was, literally, a twist. The two sequences in DNA form two molecular strands that are wound round each other in the form of a spatially compact helix (Fig. 2-1)[8]. Perhaps the most famous throwaway line ever written came at the end of Watson and Cricks' first paper [2]. Here, as an apparent afterthought, they casually noted: "It has not escaped our notice that the specific pairing we have postulated immediately suggests a possible copying mechanism for the genetic material." In other words, they were claiming not only to have discovered the fundamental structure of genetic information, but also to have discerned from that structure how the information would be faithfully replicated. When the underlying *chemistry* was understood, the *physiology* could be explained – a triumph for the"reductionist" approach. They had shown how the chemical structure of DNA provided a basis for the continuity of inherited characteristics from organism to organism, and from cell to cell within an organism. In Bateson's words, they had discovered how "the allotment of characteristics among offspring is ... accomplished." This was made explicit in a second paper [9]:

> "Previous discussions of self-duplication [of genetic information] have usually involved the concept of a template or mould. Either the template was supposed to copy itself directly or it was to produce a 'negative,' which in its turn was to act as a template and produce the original 'positive' once again. ... Now our model for deoxyribonucleic acid is, in effect, a *pair* of templates [Watson and Cricks' italics], each of which is complementary to the other. We imagine that prior to duplication ... the two chains unwind and separate. Each chain then acts as a template for the formation on to itself of a new companion chain, so that eventually we shall have two pairs of chains, where we only had one before."

Armed with this powerful clue, within a decade biochemists such as Arthur Kornberg in the USA had shown Watson and Crick to be correct, and had identified key enzymes (e.g. DNA polymerase) that catalyze DNA replication [10]. The stunning novelty of the Watson-Crick model was not only that it was beautiful, but that it also explained so much of the biology of heredity. One strand is the complement of the other, so that the text of one strand can be inferred from the text of the other. If there is an error in one strand, then there is the potential for its repair on the basis of the text of the opposite strand. When the cell divides the two strands separate. New "child" strands are synthesized from nucleotide "building blocks" corresponding to **A**, **C**, **G** and **T**. Each of these blocks, consisting of phosphate, ribose and a base (Fig. 2-2), replaces the former pairing partners of the separated strands, so that two new duplexes identical to the parental duplex are created. In each duplex one of the parental strands is conserved, being paired with a freshly synthesized child strand (Fig. 2-3).

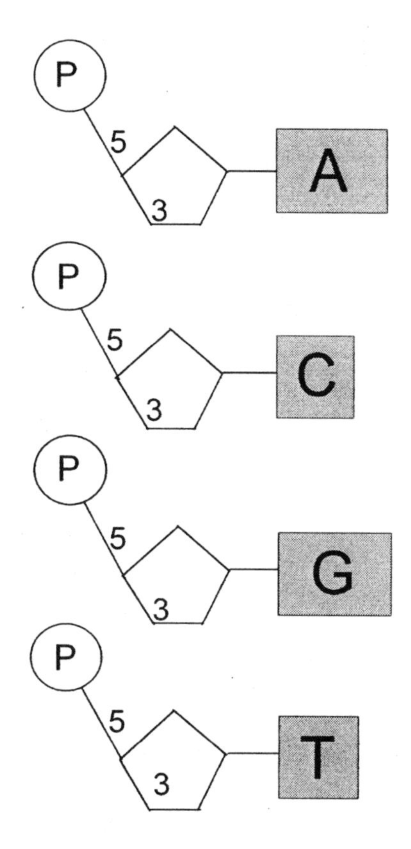

Fig. 2-2. The four nucleotide "building blocks" of which DNA is composed. Each base is connected by way of a pentose sugar (pentagon) to a phosphate (circle). The purine bases (**A** and **G**) are represented by larger boxes than the pyrimidine bases (**T** and **C**). Nucleotides have in common a pentose sugar and a phosphate, and differ in their bases. A major difference between DNA and RNA is that the pentose sugar in DNA is deoxyribose (hence "deoxyribonucleic acid" = "DNA"), whereas the sugar in RNA is ribose (hence "ribonucleic acid" = "RNA"). The carbon atoms that are part of the pentose sugar are numbered, as indicated here for the third and fifth carbon atoms

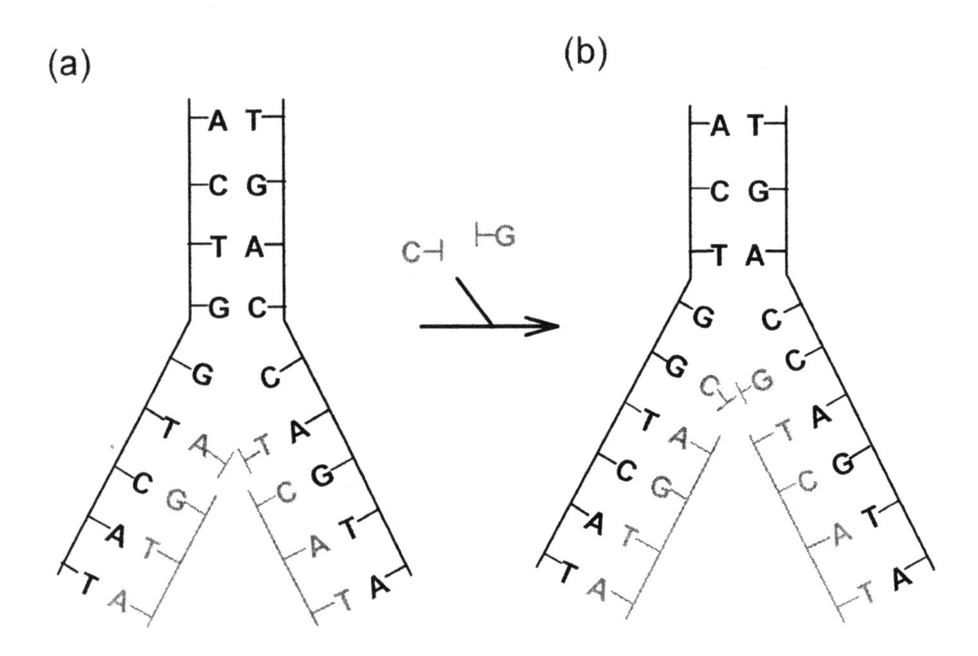

Fig. 2-3. DNA replication. Individual strands of a parental duplex partially separate, and fresh child strands are synthesized by sequential introduction and "zipping" up (polymerization) of complementary base nucleotide "building blocks" (shown in grey). Thus, DNA is a linear polymer (Greek: *poly* = many and *meros* = part) of nucleotide units (i.e. it is a polynucleotide). In *(a)*, at the point of child strand growth in the left limb of the replication fork (inverted Y), an **A** (grey) is about to be joined to a **G** (grey). This join is complete in *(b)*, where the two parental strands are further separated. The new duplexes each contain one parental strand (black), and one child strand (grey). Details of synthesis in the right limb are dealt with in Chapter 6 (Fig. 6-6)

All nucleotide "building blocks" have in common phosphate and ribose, which are essential for continuing the phosphate-ribose "medium," upon which the base "message" or "pattern" is "written." Thus, *any* nucleotide can serve to ensure continuity of the phosphate-ribose medium, and the message itself is determined only by which particular base-containing nucleotide is placed in a particular position. This, in turn, is determined by the complementary template provided by the parental DNA strands, which are recognized according to the specific base-pairing rules (Fig. 2-4).

The message you are now reading was imposed by the stepwise sequential addition of letters to a *pre-existing* medium (the paper). Each letter required a small local piece of the medium, but that medium was already in place when the letter arrived. The medium had already been generated. When DNA is synthesized each base "letter" arrives in a pre-existing association with a small piece of the medium (phosphate-ribose) that it locally requires.

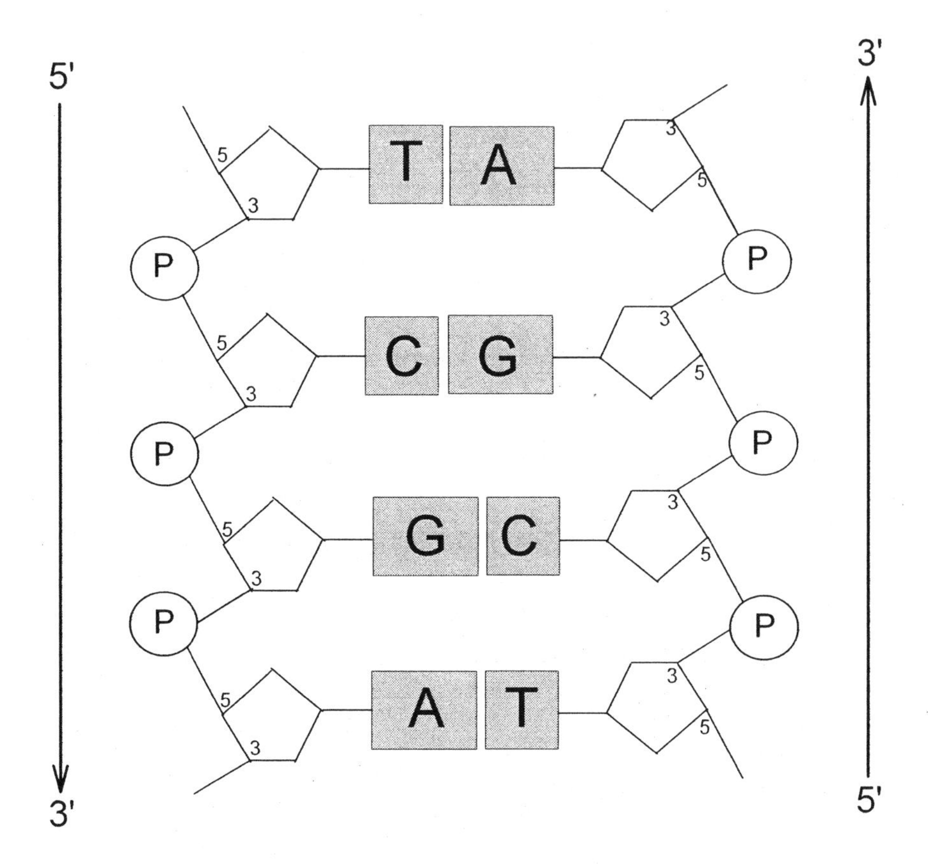

Fig. 2-4. Model of base-pairing between the two strands of a DNA duplex. Note that the larger purines pair with the smaller pyrimidines, so that the distance between the two strands remains relatively constant. Because of this size difference, the flat bases do not just "stack" (form a "pile of coins") above and below neighbouring bases in the same strand (e.g. note that the two **G**'s on separate strands overlap each other, and thus partially stack together). Rather, base-pairs "stack" with base-pairs. Numbering associated with the pentose sugars indicates that strands have distinct directionality (polarity) that, by convention, is written from 5' to 3' (see vertical arrows). Thus, the left strand reads 5'**TCGA**3' from top to bottom. The right strand also reads 5'**TCGA**3', but from bottom to top. The two strands are described as "antiparallel." Since this short duplex *as a whole* has symmetry (i.e. putting the purines first, the order is **A**-**T** base pair, **G**-**C** base pair, **G**-**C** base pair, **A**-**T** base pair), then it can be said to show palindromic properties (see Chapter 4)

Thus, the message and the medium are generated at the same time. The message and the medium increase in length simultaneously. Remarkably, all this had been sensed by Muller, who had mentored Watson in the 1940s. In 1936 he wrote [11]:

> "The gene is, as it were, a modeller, and forms an image, a copy of itself, next to itself, and since all genes in the chain do likewise, a duplicate chain is produced next to each original chain, and no doubt lying in contact with a certain face of the latter. ... There are thousands of different levels of genes, i.e. of genes having different patterns, ... and ... each of these genes has to reproduce its own specific pattern out of the surrounding materials common to them all. When, through some microchemical accident, or chance quantum absorption, a sudden change in the composition ('pattern') of the gene takes place, known to biologists as a 'mutation,' then the gene of the new type, so produced, reproduces itself according to this new type, i.e. it now produces precisely the new pattern. This shows that the copying property depends upon some more fundamental feature of gene structure [phosphate-ribose chain to the modern reader] than does the specific pattern which the gene has [base sequence to the modern reader], and that it is the effect of the former to cause a copying not only of itself but also of the latter, more variable, features. It is this fact which gives the possibility of biological evolution and which has allowed living matter ultimately to become so very much more highly organized than non-living. It is this which lies at the bottom of ... growth, reproduction, and heredity."

As we shall see in Chapter 7, the "possibility of biological evolution" occurs because, although mutations are often repaired, sometimes they are not. A change in "specific pattern" can then be passed on, by copying, to the next generation. When considering pairs of bases, care should be taken to distinguish between: (i) a Watson-Crick base-pair (i.e. two bases on separate strands, or separate parts of a strand, which are involved in the classical **A-T** and **G-C** pairings), (ii) a dinucleotide consisting of two ordered contiguous bases on the same strand (e.g. **CpG**; see Chapter 15), and (iii) the base composition of a nucleic acid segment (e.g. **(G+C)**%; see Chapter 8).

Turnover

The "microchemical accident," to which Muller referred might have a definite cause (e.g. Muller himself had noted increased mutations following X-irradiation), or might loosely be described as "spontaneous." The accident might result in one regular letter being substituted for another (e.g. "hat" rather than "mat"), or a regular letter might change to something else (e.g. "ψat" rather than "mat"), or simply be eliminated (e.g. "at" rather than "mat"). As will be discussed later (under the heading "entropy;" Chapter 12), it seems to be a general property of the universe that the elements that compose it, whatever their size, tend to become disordered and evenly distributed. This is true at the chemical level where macromolecules tend to break down to their micromolecular building blocks, and the building blocks themselves, either separate or when they are part of macromolecules, live under a constant threat of structural change and dismemberment into their constituent atoms.

Photographers are sometimes confronted with the problem that they want to photograph a busy city scene but without the people, traffic and parked cars. The solution is to use time-lapse photography. A fixed camera takes a picture once a day with a very short exposure time. The film is not wound on, so daily pictures are superimposed. The first picture, if developed, would show nothing. However, over weeks and months static objects begin to appear, whereas the transient objects are never present long enough to register. Since macromolecules tend to be transient, a magic time-lapse camera that could see individual molecules in bodies would tend to register nothing – except for molecules of DNA. From this crude metaphor one should not deduce that DNA molecules are static. Even buildings vibrate and move within their confines. So do DNA molecules.

Two cell strategies for dealing with the constant breakdown of its parts are *recycling* (so that macromolecules are degraded and then resynthesized from their component parts), and *repair*. The former strategy (turnover) applies mainly to four of the five major classes of macromolecules (lipids, carbohydrates, proteins and RNA). The latter strategy applies mainly to the fifth class, DNA. Thus, whereas a damaged amino acid in a protein (a polymer of amino acid units) leads to the protein being degraded by specific enzymes (proteases) to its constituent amino acids, a damaged nucleotide in a DNA molecule (a polymer of nucleotide units) often invokes a "rapid response team" of repair enzymes that will do its best to effect on-site repair without necessarily interrupting macromolecular continuity.

Promiscuous DNA

Sometimes there is a break in a DNA duplex. The two ends may be reconnected by various enzymes (e.g. "ligases"). However, the tendency towards disorder sometimes means that a DNA segment is incorrectly reconnected. A random "cut" followed by a "paste" may result in one segment of DNA recombining with a new segment of DNA so that the order of the information they contain is changed (transposed or inverted). To the extent that such changes are not critical for survival of the line, genomes are vulnerable to an on-going kaleidoscopic diversification, a constant shuffling, of the sequences they contain.

More than this, DNA molecules are promiscuous – meaning, literally, that DNA molecules are "pro-mixing." Place two duplex DNA molecules within a common cell wall and they will seek each other and attempt to recombine. We shall see that biological evolution became possible when DNA "learned," by adjusting sequence and structure, how to constrain and channel this tendency. Often the order of information in DNA is critical. Specific segments of DNA have specific "addresses" in their chromosomes. The ability to accurately recombine specific segments of duplex DNA, while maintaining segment order and the integrity of functional units, is a fundamental property of living organisms. Indeed, US biologist George Williams, one of those responsible for our modern "selfish gene" concept, thought it better to define genes in terms of their abilities to resist dismemberment by recombination, than in terms of their functions (see Chapter 8). The great evolutionary significance of recombination was pointed out by Crick in 1970 [12]:

> "There is also a major problem to which I believe biologists have given insufficient attention. All biologists essentially believe that evolution is driven by natural selection, but ... it has yet to be adequately established that the rate of evolution can be adequately explained by the processes which are familiar to us. It would not surprise me if nature has evolved rather special and ingenious mechanisms so that evolution can proceed at an extremely rapid rate – recombination is an obvious example."

A year later Crick presented his "unpairing postulate" to explain how the inward-looking bases in a DNA double-helix might look *outward* to recognize complementary bases in another helix (see Chapter 8).

Bits and Bats

There is a link with information theory. Since there are two main types of bases, purines (**R**) and pyrimidines (**Y**), then, disregarding the phosphate-ribose medium upon which the base message is written, a nucleic acid can be represented as a binary string such as:

YRYRRYRYYRRYRRYRYYRYR (2.4)

Electronic computers work with information in this form – represented as strings of 0s and 1s. If a **Y** and an **R** are equally likely alternatives in a sequence position, then each can be quantitated as one "bit" (binary digit) of information, corresponding to a simple yes/no answer.

Confronted with a generic base (often expressed as **N**) you could first ask if it was a purine (**R**). A negative reply would allow you to infer that **N** was a pyrimidine (**Y**). You could then ask if it was cytosine (**C**). A positive reply would allow you to infer that **N** was not thymine (**T**). Thus, by this criterion, each position in a DNA sequence corresponds to two potential yes/no answers, or two "bits" of information. By this measure, the entire single-strand information content in the human haploid genome (3×10^9 bases) is 750 megabytes (since 8 bits make a byte), which is of the order of the amount of information in an audio compact disk.

This way of evaluating DNA information has been explored [13], but so far has not been particularly illuminating with respect to DNA function. One reason for this may be that DNA is not just a binary string. In the natural duplex form of DNA, a base in one string pairs with its complementary base in another string. Each base is "worth" 2 bits, so that a base pair would correspond to 4 bits. However, even if not paired the two bases would still collectively correspond to 4 bits. Thus, the chemical pairing of bases might increase their collective information content to some value greater than 4 bits. But does this come at a price?

So breath-taking was Watson and Crick's model that some potentially major criticisms were overlooked. If every line of the present book were repeated, after the fashion of the cat sentences with which this chapter begins, then the book would be twice as long as it now is. Not only does it make sense to minimize the duplication of information in books, but there are circumstances where it would appear advantageous not to duplicate information in biological systems. Despite this, duplication is the rule. For example, one of the two forms of gamete, usually the male spermatozoon, has to be highly mobile and hence has a streamlined shape and, tadpole-like, is often equipped with a flagellum. There appears to have been a selection pressure to keep the quantity of contained information (i.e. DNA) to a minimum (Fig. 1-3). Virus genomes, which have to be packaged for transfer from organism to organism, are also very compact. Yet, the DNA of spermatozoa and viruses is always in duplex form (with a few special exceptions).

At another level (literally and otherwise) consider flying organisms – bats, birds, insects. In every case we find duplex DNA in cells. Every cell of all multicellular organisms has duplex DNA, and flying organisms are no exceptions. Bats have 5.4 picograms per cell [14], whereas equivalent mammals (mice) have 7 picograms of DNA per cell. Bats appear to have shed some of the "excess" DNA (see Chapter 12 for discussion of "junk" DNA), but that which remains is still in *duplex* form. Birds have approximately 2.5 picograms of DNA/cell. A bird that could shed half its DNA and exist with single stranded DNA would seem to have a weight advantage compare with a bird that had duplex DNA. It should be able to fly faster and farther than those with duplex DNA, a feature of particular importance for migratory birds. But again, the DNA is always in duplex form. Relative to humans, birds have shed "excess" DNA, and this has occurred more in the parts of genes that do not encode proteins (introns) than in the parts of genes that mainly encode proteins (exons; Fig. 2-5; see Chapter 10).

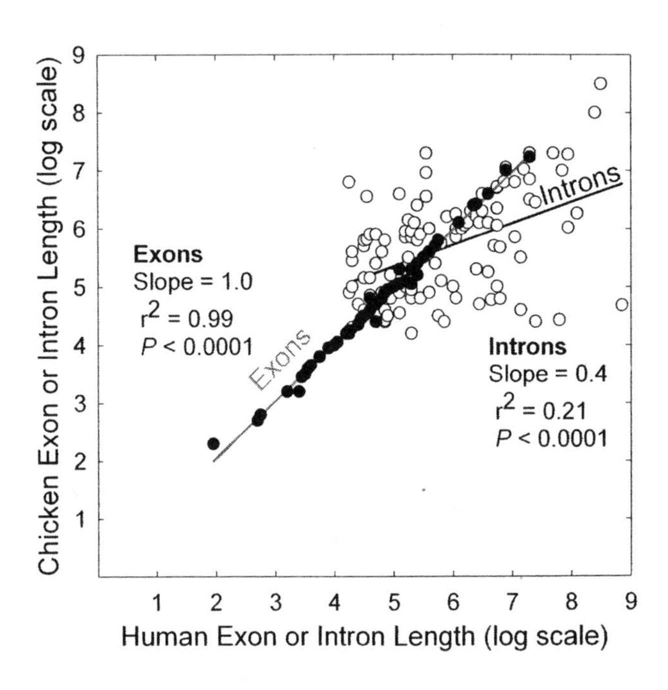

Fig. 2-5. Comparison of exon and intron lengths in a set of corresponding genes of humans and chickens. Lengths are expressed in natural logarithms, and the lines are drawn to best fit the points (see Appendix 1). In the case of exon lengths (black circles) the slope is 1.0 showing that, on average, each chicken exon is the same size as the corresponding human exon. In the case of intron lengths (open circles) the slope is 0.4 showing that, on average, each chicken intron is four-tenths the size of the corresponding human intron. Both slopes are significantly different from zero ($P < 0.0001$). Note that, in general, exons are smaller than introns (see Chapter 10), and the points for exons fit more closely to their line (SEE = 0.11; $r^2 = 0.99$) than the points for introns fit to their line (SEE = 0.80; $r^2 = 0.21$). The lesser scattering of exon length values indicates that the exons in the set of genes studied have been under strong negative selection (i.e. individuals with exon mutations have tended not to survive). Thus, since the time when humans and chickens diverged from a common ancestor, exon sequences have been less successful at varying than intron sequences (i.e. individuals with intron mutations have survived more often than individuals with exon mutations; see Chapter 7). This figure is adapted from reference [15]

From this it seems that there are compelling reasons for keeping DNA in duplex form at all stages of life. As we shall be considering, in biological systems there are conflicts and there have to be trade-offs. But abandoning the duplication of DNA information is seldom one of them.

Haploidy and Diploidy

Many organisms alternate during their life cycle between haploidy (one copy of each chromosome, containing one DNA duplex, per cell) and diploidy (two copies of each chromosome per cell). Most gametes are haploid and so contain only one copy of each DNA duplex. When male and female gametes unite, the product (zygote) is diploid with two copies of each DNA duplex-containing chromosome, one of paternal origin and one of maternal origin. Some organisms, such as the malaria parasite *Plasmodium falciparum*, quickly switch back to the haploid state, so its adult form is haploid. But for many organisms, diploidy is the adult norm. Only when new gametes are formed is there a brief flirtation with haploidy.

Thus, we are confronted by the fact that there is redundancy of information not only because DNA molecules come as duplexes, but also because many organisms "choose" for most of their life cycles to have two copies of each duplex. Since each duplex has *at least* two-fold redundancy, diploid organisms have *at least* four-fold redundancy in their content of DNA.

In-Series Redundancy

Why "*at least*"? There is only at least four-fold redundancy because we have so far considered only in-parallel redundancy. The phenomenon of in-series redundancy was discovered when measurements were made of the rate at which duplexes would reform from single strands when in solution in test tubes.

From knowledge of the length of a DNA duplex and the number of such duplexes in a solution it was possible to calculate how rapidly the duplexes should reform after the two strands have been separated from each other by heating. Like separated partners on a dance floor, to reform, each single-strand would have to find its complement. If there were just one DNA duplex present, then each single strand would have no option but to find its original complementary partner. If two identical duplexes were present it would not matter if a strand found a partner from the other duplex (i.e. it would switch dancing partners). However, in this case there would be twice the chance of finding a partner in a given space and time, compared with the situation when only one duplex was present. Thus, the more identical DNA duplexes present, the more rapidly would the strands reform duplexes (anneal) after heating.

When the experiment was carried out, it was found that for many DNA samples the rate of duplex reformation was far greater than anticipated [16]. This was particularly apparent in the case of species with very long DNA molecules. Further studies showed that within DNA there is a redundancy due to the presence of repetitive elements. There are many more copies of certain segments of DNA than the four expected from in-parallel considerations. Molecular "dancing partners" may be found in series as well as in parallel. This in-series redundancy will be discussed further in Chapter 12.

Accidents and Non-Accidents

In our lives we encounter two classes of adverse events – random and non-random. The non-paranoid designate random adverse events as "accidents." There is conflict between the random forces of disorder and the forces of order, the former being not deliberately hostile, but merely reflecting the tendency of things, if left alone, to become untidy rather than tidy. This tendency gets greater when things move faster, which usually means they get hotter, as will be considered at the molecular level in Chapter 12.

In this chapter we have considered error-generation as driven by random processes ("microchemical accidents"), and sequence redundancy as having arisen to permit error-detection, and so, possibly, error-correction. Redundancy means that the *qualitative* characteristics (e.g. sequence) of an organism's *own* DNA molecules can be compared, so allowing *quality control.* By mechanisms to be touched on in Chapter 6, the total quantity of DNA in a cell is maintained relatively constant. This is *quantity control.* Since the quantity of DNA determines the "dose" of gene-products (e.g. proteins) that a cell contains (see Chapter 14), this implies that the quantity of cellular macromolecules can be regulated, directly or indirectly, by DNA quantity-control mechanisms.

Sometimes the forces of disorder have an appreciable non-random component, as when a foreign virus (i.e. foreign DNA) deliberately enters a cell. The repertoire of "self" macromolecules (M) then is supplemented by sets of "not-self" macromolecules (VM). So the total quantity of cellular macromolecules (TM) can be written:

$$M + VM = TM \tag{2.5}$$

Under normal circumstances, the quantity of macromolecules of virus origin (VM) would be zero. As will be seen in Chapters 13, there are sophisticated host strategies for distinguishing "self" from "not-self," and to be successful (i.e. to increase VM) a virus must outsmart them. The closer the virus, in its *qualitative* characteristics, can approach to self (i.e. become "near-self" with respect to its host) the more likely it is to succeed. Host quality-control mechanisms, are then likely to be less effective. There is, however, the theoretical possibility of using *quantitative* characteristics of viruses (i.e. VM itself) as a basis for distinction by the host. The available strategies for organisms to respond internally to non-random adversities, in the forms of viruses, are somewhat similar to the available strategies for countries to respond internally to non-random adversities, in the forms of forgers of their currencies. The metaphor may be helpful.

The aim of a forger is to fool you with counterfeit currency. If successful the forger prospers, but if too successful there is the possibility that the entire monetary system would collapse. This would not serve the forger well and, so far as we know, no forger has gone to this extreme. Nevertheless, the counterfeit notes must be as like the real thing as the forger can contrive. At the qualitative level, your visual and tactile sensory mechanisms for distinguishing real notes from counterfeit notes must be evaded. Accordingly, manufacturers of a country's true currency are engaged in an "arms race" with the illegitimate manufacturers of false currency. As forgers get progressively better at counterfeiting currency that approaches progressively closer to the real thing, so the manufacturers of true currency must add embellishments that are difficult for forgers to imitate. This allows you to continue to make a *qualitative* distinctions on a note-by-note basis.

At the level of the entire currency system, however, it should *in theory* be possible to detect that forged notes (FN) are present without looking at individual notes. Designating the quantity of real notes as N, we can write:

$$N + FN = TN \qquad (2.6)$$

TN represents the total quantity of notes. Here is how it would work. Given knowledge of the initial quantity of real notes, and their rates of manufacture and of destruction when worn-out, then it should be possible to know how many real notes (N) exist. If there were a way of directly monitoring how many notes actually existed at a particular time-point (e.g. knowing the average "concentration" of notes and the area over which they were distributed), then the actual number (TN) could be compared with the calculated number (N). If the actual number exceeded the calculated number, then the presence of forged notes (FN) would be inferred, alarm bells would be rung, and appropriate corrective measures implemented. In principle, if the system were sufficiently sensitive, a small initial increase in forged notes would be immediately responded to. A forger would have difficulty opposing this form of monitoring. But, of course, in practice such monitoring is difficult for countries to implement.

Biological organisms are not so constrained. In general, "self" molecules are manufactured at rates that have been fine-tuned over millions of years of evolution. Similarly, rates of destruction have been fine-tuned. Accordingly, the concentrations of many molecules, including nucleic acids and proteins, fluctuate between relatively narrow limits. Intrusive foreign "not-self" macromolecules would tend to increase total macromolecule concentrations in ways that, in principle, should be detectable. This theme will be explored in Chapters 12 and 13.

Summary

Most cell components undergo cycles of degradation and resynthesis ("turnover"), yet their concentrations fluctuate between only very narrow limits. The DNA of a cell provides information that specifies the quality and quantity of these components. Accurate transmission of this information requires that errors be detected and corrected. If there is more than one copy of the information (redundancy) then one copy can be compared with another. For hereditary transmission of information, a "message" is "written" as a sequence of four base "letters" – **A**, **C**, **G**, **T** – on a strand of phosphate and ribose (the "medium"). In duplex DNA there is two-fold redundancy – the "top" strand is the complement of the "bottom" strand. **A** on one strand matches **T** on the other, and **G** on one strand matches **C** on the other. A check for non-complementarity permits error-detection. Thus, Chargaff's first parity rule is that, for samples of duplex DNA, the quantity of **A** (adenine) equals the quantity of **T** (thymine), and the quantity of **G** (guanine) equals the quantity of **C** (cytosine). In diploid organisms there is four-fold sequence redundancy, due to the presence of a DNA duplex (chromosome) of maternal origin, and a DNA duplex (chromosome) of paternal origin. There is also some in-series, within-strand, redundancy. Trade-offs to optimize utilization of sequence space usually do not include abandonment of duplex DNA or diploidy. Birds lighten their DNA load by decreasing sequences (introns) that do not encode proteins. DNA is promiscuous in readily acquiescing to a "cutting-and-pasting" (recombination between and within strands) that shuffles the information it contains. Indeed, George Williams thought it better to define genes in terms of their abilities to resist dismemberment by recombination, than in terms of their functions. Furthermore, a codiscoverer of DNA structure, Francis Crick, questioned the potency of natural selection of functional differences as an evolutionary force, and pointed to possible "ingenious mechanisms" involving recombination that might accelerate evolutionary processes.

Chapter 3

Information Levels and Barriers

> "All messages and parts of messages are like phrases or segments of equations which a mathematician puts in brackets. Outside the brackets there may always be a qualifier or multiplier which will alter the whole tenor of the phrase."
>
> Gregory Bateson (1960) [1]

Aristotle wrote that the form-giving essence, the *eidos*, "does not become part of the embryo, just as no part of the carpenter enters the wood he works… but the *form is imparted* by him to the material" [2, 3]. A carpenter will often hold the information for imparting form to wood in his head. If he wishes to communicate this information to other carpenters separated from himself in space and time he has to choose symbols and a code. The symbols, D, O, and W, if repeated and arranged in the sequence W, O, O, and D, would be meaningless unless the receiver knew the transmitting carpenter's code. Thus, implicit to the information concept is the idea that a grouping of symbols can have a meaning, and that there can be a linkage ("mapping") between the grouping and the meaning, which we call a code (an understood convention).

Symbols and Code

If symbols are arranged as a string in one dimension, then the corresponding code is also likely to be one-dimensional. The code can be three-dimensional, as in the way a lock, by its complementary shape, "decodes" the contours (i.e. the symbols) of a key. Indeed, it is in this three-dimensional manner that biomolecules communicate with other biomolecules within and between cells, to create the phenomenon we know as life. If the time of the intermolecular communication were also encoded, then the code would be four-dimensional, with time being the extra dimension. This would mean that the decoding process might also involve the molecular resonances or vibrations referred to in Chapter 2.

Primary and Secondary Information

Returning to our human written form of information, the sentence "Mary had a little lamb its fleece was white as snow" contains the *information* that a person called Mary was in possession of an immature sheep. The sentence is a *message*, whereas the information, the meaning of the message, is itself an abstract entity, which can only exist as a message. The message can be in the form of an English text as above, or a French text, or a German text, or in some (at the time of this writing) unknown molecular form in our brains. The determination that the information was the same in these sources would be made when it was extracted from the corresponding messages by applying a decoding process and shown to be convertible to a common message form (e.g. the English text).

The information about Mary and the lamb can be conveyed in different languages. So the message itself can contain not only its *primary* information, but also *secondary* information about its source – e.g. it is likely that an author is more familiar with one language than others. Some believe that English is on the way to displacing other languages, so that eventually it (or the form it evolves to) could constitute the only language used by humans on this planet. Similarly, in the course of early evolution it is likely that a prototypic nucleic acid language, perhaps initially in a two-letter form (R and Y), displaced contenders.

Because languages diverged from primitive root languages many thousands of years ago, it would be difficult to discern directly a relationship between the English, French and German versions of "Mary had a little lamb its fleece was white as snow." However, in England, if a person with a Cockney accent were to speak the sentence it would sound like "Miree ader liawl laimb sfloyce wors woyt ers snaa." Cockney English and "regular" English diverged more recently (they are "allied languages") and it is easy to discern similarities. Now look at the following:

yewas htbts llem ws arifea ac Mitte alidsnoe lahw
irsnwwis aee ar lal larfoMyce b sos woilmyt erdea

One line of text is the regular English version with the letters shuffled. The other line is the cockney version with the letters shuffled. Can you tell which is which? If the shuffling was thorough, the primary information has been destroyed. Yet, there is still *some* information left. With the knowledge that cockneys tend to "drop" their Hs, it can be deduced that the upper text is more likely to be from someone who spoke regular English. With a longer text, this could be more precisely quantitated.

Thus, languages have characteristic letter frequencies. You can take a segment ("window") and count the various letters in that segment. In this way you can identify a text as English, Cockney, French or German. We can call this information "secondary information." There may be various other *levels* of information in a sequence of symbols. To evaluate the secondary information in DNA (with only four base "letters"), you can select a window (say 1000 bases) and count each base in that window. You can apply the same window to another section of the DNA, or to another DNA molecule from a different biological species, and repeat the count. Then you can compare DNA "accents" (see Appendix 2).

Primary Information for Protein

The best understood type of primary information in DNA is the information for proteins. The DNA sequence of bases (one type of "letter") encodes another type of "letter," the "amino acids." There are 20 members of the amino acid alphabet, with names such as aspartate, isoleucine, serine and valine. These are abbreviated as **Asp**, **Ile**, **Ser** and **Val** (and sometimes as single letters; i.e. D, I, S, V). Under instructions received from DNA, the decoding and assembling machinery of cells joins amino acids together in the same order as they are encoded in DNA, to form proteins. The latter, amino acid chains that fold in complicated ways, play a major role in determining how we interact with our environment. Proteins constitute a major part of, for example, the enzymes, which interact with other molecules (substrates) to catalyze chemical changes. The proteins largely determine our phenotype (Figs. 1-1, 1-4).

In an organism of a particular species ("*A*"), consider the twenty one base DNA sequence:

TACGACGCCGATAGCGTCGTA (3.1)

This, read in sets of three bases ("codons"), can convey primary information for a seven amino acid protein fragment (**TyrAspAlaAspSerValVal**). All members of the species will tend to have the same DNA sequence, and differences between members of the species will tend to be rare and of minor degree. If the protein is fundamental to cell function it is likely that organisms of *another* species ("*B*") will have DNA that encodes the *same* protein fragment. However, when we examine their DNA we might find major differences compared with the DNA of the first species (the similarities are underlined):

TATGATGCTGACAGTGTTGTT (3.2)

This sequence *also* encodes the above protein fragment, showing that the DNA contains the *same* primary information as DNA sequence 3.1 above, but it is "spoken" with a different "accent." Such *secondary* information could have a biological role. It is theoretical possible (but unlikely) that all the genes in an organism of species *B* would have the "accent," yet otherwise encode the *same* proteins as species *A*. In this case, organisms of species *A* and *B* would be both anatomically and functionally (physiologically) *identical*, while differing dramatically with respect to secondary information. Biologists sometime refer to species that show this tendency as "sibling species" (*espèces jumelles*, *Geschwisterarten*).

On the other hand, consider a single change in the sequence of species *A* to:

TACGACGCCGTTAGCGTCGTA (3.3)

Here the difference (underlined) would change one of the seven amino acids. It is likely that such *minor* changes in a *very small* number of genes affecting development would be sufficient to cause anatomical differentiation *within* species *A* (e.g. compare a bulldog and a poodle, as "varieties" of dogs). Yet, in this case the secondary information would be hardly changed.

A view to be explored in this book is that, like the Cockney's dropped H's, the role of secondary information is to *initiate,* and, for a while, *maintain*, the reproductive isolation (recombinational isolation) that is essential for the process by which biological species are formed. Secondary information is information that *preserves.* Indeed, such is the promiscuity of DNA that, even *within* the genome of a species, secondary information plays a similar role in maintaining the recombinational isolation of individual genes. Secondary information *preserves* genes (Chapter 8).

Secondary information can exist because the genetic code is a "redundant" or "degenerate" code, as we will see in Chapter 7; for example, the amino acid serine is not encoded by just one codon; there are six possible codons (**TCT**, **TCC**, **TCA**, **TCG**, **AGT**, **AGC**). In DNA sequence 3.1 (species *A*), serine (**Ser**) is encoded by **AGC**, whereas **AGT** is used in DNA sequence 3.2 (species *B*). On the other hand, the change in species *A* from **GAT** (sequence 3.1) to **GTT** (sequence 3.3) changes the encoded amino acid from aspartic acid (**Asp**) to valine (**Val**), and this might suffice to change the properties of the corresponding protein, and hence to change one or more of the anatomical or physiological characters of an organism (its phenotype) that depend on that protein. As a consequence of the degeneracy of the genetic code, given a nucleic acid sequence one can deduce the corresponding protein, but not the converse. Thus, in this context, information is doomed to flow one way from nucleic acid to protein.

Information Barriers

That it might be appropriate to draw an analogy between the evolution of languages and of species has long been recognized [4]. A biological interest of linguistic barriers is that they also tend to be reproductive barriers. Even if a French person and an English person are living in the same territory (biologists call this living "sympatrically"), if they do not speak the same language they are unlikely to marry. Of course, being of the same species, if there were not this early barrier to reproduction, then the union would probably be fruitful in terms of the number and health of children. Thus, if this early barrier were overcome, there would probably be no later barriers.

However, the French tend to marry French and produce more French. The English tend to marry English and produce more English. Even in England, because of the class barriers so colorfully portrayed by George Bernard Shaw in *Pygmalion*, Cockneys tend to marry Cockneys, and the essence of the barrier from people speaking "regular" English is the difference in accent [5]. Because of other ("blending") factors at work in our society, it is unlikely that this linguistic "speciation" will continue to the extent that Cockney will eventually become an independent language.

The point is that when there is incipient linguistic "speciation," it may be the *secondary* information (dropped H's), rather than the primary information, which constitutes the barrier. Similarly, secondary information in DNA may provide a species barrier that keeps organisms reproductively separate as distinct quantal "packages." Organisms are reproductively isolated (i.e. only able to reproduce sexually with members of their own species). For purposes of reproduction, members of a species assort with their own type. An organism that breeds with a member of an allied species may sometimes produce healthy offspring, but these offspring are usually sterile ("hybrid sterility").

Secondary information in DNA can be considered as part of the "genome phenotype," which in the context of reproduction, we may refer to as the "reprotype." In contrast, primary information in DNA is concerned with conventional phenotypic attributes, often dependent on proteins, such as the color of a flower, or the speed at which a horse can run. It is with such attributes that classical Darwinian natural selection is mainly concerned.

Barriers Preserve Discontinuity

If you fully understanding a concept then you should readily be able to deduce its implications. While it can be argued that the Victorians understood the information concept (Chapter 1), with the exception of Butler they did not explore what was implicit in the language they spoke and the texts they read and wrote. For example, information can be inaccurate; it may be accurate at one moment at one place, but at a moment later, at the same or a different place, it may be inaccurate. A concept of information that does not include the concept that information is subject to error is an incomplete concept of information.

Galton in 1876 compared gametes to "mailbags" containing "heaps of letters," and regretted that, with current technology biologists were like persons looking through the windows of a post office; they "might draw various valuable conclusions as to the postal communications generally, but they cannot read a single word of what the letters contain." Noting that the "germs" from which bodily units were "derived" should have independently varied in each parent, and hence become "deficient" in different ways, Galton saw an advantage of sexual reproduction [6]:

> "When there are two parents, and therefore a double supply of material, the chance deficiency in the contribution of either of them, of any particular species of germ [gene], tends to be supplied by the other. No doubt, cases must still occur, though much more rarely than before, in which the same species of germ is absent from the contribution of both, and a very small proportion of families will thereby perish."

Butler said this more colorfully in 1877 when pondering what sex was for [7]:

> "We should expect to find a predominance of sexual over asexual generation, in the arrangements of nature for continuing her various species, inasmuch as two heads are better than one, and a *locus poenitentiae* is thus given to the embryo – an opportunity of correcting the experience of one parent by that of the other. And this is what the more intelligent embryos may be supposed to do; for there would seem little reason to doubt that there are clever embryos and stupid embryos, with better or worse memories, as the case may be, of how they dealt with their protoplasm before, and better or worse able to see how they can do better now."

The possibility of error did not escape Miescher (Fig. 3-1) to whom we will be returning in Chapter 14. In 1892 he suggested that sex might have arisen to correct structural defects in molecules [8]:

> "To me the key to the problem of sexual reproduction is to be found in the field of stereochemistry. The 'gemmules' of Darwin's pangenesis are no other than the numerous asymmetrical carbon atoms of organic substances. As a result of minute causes and external conditions these carbon atoms suffer positional changes and thus gradually produce structural defects. Sexuality

is an arrangement for the correction of these unavoidable stereometric architectural defects in the structure of organized substances. Left handed coils are corrected by right-handed coils, and the equilibrium restored."

Fig. 3-1. Johann Friedrich Miescher (1844-1895)

As noted in Chapter 2 when considering Muller's "microchemical accidents," error detection usually requires some form of redundancy, so that one copy can be compared with another. To correct a play by Shakespeare you need another copy of the same play. You cannot correct a copy of *Macbeth* with a copy of *The Tempest*. Furthermore, a blended form, – a "*Macpest*" or a "*Tembeth*" – also would not suffice. Thus, like biological species, the plays must be discrete, or "discontinuous." They must come as packets, but not clonal packets. If you do not believe something in your newspaper you do not go out and buy another copy of the same newspaper which, having been produced as part of the same print run, contains identical information with identical errors.

The property of discreteness implies that there are limits. Shakespeare was able to explore the theme of personal ambition in *Macbeth*. A separate play was required to explore the theme of shipwreck on an enchanted island. The diversity of Shakespeare's plays can be compared with the diversity of biological species. Within each species there is room for change, but within limits. A species can attain a certain type and level of complexity, but for further change barriers must be breached and new barriers established. There must be an origin of species. Butler related this to information [7]:

"Many plants and animals do appear to have reached a phase of being from which they are hard to move – that is to say, they will

> die sooner than be at pains of altering their habits – true martyrs to their convictions. Such races refuse to see changes in their surroundings as long as they can, but when compelled to recognize them, they throw up the game because they cannot and will not, or will not and cannot, invent. This is perfectly intelligible, for a race is nothing but a long-lived individual, and like any individual, or tribe of men whom we have yet observed, will have its special capacities and its special limitations, though, as in the case of the individual, so also with the race, it is exceedingly hard to say what those limitations are, and why, having been able to go so far, it should go no further. Every man and every race is capable of education up to a point, but not to the extent of being made from a sow's ear into a silk purse."

In Scotland in 1867 a professor of engineering, Fleeming Jenkin, saw members of a biological species as enclosed within a sphere (Fig. 3-2). For a particular character, say potential running speed, most members would be at, or near, the centre of the sphere, since most members are capable of average speed. Jenkin noted that, under the then prevailing view that parental characters were blended in offspring, there would always be a tendency for individuals to vary centripetally, rather than centrifugally [9]:

> "A given animal ... appears to be contained ... within a sphere of variation; one individual lies near one portion of the surface; another individual, of the same species, near another part of the surface; the average animal at the centre. Any individual may produce descendents varying in any direction, but is more likely to produce descendents varying towards the centre of the sphere, and the variations in that direction will be greater in amount than the variations towards the surface. Thus a set of racers of equal merit indiscriminately breeding will produce more colts and foals of inferior than of superior speed, and the falling off of the degenerate will be greater than the improvement of the select."

This "regression to the mean," given much weight by Galton, was invariably observed when individuals of mixed pedigree were crossed. To maintain the range of variant forms (i.e. to maintain racers of superior speed), ongoing variation would have to replace those lost through regression. Johannsen was later to show that if an individual of unmixed pedigree ("pure line") were crossed with one of the same pedigree, then a position near the surface of the sphere could be sustained. This was in keeping with Mendel's laws (see Chapter 7). However, even pure lines could not escape beyond the surface of the sphere.

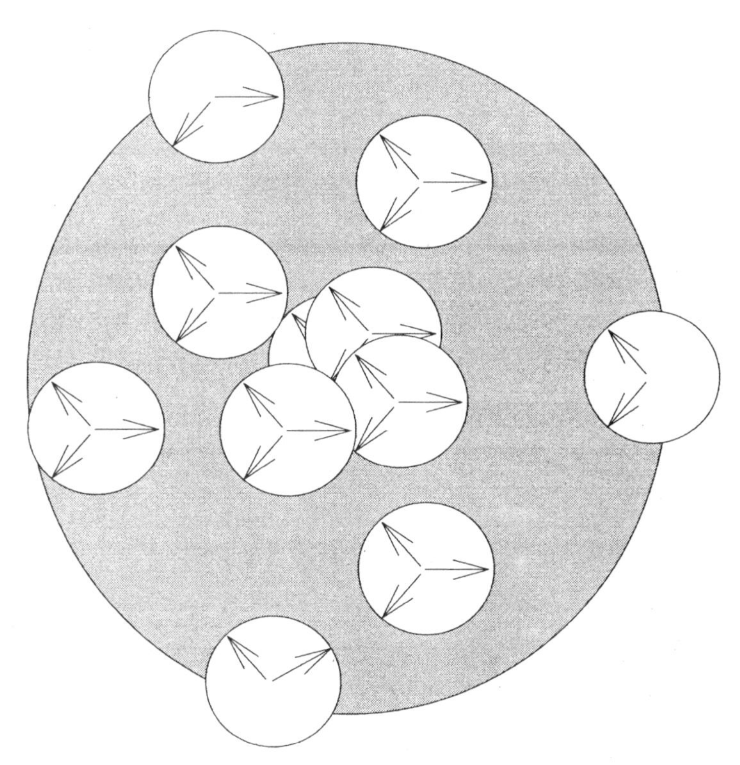

Fig. 3-2. Jenkin's sphere of species variation (grey). Members of a species and the ranges of variation seen in their children are represented as white circles, with arrows representing the directions of variations. Most individuals are clustered at the centre (average phenotypes) where they are free to vary in any direction. Those at the precise center can only vary away from the center (i.e. centrifugally). The few individuals at the periphery (rare phenotypes with exceptionally positive or negative qualities), tend to vary towards the center (i.e. centripetally). Thus their children cannot escape the confines of the species (i.e. their exceptional qualities cannot develop further). Note that the variations Jenkin refers to reflect differences in nature, not nurture. He is not referring to variations acquired through interaction with the environment within an individual lifetime (e.g. increased muscular strength in an individual that exercises)

To Select is Not To Preserve

Johannsen in 1909 suggested the word "gene" as a *Rechnungseinheiten*, or accounting unit, that symbolized one of Bateson's "character units." In most definitions of the "gene" there is a loose or explicit reference to function. Biologists talk of a gene encoding information for tallness in peas. Biochemists talk of the gene encoding information for growth hormone, and relate this to a segment of DNA. One popular biochemical text states that "each region of the DNA helix that produces a functional RNA molecule constitutes a gene" [10]. Another states that "a gene is defined biochemically as that segment of DNA (or in a few cases RNA) that encodes the information required to produce a functional biological product" [11]. However, before it can function, information must be preserved. Classical Darwinian theory proposes that function, through the agency of natural selection, is itself the preserving agent (see Chapter 7). Thus, function and preservation go hand-in-hand, but function is held to be more fundamental than preservation.

There is nothing so strange in this. Shakespeare's plays entertained far better than plays by other authors, and by virtue of this function they are with us to this day. But the physical preservation of the plays, the maintenance of their physical discontinuity from their surroundings, required much more – durable ink and paper, and the bindings and cover that make a folio or book. This was not all. The folios and books had to be carefully housed. The selection and reproduction of ancient texts by generations of priests and scribes came to nothing when the great library of Alexandria was burned to the ground. It is only by references from other sources that we now know the magnitude of our loss. The epic cycle poems of Homer's contemporaries ("The Little Iliad," "The Destruction of Troy"), some of the plays of Sophocles, and the works of ancient historians such as Hecataeus, Hellanicus, and Ephorus, are preserved only in fragments by later authors.

Bringing the metaphor closer to home, Mendel's ideas lay fallow for 35 years until their discovery in 1900. Initially, the ideas were in Mendel's head and in a certain issue of the obscure journal *Verhandlungen des naturforschended Vereines in Brunn*, copies of which were distributed to many academic centers. Some copies were lost. Mendel died in 1884. But some of the copies were bound into journals. For 35 years Mendel's ideas were held between pages 3 to 47 in volume four of the journal, not because their value was recognized, but because the volumes were securely bound and stably housed. The preservation of the journal was a *precondition* for the positive selection of its contents.

Eventually Mendel's ideas ("mnemes," or "memes") [12] gained a life of their own. The texts that contained them were sought out, reproduced, and translated into many languages, and the ideas began to spread from head to head. An interesting side-product was that the *Verhandlungen des naturforschended Vereines in Brunn*, and in particular the issue containing Mendel's paper, gained increased attention. Consequently, other papers in that issue may have got a second reading, but not by virtue of their own intrinsic merits. In essence, they, like the journal, were able to "hitch-hike" on the success of Mendel's paper (see Chapter 7).

So, in biological systems, which is more important, continuing demand for function, or actual physical preservation? This book argues that preservation is often more fundamental than function. Furthermore, secondary information (context) is an agency, distinct from natural selection, which brings this about. Once biological information has been preserved then its meaning can be displayed and its accuracy checked.

Another Order of Variation

William Bateson was concerned with the discreteness of biological species as distinct entities and the relationship of this discreteness to that of the environments they occupied. Thus fish species live in water and non-fish species live on land. The two species groups are discrete, as are their respective environments. So the discontinuity of the species matches the discontinuity of their environments. But in the case of allied species – likely to have diverged relatively recently from a common ancestral species – often differences in environments do not match the observed differences in species. Allied species are often found in the same territory (i.e. they are sympatric). Conversely, members of a single species sometime thrive in a wide range of environments. In 1894 the 33-year-old Bateson noted in his book *Materials for the Study of Variation Treated with Especial Regard to Discontinuities in the Origin of Species* [13]:

> "The differences between Species on the whole are Specific, and are differences of kind, forming a discontinuous Series, while the diversities of environment to which they are subject are on the whole differences of degree, and form a continuous Series; it is therefore hard to see how the environmental differences can thus be in any sense the directing cause of Specific differences, which by the Theory of Natural Selection they should be." [Bateson's capitalization]

The preservation of a species did not seem readily explicable on the basis of conventional natural selection, often a relatively continuous environmental force capable of bringing about profound changes, but usually *within the confines of a species*. For the discontinuity of species, Bateson believed there should be a discontinuity in the way in which members of a species vary from one another.

His use of the term "discontinuity" can perhaps be best understood if we first consider the term "continuity." If a road is continuous then *one process* – walking, running or riding – should suffice to reach a destination. If a road is discontinuous, perhaps because there is an unbridged river or ravine, then *another process* – wading, swimming or climbing – must intervene. If variation is continuous then, in principle, every possible intermediate between two forms can now exist, and one process that has removed intermediates that do not match their environment suffices to explain the distinctness of diverse of types we see around us. In short, Darwin's natural selection reigns. If variation is discontinuous then each of the two forms is limited in the extent to which it can now vary. Every possible intermediate between the two forms cannot now exist (although intermediates may have existed in the past when the forms were closer, having just begun to diverge from a common ancestor). So, in addition to natural selection, another process that has something to do with the gap between forms, is needed. However, at the time Bateson could describe this only in very abstract terms:

> "Upon the received hypothesis it is supposed that Variation is continuous and that the Discontinuity of Species results from the operation of Selection. ... There is an almost fatal objection in the way of this belief, and it cannot be supposed both that all variation is continuous and also that the Discontinuity of Species is the result of Selection. With evidence of the Discontinuity of Variation this difficulty would be removed." [Bateson's capitalization]

The need for some other order of variation had been perceived earlier by Jenkin [9]:

> "The theory advanced [by Darwin] appears rather to be that, if owing to some other qualities a race is maintained for a very long time different from the average or original race (near the surface of our sphere), then it will spontaneously lose the tendency to relapse, and acquire a tendency to vary outside the sphere. What is to produce this change? Time simply, apparently. ... Not only do we require for Darwin's theory that time shall first permanently fix the variety near the outside of the assumed sphere of variation, we require that it shall give the power of varying beyond that sphere."

In his 1869 book *Heredity Genius*, Galton likened this power of varying beyond the sphere to the effort needed to roll a rough stone from one position of stability to another [14]. Initially, Bateson examined the observable characteristics of a wide range of plants and animals to catalogue the range of variations that could occur. With the emergence of Mendel's laws on quantitative aspects of variation, which demonstrated the discontinuity of genes (i.e. the discreteness of the characters they encoded; see Chapter 7), it was tempting think that this genic discontinuity might be the evidence he sought. However, although the discontinuity of genes clarified many issues, Bateson came to the view that a different type of discontinuity was needed.

From the discontinuity of genes it followed that if a *rare* new mutant organism crossed with one of the *abundant* non-mutant forms in its species (the most likely occurrence), then the emergent mutant character would not immediately be diluted due to the blending of parental characters. The character's increase or decrease in the population would be statistically determined, depending on (i) its degree of dominance (see Chapter 7), (ii) whether it was seen by natural selection as beneficial or detrimental, and (iii) whether at some point, perhaps after preservation for many generations, there was a cross with another organism with the same mutation. This latter point was emphasized by Jenkin [9]:

> "Any favourable deviation must ... give its fortunate possessor a better chance of life; but this conclusion differs widely from the supposed consequence that a *whole* species may or will gradually acquire some one new quality, or *wholly* change in *one* direction and in the *same* manner."

Easy access to a mate with the same mutation had been a strength of the Lamarckist doctrine of the inheritance of acquired characters. According to Lamarck (and Butler) organisms subjected to the same environmental provocation would *simultaneously* adapt by mutation. Being colocalized in the same environment there would then be little difficulty in their finding a mate with the same adaptation. In this respect, the discovery of genic discontinuity, raising the possibility of preservation for many generations, was another nail in the Lamarckist coffin.

Context as Barrier

Some mutations bring about dramatic, sometimes large scale, changes in a single generation. For example, dwarfism, extra fingers, and a leg where a fly's antenna should be (see Chapter 7). Although initially misled into believing that these "monstrosities" or "sports" might be examples of the discontinuity he sought, William Bateson came to realized that the process of variation had to generate a *non-genic* discontinuity as a barrier that would preserve species (see Chapter 8). Yet he did not see this discontinuity as a barrier that might relate to error detection and correction. His thoughts on evolution greatly influenced his son Gregory (named after Mendel; Fig. 3-3), who became an anthropologist/psychologist much interested in the hierarchy of information levels in human communication.

In *Steps to an Ecology of Mind*, Gregory Bateson claimed his father's heros (Voltaire, Butler) as his own, and pointed to the all-important role of context [15]. He noted that schizophrenic patients often appear unable to recognize the context in which primary information is conveyed. More simply stated, there is no smile or wink acting as a "classifying metamessage." This work of the younger Bateson and his colleagues is well summarized by his biographer [16]:

"Assuming the view that in communication there are primary messages which are verbal and classifying messages which express the on-going relationship of the actors, they proposed that the discourse of schizophrenics especially distorts or omits the signals about relationship. ... The ability to discriminate the various contextual cues which specify both the semantic and personal meaning of a message seemed to be impaired. Human communication was construed in terms of multiple levels of reference, ... The problem which they posed for the schizophrenic was how does a simple message *mean* if it is not pinned down by normal metamessages."

Figure 3-3. Gregory Bateson (1904-1980)

Gregory Bateson extrapolated this to biological information transfer, writing [15]:

"Both grammar and biological structure are products of communicational and organizational processes. The anatomy of the plant is a complex transform of genotypic instructions, and the 'language' of the genes, like any other language, must of necessity have contextual structure. Moreover, in all communication there must be a relevance between the contextual structure of the message and some structuring of the recipient. The tissues of the plant could not 'read' the genotypic instructions carried in the chromosomes of every cell unless cell and tissue exist, at that given moment, in a contextual structure."

Gregory Bateson also equated a change in contextual structure to a special type of change in the genomic information required for species divergence that his father had postulated [17]:

> "If evolution proceeded in accordance with conventional theory, its processes would be blocked. The finite nature of somatic changes indicates that no ongoing process of evolution can result only from successive externally adaptive genotypic changes, since these must, in combination, become lethal, demanding combinations of internal somatic adjustments of which the soma is incapable. We turn therefore to a consideration of *other classes of genotypic change*. What is required to give a balanced theory of evolution is the occurrence of genotypic changes, which shall *increase* [Bateson's italics] the available range of somatic flexibility. When the internal organization of the organisms of a species has been limited by environmental or mutational pressure to some narrow subset of the total range of living states, further evolutionary progress will require some sort of genotypic change which will compensate for this limitation."

In Germany, Richard Goldschmidt (Fig. 3-4) had termed this "macroevolution."

Fig. 3-4. Richard Goldschmidt (1878-1958)

In Chapter 7 we will consider Goldschmidt's proposal that, within the confines of a species, the plasticity of the phenotype is limited since changes in genotypic "pattern" (microevolution) are limited. A chromosomal "repatterning" (macroevolution) is required if species are to diverge into new species, so expanding the scope of possible phenotypes. The relationships of some of the dichotomies considered here to Goldschmidt's terminology are indicated in Table 3-1

	Dichotomy	
Information Type	Primary	Secondary
DNA involved	Genic	Non-Genic
Selection Type	Natural	Physiological
Phenotype	Conventional	Genome
Adaptation for:	Function	Preservation
Scope of Variation	Continuous	Discontinuous
Evolution Type	Non-branching	Branching
Goldschmidt Terminology	Microevolution	Macroevolution

Table 3-1. Some of the dichotomies considered in this chapter. "Physiological selection" is a term introduced by Romanes to contrast with "Natural selection." When a particular *genome phenotype* ("reprotype") confers reproductive success or failure, then physiological selection has operated. Similarly, when a particular *conventional phenotype* confers reproductive success or failure, then natural selection has operated (see Chapters 7 and 8)

Not a Palimpsest

Today's metaphors may be tomorrow's truths. As we saw in Chapter 1, Romanes could tolerate Butler's relating heredity with memory *as a metaphor*, but at that time the thought that within cells there might be something akin to a written text was going too far. Metaphors can help us understand our genomes. But they have to be the right metaphors. Some metaphors are just not helpful. Others can actively mislead.

The idea of levels of information invites comparison with palimpsests (Greek: *palimpsestos* = scraped again). In ancient times messages on clay tablets could be scraped off to allow the tablet to be reused. The scrapings may have been incomplete, but sufficient to allow reuse. Likewise, the memory discs of today's computers can have their contents erased so allowing fresh messages to be entered. Again, the erasures may be incomplete. Computers can "crash" and the messages on their discs can be accidentally erased. There are processes by which erased messages can sometimes be restored.

In 1906 in Constantinople a faded tenth century copy of Archimedes' long lost *Method of Mechanical Theorems* was discovered on a parchment beneath the lettering of a twelfth century Greek prayer book [18]. Such a document is also referred to as a palimpsest. With modern technologies some palimpsests are found to have multiple layers of text. Clearly, the writers of the top layers of text had no use for the preceding layers. For their purposes they were garbage and the only lament, if any, would be that they had been insufficiently erased. Translating this mode of thought to the layers of information that have been discovered in genomes, it was easy to assume that, while these may be of interest as possible remnants of ancient messages laid down over billions of years of evolution, they are not of relevance to genomes as they now exist. Thus, the term "junk DNA" did not seem inappropriate.

This view is challenged in Chapter 12. Because we are unable to conjure up a functional explanation for something, it does not follow that it has no function. Following the admonition attributed to William of Occam, we should accept a simple explanation if it fits in with the facts as well as, or better than, a more complex explanation. But we should also heed the admonition attributed to Albert Einstein: "Make everything as simple as possible, but not simpler."

Certainly, from modern DNA sequences we can infer some of the characteristics of ancient sequences, but it does not therefore follow that modern sequences are not strictly functional, serving the needs of modern organisms. A popular form of "molecular archaeology" is to compare the sequences of two species that are considered to have derived from a common ancestral species (phylogenetic analysis). What they have *in common* may have been in the ancestral sequence (i.e. has been conserved). What they do not have in common may have been acquired since their divergence into separate species. For example, modern passenger automobiles and airplanes both serve the practical and aesthetic needs of modern humans. They have certain features in common, such as wheels, windows and at least one forward-looking seat at the front. It is not unreasonable to infer that many "ancient" horse-drawn carriages had these features.

Hybrid Disciplines and Hybrid Sterility

Barriers protect and preserve. People join clubs and societies in order to associate with groups of like-minded types. This is also true for academic disciplines. To become a member of a club, society or discipline, various entry criteria must be satisfied (i.e. barriers must be breached). Here again, if not used with care, metaphors can mislead. When emphasizing the importance of chemistry for understanding genetic function, Crick noted [19]:

> "Classical genetics is ... a black-box subject. The important thing was to combine it with biochemistry. In nature hybrid species are usually sterile, but in science the reverse is often true. Hybrid subjects are often astonishingly fertile, whereas if a scientific discipline remains too pure it usually wilts. In studying a complicated system, it is true that one cannot even see what the problems are unless one studies the higher levels of the system, but the *proof* of any theory about higher levels usually needs detailed data from lower levels." [Crick's italics]

Hybrid subjects are indeed astonishingly fertile, but not quite in the way Crick implied. Horses and donkeys (asses) have been artificially crossed for millennia because the mule that results, although sterile, is endowed with great energy and vigor. Thus, hybrid animals and hybrid disciplines have *in common* the feature that they tend to be *more vigorous* than the corresponding parental animals or disciplines. They are both highly productive, but not highly reproductive. This is because they also have *in common*, a tendency towards sterility, and for much the same reasons.

Within the barrier afforded by an academic discipline members are likely to thrive, provided their activities remains within the confines of the group. Here they are likely to be best understood, so their ideas become subject to constructive criticism and error-correction. However, those who stray beyond the confines of one discipline (in Mendel's case, Biology) and engage with another discipline (in Mendel's case, Physics), may be misunderstood by those in both parental disciplines. Barrier breachers may be vigorously productive, but the communication gap may imperil both the novel ideas they wish to convey and their personal fortunes. They run the risk of an alienation that can disrupt their lives. At least in the short term, they are academic:ally sterile (i.e. their ideas are not reproduced; see Epilogue).

Again, Butler, unique among the Victorians in his articulation of evolution in informational terms, had some sense of this [7]. First he considered the case of a cross between two types so disparate that the embryo did not develop correctly ("hybrid inviability"):

> "We should expect to find that all species, whether of plants or animals, are occasionally benefited by a cross; but we should also expect that a cross should have a tendency to introduce a disturbing element, if it be too wide, inasmuch as the offspring would be pulled hither and thither by two conflicting memories or advices, much as though a number of people speaking at once were without previous warning to advise an unhappy performer

> to vary his ordinary performance. ... In such a case he will ... completely break down, if the advice be too conflicting, ... through his inability to fuse the experiences into a harmonious whole."

Then, Butler considered the reproductive events leading, in turn, to a parental male horse, to a parental female donkey, and finally to their child, a mule ("hybrid sterility"):

> "[In] the case of hybrids which are born well developed and healthy, but nevertheless perfectly sterile, it is less obvious why, having succeeded in understanding the conflicting memories of their parents, they should fail to produce offspring. ... The impregnate ovum from which the mule's father was developed remembered nothing but its horse memories; but it felt its faith in these supported by the recollection of *a vast number* of previous generations, in which it was, to all intents and purposes, what it now is. In like manner, the impregnate ovum from which the mule's mother [donkey] was developed would be backed by the assurance that it had done what it is going to do now a hundred thousand times already. All would thus be plain sailing. A horse and a donkey would result. These two are brought together; an impregnate ovum is produced which finds an unusual conflict of memory between the two lines of its ancestors, nevertheless, being accustomed to *some* conflict, it manages to get over the difficulty, *as on either side it finds itself backed by a very long series of sufficiently steady memory.* A mule results – a creature so distinctly different from either horse or donkey, that reproduction is baffled, owing to the creature's having nothing but its own knowledge of itself to fall back upon." [Butler's italics]

Sadly, Butler himself turned out to be an academic mule. With a few late-in-the-day exceptions (e.g. William Bateson and Francis Darwin), his ideas were scorned by the scientific establishment (see Epilogue) [20].

Summary

Information is an abstract entity that we recognize in the form of a message conveyed in a medium. Messages are read (decoded) to extract their meanings (information content) and are compared with copies of the same messages to detect and correct errors. These functions require that a message contain both its primary information and a barrier (secondary information) that preserves the context within which the primary information is read both for meaning and for errors. There must be discontinuity between different messages. Just as the themes that can be developed within the context of a single play (e.g. *Macbeth*) are limited, so that new themes require new plays (e.g. *The Tempest*), evolutionary themes that can be developed within the context of a biological species are limited, so that further evolutionary progress requires a barrier-breaching change that establishes a new species. Just as error-correction of the text of *Macbeth* requires another text of *Macbeth* from an independent source, so error-correction of a genome text is best carried out with a genome text from an unrelated member of the same species. Sex is primarily a within-species error-correcting device requiring that species be discontinuous and reproductively isolated from each other. Members of allied species that fail to recognize their context and breach the reproductive barrier may produce sterile offspring (hybrid sterility). Similarly, humans who fail to recognize the context of cognitive discourse may be schizophenic.

Part 2 Parity and Non-Parity

Chapter 4

Chargaff's Second Parity Rule

> "Poetry's unnat'ral; no man ever talked poetry 'cept a beadle on boxin'day, or Warren's blackin' or Rowland's oil, or some o' them low fellows."
>
> Mr. Weller. *Pickwick Papers* [1]

Information has many forms. If you turn down the corner of a page to remind you where you stopped reading, then you have left a message on the page ("bookmark"). In future you "decode" (extract information from) the bookmark with the knowledge that it means, "continue here." A future historian might be interested in where you paused in your reading. Going through the book, he/she would notice creases suggesting that a flap had been turned down. Making assumptions about the code you were using, a feasible map of the book could then be made with your pause sites. It might be discovered that you paused at particular sites, say at the ends of chapters. In this case pauses would be correlated with the distribution of the book's "primary information." Or perhaps there was a random element to your pausing – perhaps when your partner wanted the light out. In this case pausing would be influenced by your pairing relationship. While humans not infrequently get in a flap about their pairing relationships, we shall see that flaps are useful metaphors when we observe DNA molecules in a flap about their pairing relationships.

Information Conflict

The primary information in this book is in the form of the linear message you are now decoding. If you turn down a large flap it might cover up part of the message. Thus, one form of information can interfere with the transmission of another form of information. To read the text you have to correct (fold back) the "secondary structure" of the page (the flap) so that it no longer overlaps the text. Thus, there is a conflict. You can either retain the flap and not read the text, or remove the flap and read the text.

In the case of a book page, the text is imposed on a flat two-dimensional base – the paper. The message (text) and the medium (paper) are different. Similarly, in the case of our genetic material, DNA, the "medium" is a chain of two units (phosphate and ribose), and the "message" is provided by a sequence of "letters" (bases) attached at intervals along the chain (see Fig. 2-4). It will be shown in chapter 10 that, as in the case of a written text on paper, "flaps" in DNA (secondary structure) can conflict with the base sequence (primary structure). Thus, the pressures to convey information encoded in a particular sequence, *and* to convey information encoded in a "flap", may be in conflict. If it is possible to recognize that one form of information (e.g. flap) is in some way higher than another form (e.g. text), then hierarchical levels can be assigned, and it can then be said that there is a conflict not only between different *forms* of information, but also between different *levels* of information. In biological systems where there is competition for genome space, the "hand of evolution" has to resolve these *intrinsic* conflicts while dealing with other pressures (*extrinsic*) from the environment.

Prose, Poetry and Palindromes

Primary information can be conveyed in pidgin English. "Cat sat mat" can convey essentially the same primary information as the more formal "the cat sat on the mat." It is known that the act of sitting is something cats frequently do, and that mats usually occupy lowly places where they are vulnerable to being sat upon. An important function of the syntax of formal prose – the form the primary information would take in normal human discourse – is that it can suggest the possibility that there has been an error in the primary information. The sentence "The cat sat on the mat" is syntactically correct, and on this basis we might judge that the author had really written what he/she had meant to write. However, this comes at a price. "Cat sat mat" takes less time to write or speak, and occupies less space. Definite articles and prepositions can be considered as redundant information. Provided we know that it is cats that do the sitting, the less formal prose of pidgin English has certain efficiencies.

To those whose first language is English this may not be obvious. Like Monsieur Jourdain in Moliere's *Le Bourgeous Gentilhomme*, we discover that we have been speaking and writing a high level formal prose quite effortlessly all our lives [2]. But, in fact, it has not been all our lives. As infants we began by stringing single words together in forms not too different from "cat sat mat," and had to learn, slowly and laboriously (as most parents know), the modifications that led to a higher level of discourse.

An even higher level of discourse would be poetry. The error-detecting function would then be considerably increased. In this case the ear and eye would, from childhood, have been so accustomed to meter and rhyme that, if these elements were missing, the listener or viewer would then be alerted to the possibility of error in the primary information. Yet, as Mr. Weller notes, we do not naturally speak poetry. The advantage of this extra layer of error-detecting ability does not appear to have outweighed its disadvantages.

The same can be said of the writings of Christian Bök [3]. You should be able to spot the potential error-detecting device:

> "Relentless, the rebel peddles these theses, even when vexed peers deem the new precepts mere dreck. The plebes resent newer verse; nevertheless, the rebel perseveres, never deterred, never dejected, heedless, even when hecklers heckle the vehement speeches."

Bök constrains himself to use only the vowel "e". Georges Perec also tried this in 1972 [4], and in an earlier novel he employed all the vowels with the exception of "e" [5]. That an error had occurred in Bok's text, and in Perec's 1972 text, would become detectable if a vowel other than "e" appeared. That an error had occurred in Perec's earlier text would become detectable if "e" appeared.

The constraints become greater when discourse is palindromic (Greek: *palindromos* = running back again):

"Madam I'm Adam."

The sentence reads the same forwards and backwards. If we count letters we get: A = 4, D = 2, I = 1, M = 4. The palindrome is asymmetric with a lone central "I." Otherwise, all the letters correspond to even numbers. Palindromes obey a parity rule.

With sufficient ingenuity you can come up with quite long palindromes. Indeed, they have been known since ancient times. Inscribed in Greek on medieval fountains there is sometimes the admonition "Wash sins, not only your face:"

"ΝΙΨΟΝ ΑΝΟΜΗΜΑΤΑ ΜΗ ΜΟΝΑΝ ΟΨΙΝ"

Here, A = 4, H = 2, I = 2, M = 4, N = 2, O = 4, Ψ = 2, and T = 1. The following, by geneticist Susumo Ohno, is symmetric:

> "Doom! Sad named was dog DNA, devil's deeds lived, and God saw demand as mood."

Here again the parity rule is obeyed (A = 8, D = 10, E = 6, G = 2, I = 2, L = 2, M = 4, N = 4, O = 6, S = 4, V = 2). A check for even parity, which disregarded sentences with lone central letters, should be capable of detecting errors quite powerfully, but at a very high cost. Normal human discourse would be severely constrained if our communication were so limited.

Yet we shall see that this form of human language serves as a good metaphor for the language of our genomes. In Ohno's words: "All DNA base sequences, regardless of their origins or functions (coding versus non-coding) are messages written in palindromic verses" [6]. This indicates that the error-detecting role of the genome language (involving various forms of information that can be referred to as secondary) may be of *more* importance than the immediate efficiencies of communicating primary genetic messages (primary information).

Intrastand Parity

Chargaff's second parity rule is that his first parity rule, which applies to duplex double-stranded DNA, also applies, to a close approximation, to single-stranded DNA. If the individual strands of a DNA duplex are isolated and their base compositions determined, then **A%** ≅ **T%**, and **G%** ≅ **C%** [7]. This also means that the "top" and "bottom" strands of duplex DNA have approximately the same base composition. For example, if we arrange for base parity in a single "top" strand:

5' **AAAAACCCGGGTTTTT** 3' (4.1)

Then, following Chargaff's first parity rule, the complementary bottom strand must have the same base composition:

3' **TTTTTGGGCCCAAAAA** 5' (4.2)

The validity of the second parity rule became clearer when complete genomic sequences became available. For example, the cowpox virus (Vaccinia virus; Latin, *vacca* = a cow) that is used to vaccinate humans against smallpox, has in its "top" strand, 63,921 **A**s, 63,776 **T**s, 32,030 **G**s, and 32,010 **C**s, for a total of 191,737 bases. Cut the strand into two equal halves and the rule still holds – **A%** ≅ **T%** and **G%** ≅ **C%**. Cut each of these halves into halves, and again the rule holds. Indeed, it holds, albeit less precisely, even when the divisions continue down to segments of a few hundred bases [8].

To simplify, the relationship can be expressed as **(A+G)%** = **(C+T)%**, or **R%** = **Y%**. Since the proportions of purines (**R**) and pyrimidines (**Y**) are reciprocally related (i.e. if a base is not an **R** it must be a **Y**, and vice-versa), then the degree of parity can be expressed as **R%** or **(A+G)%**, with a value of 50% representing precise parity (i.e. **R%** = **Y%**). This will be considered further in Chapters 6 and 9.

It is easy to dismiss the second parity rule as a product of random forces. For example, in many species there are approximately equal numbers of males and females. This sexual parity appears to follows from the fact that in one of the sexes there are usually two types of gamete (e.g. in human males there are X-chromosome bearing spermatozoa and Y-chromosome bearing spermatozoa; see Chapter 14). These have approximately equal opportunities of encountering a gamete of the opposite sex (e.g. the female X-chromosome-bearing ovum; see Fig. 1-3). Thus, on a chance basis, there should be equal numbers of males (XY) and females (XX). However, chance is supplemented by powerful evolutionary forces [9, 10].

Imagine that, say, after a war, there were more females than males. Then there would be more potential partners for the rare sex (male). Each male would, on average, come to leave more offspring than each female (polygamy being permitted). Biologists call this "frequency-dependent selection." Now, whereas most families have a genetic tendency to produce equal numbers of males and females, some families have a genetic tendency to produce more males and some have a genetic tendency to produce more females (for an example, see Chapter 14). When males are the rare sex, families with a genetic tendency to produce more males would leave more offspring. Hence the proportion of males in the population would increase. When the male/female ratio was restored, then families tending to produce more males would no longer have this advantage, and their genetic characteristics, *including the tendency to produce more males*, would no longer gain a greater representation in future populations. A similar converse argument would apply if there were a population fluctuation causing excess females. We shall see that, as in the case of sexual parity, Chargaff's second parity rule is the result of randomness supplemented by powerful evolutionary forces.

If the second parity rule holds for the two pairs of complementary bases (**A** and **T**, **G** and **C**), what about the ten pairs of complementary base duplets, or the 32 pairs of complementary base triplets, or the 128 pairs of complementary base tetruplets, or ... ? But we are getting a little ahead of ourselves. Let us first sort out what we mean by complementary base duplets (dinucleotides), triplets (trinucleotides), tetruplets (tetranucleotides), quintruplets (pentanucleotides), etc. (Table 4-1).

Polarity and Complementarity

Sentences in English show polarity (i.e. they have a beginning and an end), and are read sequentially from left to right. Following this, by convention we write the "top" strand of a DNA duplex from left to right. The strand has polarity beginning on the left at what, for chemical reasons, is referred to as the 5' ("five prime") end, and terminating at the right at what, for chemical reasons, is referred to as the 3' ("three prime") end (see Fig. 2-4). The top strand is the strand that is recorded in databases, such as the GenBank database maintained at the National Center for Biotechnological Information (NCBI) in Washington, DC. It should be noted that although only one strand is recorded in Washington there is global redundancy, since the same sequence is stored at equivalent sites in Europe and Japan.

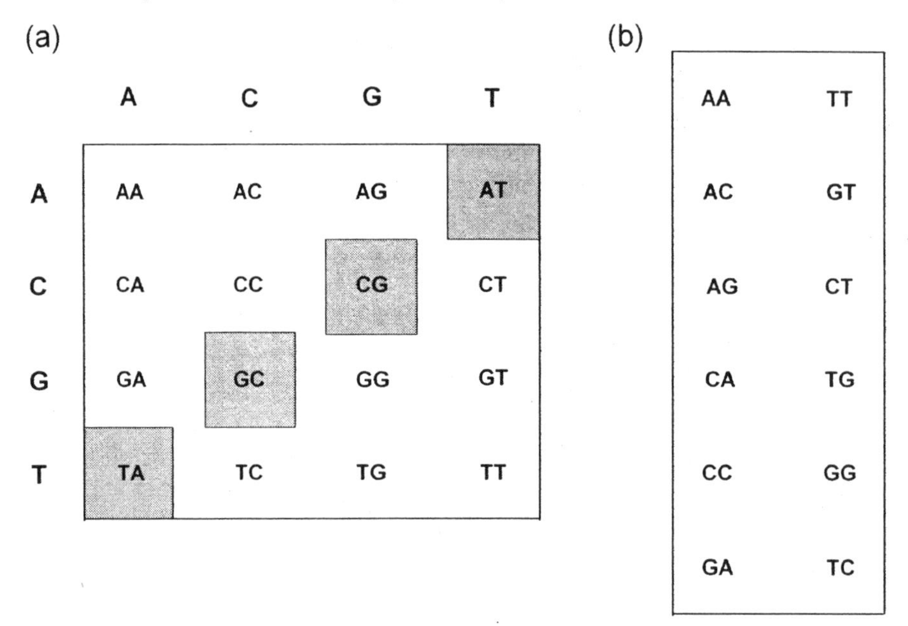

(a)

	A	C	G	T
A	AA	AC	AG	AT
C	CA	CC	CG	CT
G	GA	GC	GG	GT
T	TA	TC	TG	TT

(b)

AA	TT
AC	GT
AG	CT
CA	TG
CC	GG
GA	TC

Table 4-1. Given that any base (**N**) can accompany any of the four bases (**A**, **C**, **G**, **T**) to form a dinucleotide (i.e. **NA**, **NC**, **NG**, **NT**), then there are 16 possible dinucleotides (a), which include 6 pairs of complementary dinucleotides (b). The four dinucleotides **TA**, **GC**, **CG** and **AT** (in boxes) have themselves as complement (e.g. the reverse complement of **TA** is **TA**, and so **TA**s on one strand have **TA**s opposite them on the other strand). Thus, there are 10 complementary dinucleotide pairs, of which four are "self-complementary"

Watson and Crick found that the chemical constraints on their double helix model were best met if the bottom strand was in opposite orientation to the top strand. So a duplex is like a two-lane highway. The duplex sequence written in Chapter 2 can be labeled to show "antiparallel" strand polarities:

"top" 5'**TACGACGCCGATAGCGTCGTA** 3' (4.3)
"bottom" 3'**ATGCTGCGGCTATCGCAGCAT** 5'

Under certain circumstances, this double helix can form "flaps." The sequences of these two strands have been pre-arranged so that they each can form stem-loop secondary structures due to pairing with complementary bases in the *same* strand. For example, if we unpeel the top strand away from the above two-stranded duplex, it can fold into the following form with two major elements, a stem and a loop:

```
                 C
5'TACGACGC         G        (4.4)
                    A
3'ATGCTGCG         T
                 A
```

For this "flap" in DNA to occur there have to be matching (complementary) bases. The stem consists of paired bases. Only the bases in the loop (**CGATA**) are unpaired in this structure. In the loop there happen to be equal numbers of **C**s and **G**s, but there is an extra **A**. It can be said that, in this case Chargaff's parity rule applies, to a close approximation, to a *single strand* of DNA (i.e. the second parity rule; in this case, **A**=5, **C**=6, **G**=6 and **T**=4).

When one examines single strands of duplex DNA from whatever biological source, one invariably finds that the second parity rule applies. It turns out that, just as I arranged with my word-processor for the above sequence to have regions where the base sequences complement, so throughout evolutionary time the "hand of Nature" has arranged for biological sequences to have similar base arrangements. Look again at the sequence:

```
  "top"    5'TACGACGCCGATAGCGTCGTA 3'     (4.5)
"bottom"   3'ATGCTGCGGCTATCGCAGCAT 5'
```

Here the trinucleotide **GTA** in the top strand (underlined) has as its complement **TAC** in the bottom strand (underlined; Table 4-2). Both are, by convention, written in the sentence preceding this one with 5' to 3' polarity. The strands of duplex DNA being antiparallel (see Fig. 2-4), whenever there may be any doubt, the complement is referred to as the "*reverse* complement." Note that **CAT** in the bottom strand (underlined), written *in this particular instance* (i.e. in the sentence you are now reading) with 3' to 5' polarity, would be the "*forward* complement" of **GTA** in the top strand, if it were also in the top strand and were hence written with 5' to 3' polarity.

Considering the top strand alone, **TAC** at the beginning of the sequence is the *reverse* complement of **GTA** at the end of the sequence, meaning that its three bases could potentially base pair with the complementary three bases in **GTA** to form the stem in a stem-loop secondary structure (see sequence 4.4 above). Given the tendency of DNA sequences to form palindromes (see Chapter 5), single-stranded DNAs from most biological sources tend to have equifrequencies of inverse complements (e.g. **GTA** and **TAC**), but equifrequencies of forward complements (e.g. **GTA** and **CAT**) are much less apparent [11].

This was formally described as a "symmetry principle" by Indian physicist Vinayakumar Prabhu in 1993 [12]. Referring to dinucleotides, trinucleotides, tetranucleotides, etc., as "2-tuples," "3-tuples," "4-tuples," etc., he wrote:

	A	C	G	T
AA	AAA	AAC	AAG	AAT
AC	ACA	ACC	ACG	ACT
AG	AGA	AGC	AGG	AGT
AT	ATA	ATC	ATG	ATT
GA	GAA	GAC	GAG	GAT
GC	GCA	GCC	GCG	GCT
GG	GGA	GGC	GGG	GGT
GT	GTA	GTC	GTG	GTG
CA	CAA	CAC	CAG	CAT
CC	CCA	CCC	CCG	CCT
CG	CGA	CGC	CGG	CGT
CT	CTA	CTC	CTG	CTT
TA	TAA	TAC	TAG	TAT
TC	TCA	TCC	TCG	TCT
TG	TGA	TGC	TGG	TGT
TT	TTA	TTC	TTG	TTT

Table 4-2. Given that any of the 16 dinucleotides can accompany any of the four bases to form a trinucleotide, there are 16 x 4 = 64 trinucleotides (3-tuples). Each of these can pair with one of the others (e.g. **GTA** and **TAC**, in boxes), so there are 32 possible pairs of reverse complementary trinucleotides. Similar tables can be generated for the 64 x 4 = 256 tetranucleotides (4-tuples), and for the 256 x 4 = 1024 pentanucleotides (5-tuples), etc

> "A study of all sequences longer than 50000 nucleotides currently in GenBank reveals a simple symmetry principle. The number of occurrences of each n-tuple of nucleotides in a given strand approaches that of its complementary n-tuple in the same strand. This symmetry is true for all long sequences at small n (e.g. n = 1,2,3,4,5). It extends to sets of n-tuples of higher order n with increase in length of the sequence."

The demonstration of symmetry depended on the length (n) of the sequence under study. That this should be so is evident when one thinks about our written language. Members of the word-pairs "rat"–"tar," and "dog"–"god," can, for present purposes, be considered as complementary (i.e. if there were sense-antisense pairing, one would match the other, letter by letter). If one took a text of only a few paragraphs one might, on a chance basis, come across a disproportionate number of instances of one of these (say 95% rat and 5% tar). In a longer text one would arrive at a more proportionate number (say 60% rat and 40% tar), which might hold for similar texts of similar lengths. For "dog" and "god" it is likely that the proportionate number in the same texts would differ (say 75% dog and 25% god). In a veterinary text the proportions might be 90% dog and 10% god, and the reverse in a religious text.

The amazing thing about the DNA language is that the proportions are often close to 50% and 50% (i.e. Chargaff's second parity rule), and this holds in *all* "texts" examined for *all* pairs of complementary "words." Tuple pairs of higher order than "rat" and "tar,"are not easy to find in our language (e.g. "lager" and "regal"). One would have to search extremely long texts for these words to arrive at a proportionate number that might hold for other texts of similar length. Similarly, one has to search very long DNA texts to approximate 50%:50% symmetry for high order n-tuples. But if a long enough sequence is available, this is found.

Duplets and Triplets

It follows that if the frequency of each n-tuple in a given sequence is plotted against the frequency of its complementary n-tuple in the same sequence, then the data points fall close to the diagonal. Figure 4-1a shows this for single bases (1-tuples) in Vaccinia virus (the values used are in the text above). The four bases form two complementary base pairs, so there are only two points on the graph. A line through the two points extends back to zero. More points could be produced for the graph by taking base composition values from other organisms. However, since different organisms, or segments from different organisms, often have different absolute numbers of bases, it is better to present values as percentages of the total number of bases (Fig. 4-1b).

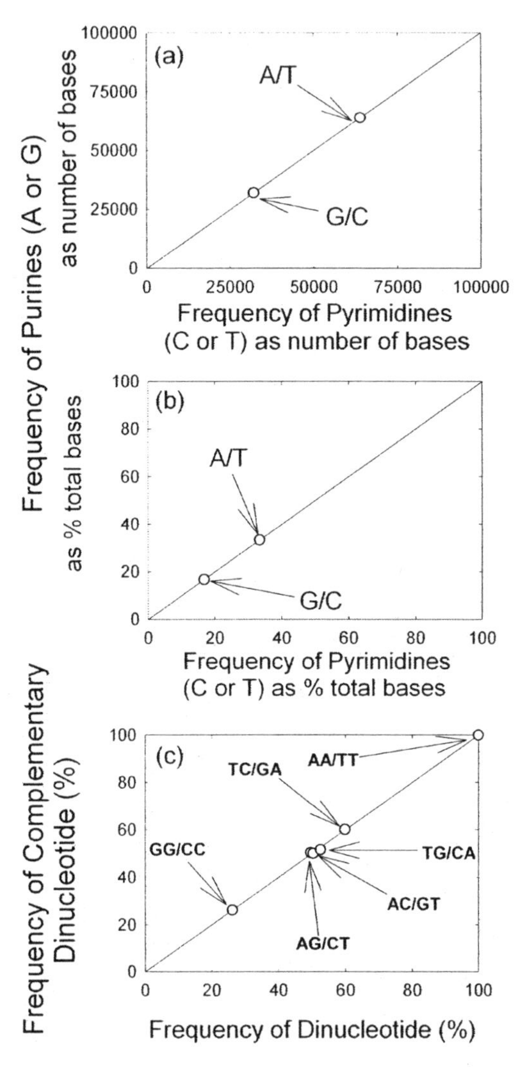

Fig. 4-1. Equifrequencies of complementary bases *(a, b)*, and dinucleotides *(c)* in the "top" strand of the Vaccinia virus genome (191737 bases). Since there are only 2 pairs of complementary bases, there are only two points in *(a)* and *(b)*. Complementary dinucleotide pairs (see Table 4-1b) generate six points in *(c)*. Location on diagonals signifies equifrequency. In *(c)* dinucleotide frequencies are expressed as a percentage of the frequency of the most abundant dinucleotide (**AA**). Note that in *(a)* and *(b)* pyrimidines are assigned to the X-axis, and purines are assigned to the Y-axis. This is not possible with dinucleotides since some dinucleotides contain both a pyrimidine and a purine. In this case assignment to one or other axis is arbitrary

Even better, the values can be presented as percentages of the frequency of the most abundant base. This is shown for the frequencies of six complementary dinucleotide pairs of Vaccinia virus in Figure 4-1c. Again, a line through the data points extends (extrapolates) back to zero. Thus, complementary dinucleotides (2-tuples) are present in approximately equal quantities. This also applies to complementary trinucleotides (3-tuples; Fig. 4-2a).

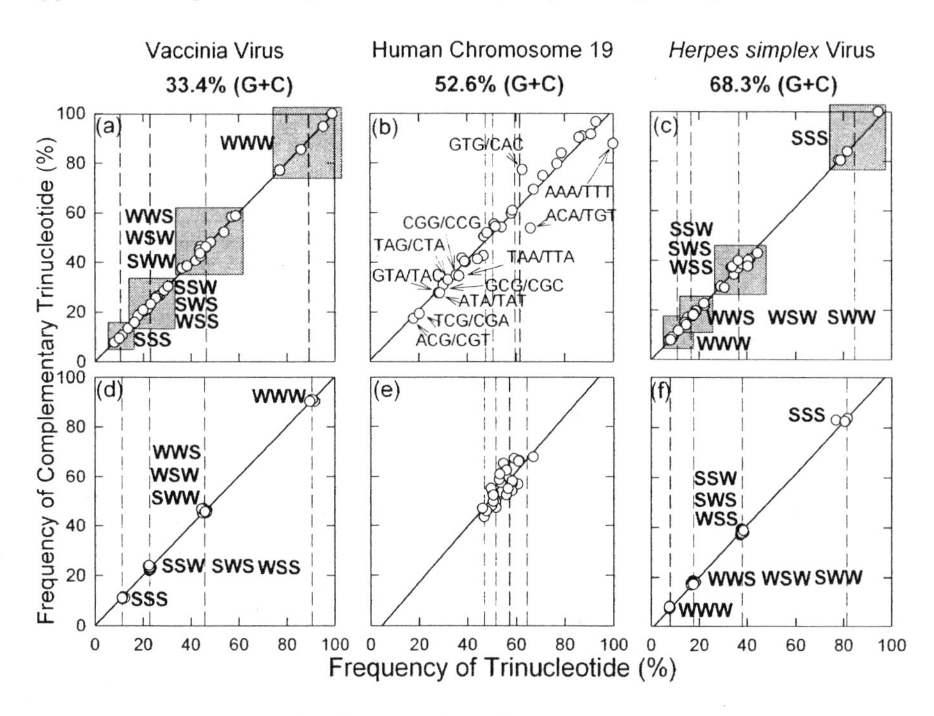

Fig. 4-2. Approximate equifrequencies of complementary trinucleotides in the natural *(a, b, c)* and randomized versions *(d, e, f)* of the "top" strand of the Vaccinia virus genome *(a, d)*, a human chromosome 19 segment *(b, e)*, and the *Herpes simplex* virus genome *(c, f)*. Frequencies are expressed as percentages of the most abundant trinucleotide in the natural sequence. Each member of the **32** sets of trinucleotide pairs is assigned to either the X-axis or the Y-axis in a standardized way (e.g. in the case of the point marked "**GTG/CAC**" the frequency of **GTG** refers to the Y-axis, and the frequency of **CAC** refers to the X-axis). "**W**" represents either **A** or **T**, and "**S**" represents either **G** or **C**. In *(a)* and *(c)* boxes surround points corresponding to the **8** members of the $\mathbf{W_3}$ group of trinucleotides (4 complementary pairs), the **24** members of the $\mathbf{W_2S}$ group, the **24** members of the $\mathbf{S_2W}$ group, and the **8** members of the $\mathbf{S_3}$ group. Vertical dashed lines indicate the average frequency among members of each group. The best line fitting the 32 points (regression line) forms the diagonal

Figure 4-2 shows that the relationship holds for 3-tuples in a low **(G+C)%** genome (Vaccinia virus), in an intermediate **(G+C)%** genome (a segment of human chromosome 19), and in a high **(G+C)%** genome (*Herpes simplex* virus). In the low **(G+C)%** genome there are many **AT**-rich 3-tuples (i.e. rich in the **W**-bases) and few **GC**-rich 3-tuples (i.e. rich in the **S**-bases; Fig. 4-2a). So, 3-tuples of general formula **WWW** ($\mathbf{W_3}$) are abundant (e.g. **ATA** with its complement **TAT**), and 3-tuples of general formula **SSS** ($\mathbf{S_3}$) are not abundant (e.g. **GCG** with its complement **CGC**). The converse applies to the high **(G+C)%** genome (Fig. 4-2c). In the case of the intermediate **(G+C)%** genome, 3-tuples corresponding to the extremes ($\mathbf{W_3}$ and $\mathbf{S_3}$) are more generally distributed (Fig. 4-2b).

2-tuples (dinucleotides) and 3-tuples (trinucleotides), are examples of oligonucleotides (Greek: *oligo* = few). The presence of a given oligonucleotide (n-tuple) in a sequence is partly a reflection of the overall **(G+C)%** composition of the sequence, and partly a reflection of the order of the bases. Thus, **GCG** would satisfy a pressure for a high **(G+C)%** genome, but the pressure would be equally satisfied by other combinations of **G** and **C** (e.g. by **CGG**, **GGC**, **CGC**; see Appendix 2).

To distinguish pressures on base *composition* (i.e. pressures determining mononucleotide frequency) from pressures on base *order* (i.e. pressures determining oligonucleotide frequencies), sequences can be shuffled (randomized). This maintains the base composition (mononucleotide frequency), but disrupts the natural order of the bases, hence randomizing oligonucleotide frequencies, (which can then be directly predicted from mononucleotide frequency; see later).

When this is done the data points become much less scattered (Figure 4-2d, e, f). Furthermore, as DNA segments depart from 50% **G+C**, there is an increasing influence of base composition (1-tuple frequency) on n-tuple frequency (i.e. oligonucleotide frequency becomes more, but not entirely, predictable from mononucleotide frequency; see later). 3-tuples are distributed among four distinct groups (centred at X-axis values marked by vertical dashed lines in Figure 4-2). These are the $\mathbf{W_3}$, $\mathbf{W_2S}$, $\mathbf{S_2W}$, and $\mathbf{S_3}$ groups [13].

A visual measure of the influence of base order is provided by comparing the degree of scatter along the diagonal of the 32 points for natural sequences (Fig. 4-2 a, b, c), with the degree of scatter along the diagonal of the 32 points for the corresponding shuffled sequences (Fig. 4-2 d, e, f). A clearer measure is provided by plotting the 64 frequencies of trinucleotides of the shuffled sequences against the corresponding 64 frequencies of trinucleotides of the natural sequences (Fig. 4-3). Here the degree of horizontal dispersion provides a measure of the role of base order. There is much more scope for base order in the DNA segment with an intermediate **(G+C)%** value (52.5%;

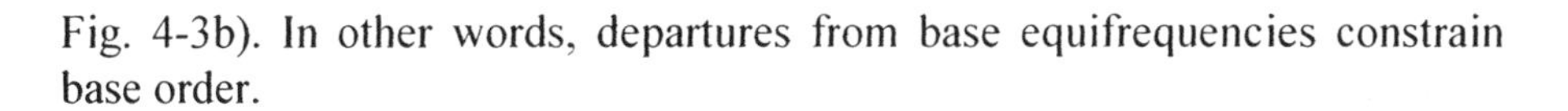

Fig. 4-3b). In other words, departures from base equifrequencies constrain base order.

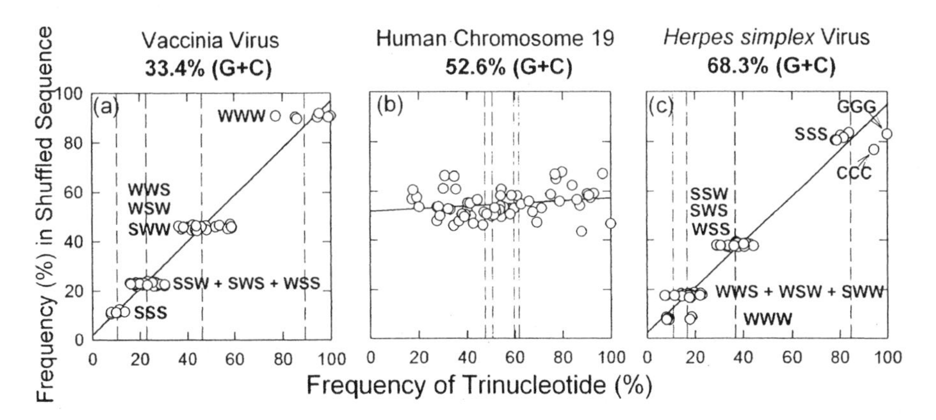

Fig. 4-3. Comparison of frequencies of the 64 trinucleotides in randomized ("shuffled") sequences with their frequencies in the corresponding natural, unshuffled, sequences. Trinucleotide frequencies, expressed as a percentage of the most abundant trinucleotide in the natural sequence, are plotted against each other. Other details are as in Figure 4-2. Note that this procedure allows us to begin to dissect out one "principle component" of the plot, base order, from another "principle component," base composition (see Appendix 1)

Frequencies of n-tuples from one DNA source can also be plotted against the frequencies of the same n-tuples in DNA from another source. In the case of trinucleotides (3-tuples) there are 64 data points. Figure 4-4a shows that when one segment of human DNA (from chromosome 4) is plotted against another segment of human DNA (from chromosome 19) that has about the same **(G+C)**%, the points again fall close to the diagonal. Yet, a plot of human DNA (chromosome 19) against a segment of DNA from the gut bacterium *Escherichia coli*, which has about the same **(G+C)**%, reveals no correlation (Fig. 4-4b). The *E. coli* sequence behaves much as the shuffled human sequence shown in Figure 4-3b.

Thus, with respect to human DNA, *E. coli* DNA would appear as randomized sequence. But two segments of DNA from different parts of the human genome, which encode entirely different functions, have very similar base order-dependent contributions to n-tuple frequencies (Fig. 4-4a). Given a particular base-composition, and other pressures being equal, the genome of a given species will opt for particular sets of n-tuples, rather than others. Genome n-tuple frequency values tend to be species specific [14]. One cause of this is a pressure in organisms considered higher on the evolutionary scale (e.g. humans, fruit fly) to loose the self-complementary dinucleotides **CpG** and **TpA** (see Chapter 15). This is not so strange. N-tuple frequencies also show specificity in spoken languages. For example, the 2-tuple "sz" is found in Polish names (e.g. Szybalski), but infrequently in English names.

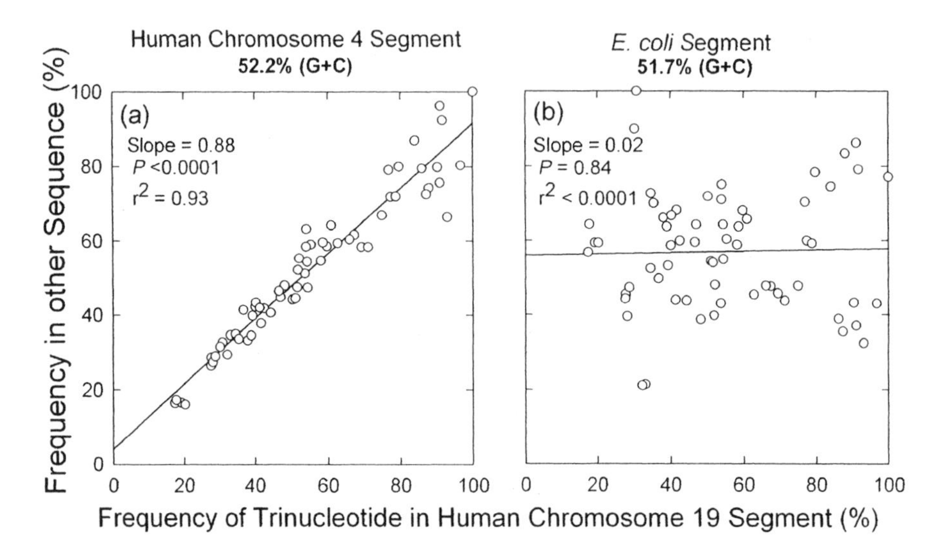

Fig. 4-4. Comparison of frequencies of 64 trinucleotides in different sequences each with **G+C** percentages around 50%. Trinucleotide frequencies in either a segment from human chromosome 4 (a), or a segment from the bacterium *E. coli* (b), were plotted against the corresponding trinucleotide frequencies in a segment from human chromosome 19 [13]

In Figure 4-4a there is a high slope (0.88) and the probability that the difference from zero slope had occurred by chance is very low ($P<0.0001$). In contrast, in Figure 4-4b the slope is low (0.02) and the probability that the difference from zero slope had occurred by chance is very high ($P = 0.84$). Thus, the former slope is highly significant, but the latter is not (see Appendix 1). By comparing the slopes for DNA segments from different species their degree of similarity can be measured, and this generally fits in with evolutionary expectations [13].

Another measure is the square of the correlation coefficient. In Figure 4-4a 93% of the variation between the two human segments can be explained by the correlation ($r^2 = 0.93$). Only 7% remains unaccounted for. In contrast, in Figure 4-4b there is no correlation ($r^2 <0.0001$). 100% of the variation remains to be accounted for. Again, by comparing r^2 values (or just r values) for DNA segments from different species, the degree of similarity can be measured, and this generally fits in with evolutionary expectations [13, 15].

Normally, before the DNA of an organism is sequenced the organism is isolated from other organisms in its environment and its DNA purified. Eventually a set of sequenced DNA fragments is obtained. These can be matched end-to-end by virtue of sequence overlaps, to generate an entire genomic sequence. However, because each species has a distinctive hierarchy of oligonucleotide frequencies it is possible to identify species-specific fragments from the total DNA obtained directly from a particular environment, *without a prior isolation of organisms*. For example, there are about 160 different species of microorganisms in one milliliter of sea water. DNA can be prepared directly from a sample of sea water and the resulting fragments then sequenced. Assignment of a particular fragment to a particular species can be made *retrospectively* based on differences in oligonucleotide hierarchies. Thus, in principle, the genomes of multiple species occupying a particular habitat can be simultaneously sequenced without identifying each species in advance. This signifies the existence of a level of information in DNA sequences that may be used to distinguish species. Whether such species-identifying information appears *after* two species have diverged from a common ancestral species, or is an agency *initiating* the process of divergence (speciation), is considered in Chapters 7 and 8.

For some purposes you can confine your study to species with similar **(G+C)**% values as in Figure 4-4, so avoiding the effects on oligonucleotide frequencies of differences in base composition (i.e. differences in mononucleotide frequencies). Alternatively, various software packages provide "principle component analysis," which allows you to dissect out the role of **(G+C)**% differences, so that the contributions of other factors (i.e. base order) to the variation can be assessed.

Introns Obey

In Figure 4-4a, two human DNA segments are compared. These segments each contain genic DNA and non-genic DNA. Furthermore, the genic DNA consists of segments that appear in the cytoplasmic RNA derived from those genes (exons), and segments that do not appear in the cytoplasmic RNA derived from those genes (introns; see Chapter 10). It is possible to compare the oligonucleotide frequencies between these distinct segments within a species, using either the above graphical approaches (Figs. 4-2 to 4-4), or principle component analysis. In this way Emanuele Bultrini and her Italian colleagues showed that intronic DNA and non-genic DNA have similar oligonucleotide frequencies, which differ from the oligonucleotide frequencies of exonic DNA. Furthermore, introns and non-genic DNA obey Chargaff's second parity rule *more closely* than exons. This provides a way of locating exons in uncharted DNA sequences, and has implications for the differing potential of these segments to adopt secondary structures. Intronic and non-genic DNA sequences should more readily form stem-loop structures than exonic sequences [15]:

> "A symmetrical trend is apparent on a scale of a few kilobases in individual ... introns. This short-range property of introns is not simply due to their symmetrical base composition, since it is drastically reduced in randomized introns. Rather, it results from the *preferred* use of reverse complementary oligomers.... It would be tempting to link the above symmetry properties of introns to formation of stem-loop structures."

Level of Selection?

All this bears on the question as to whether Nature *first* "wrote" sequences with single base parity, from which the observed parity of complementary dinucleotides, trinucleotides, etc., is an automatic consequence? Phrased evolutionarily, the question is, do selective pressures, be they extrinsic or intrinsic to the organism, operate such as to promote directly single base parity? Alternatively, do the pressures operate at another level so that single base parity is an automatic consequence of a primary pressure for parity at another (higher) level? These horse-and-cart questions were posed in 1995 as follows [13]:

> "Did evolutionary forces select for the Chargaff ratios in single DNA strands, with equality of complementary oligonucleotide frequencies being an automatic consequence? Alternatively, did evolutionary forces select for equality of complementary oligonucleotide frequencies, with the Chargaff ratios being an automatic consequence?"

If Nature "writes" a sequence with parity at the oligonucleotide level, then parity at the single base level is an automatic consequence. However, the converse does not apply. In Table 4-3 it is shown that if evolutionary forces cause a sequence to be "written" with parity at the level of single bases, then, however long the sequence, parity at the oligonucleotide level does not necessarily follow. Since biological sequences generally show parity at the oligonucleotide level, this suggests that they were initially "written" at that level.

Thus, it is likely that sequences that are of biological origin, and are of sufficient length, demonstrate parity at the single base level because "Nature" initially "wrote" them under the influence of evolutionary forces that required parity at the oligonucleotide level (e.g. to form stems in stem-loop secondary structures). Nature "writes" with parity primarily at the oligonucleotide level and, by default, there is parity at the mononucleotide level [16]. This view has been supported by Pierre-François Baisnée and his coworkers using a statistical approach [17].

(a) A A A A A A A A T T T T T T T T

(b) A A A A A A T T T T T A
A A A A A T T T T T
A A A A T T T T T T

(c) A A A A A T T T T
A A A A T T T T T
A A A T T T T T A
A A A T T T T A A
A A A T T T
A A A T T T

(d) T C A C T A G G G A T A C A T T

(e) T C C T G G A T C A T T
C A T A G G T A A T
A C A G G A A C T T

(f) T C A G G G C A T
C A C G G A A T T
A C T G A T T T T
C T A A T A T T C
T A G T A C
A G G A C A

Table 4-3. Demonstration that, if Chargaff's second parity rule applies for complementary mononucleotides (single bases), it does not always extend to complementary dinucleotides, and higher order complementary oligonucleotides, when they are arranged in a circular single-strand sequence. For simplicity, first take just two bases, **A** and **T**, which pair with each other. Write out a sequence with eight **A**s and eight **T**s. Although this sequence *(a)* is for convenience written as a linear sequence, imagine it is circular with the last base (**T)** being connected to the first base (**A)**. Does it obey Chargaff's second parity rule? Yes, there are 8 **A**s and 8 **T**s. Does the rule extend to complementary dinucleotides? Remembering that the sequence is circular, and that base pairs can overlap, score by writing dinucleotides below the sequence and then counting, as in *(b)*. Note that **AA** = 7 and its complement **TT** = 7; **AT** = 1 and its complement **AT** = 1; **TA** = 1 and its complement **TA** = 1. The similar decomposition into trinucleotides *(c)* shows **AAA** = 6 and its complement **TTT** = 6; **AAT** = 1 and its complement **ATT** = 1; **TTA** = 1 and its complement **TAA** =1. In this particular sequence there is no numerical distinction between reverse and forward complements in the same strand [11]. Note that the 16 base sequence could be repeatedly copied and the resulting copies joined together end-to-end (concatenated) to produce one large circular single strand. Whatever the extent of this replication and concatenation, the parity relationship would be retained at the levels both of single bases and of the above oligonucleotides. Thus, if this were the concatenated sequence of some organism, it would not be possible to determine whether "Nature" had first "written" the sequence at the single base level, with parity at the oligonucleotide level being an automatic consequence, or the converse. Sequence *(d)*, however, also obeys the parity rule (**A** = 5, **T** = 5, **G** = 3, **C** = 3). Here complementary dinucleotides, trinucleotides, etc., are not necessarily present in equal frequencies. For example, there is no reverse complement (**GT**) for the two **CA**s. The trinucleotide **TTT** is not matched by an **AAA**. The trinucleotide **GGG** is not matched by a **CCC**. Again, the 16 base sequence could be repeatedly copied and the replicates concatenated to produce one large circular single strand. Disparities at the oligonucleotide level would still be present. Parity at the level of single bases does not necessarily imply parity at the oligonucleotide level. Since biological sequences generally show parity at the oligonucleotide level, this suggests that they were initially "written" at that level

For the mathematically inclined we can say that, on a random basis, given that %**A** = %**T** in a sequence, then the probability of **A** (written as $P(\mathbf{A})$) at any position will equal the probability of **T** (written as $P(\mathbf{T})$) at any position. Thus, $P(\mathbf{A}) = P(\mathbf{T})$. On a random basis, the probabilities of finding the corresponding dinucleotides would be written as $P(\mathbf{AA})$ and $P(\mathbf{TT})$. Then, $P(\mathbf{AA}) = P(\mathbf{A}) \times P(\mathbf{A}) = P(\mathbf{A})^2 = P(\mathbf{T})^2 = P(\mathbf{T}) \times P(\mathbf{T}) = P(\mathbf{TT})$.

In 1995 in the USA, biochemist Noboru Sueoka argued that parity follows from mutational biases acting at the mononucleotide level [18]. However, the actual observed frequencies of higher oligonucleotides (e.g. **AA** or **TT**), cannot be predicted simply from the frequencies of the corresponding mononucleotides. Thus, if you accept values for p(**A**) and p(**T**) based on the actual observed frequencies of these single bases, you cannot then predict the frequencies of **AA** and **TT** by multiplying the probabilities of the single bases. Similarly, if you accept probability values for the four mononucleotides $P(\mathbf{A})$, $P(\mathbf{C})$, $P(\mathbf{G})$, $P(\mathbf{T})$, and for the dinucleotide $P(\mathbf{AA})$, based on actual observed frequencies, you cannot then predict the observed frequencies of the various trinucleotides containing **AA** (i.e. **AAA**, **AAC**, **AAG**, **AAT**, **CAA**, **GAA**, **TAA**). Blaisnée and his colleagues argued persuasively that parity (i.e. symmetry of base composition between two strands in a duplex) derives largely from forces acting at higher oligonucleotide levels [17]:

> "The underlying assumption is that base composition symmetry results from single point mutations that equally affect complementary strands. ... In addition higher-order symmetry is widely considered, implicitly or explicitly, as the consequence of first-

> order symmetry. ... [However], reverse-complement symmetry does not result from a single cause, such as point mutation [i.e. mutational bias], ... but rather emerges from the combined effects of a wide spectrum of mechanisms operating at multiple orders and length scales."

At the heart of the problem is the need to identify a distinct selective force operating at one or other level. We shall see in the next chapter that a primary pressure operates at least at the level of complementary dinucleotides (which include the self-complementary dinucleotides; Table 4-1). Indeed, Israeli biochemist Ruth Nussinov has suggested that dinucleotide frequencies are more fundamental than trinucleotide frequencies, so that we should seek to explain the latter in terms of the former, and not *vice versa* [19].

Rule Too Precise?

We have seen that, at least by virtue of the stems in DNA stem-loop structures, there would have been an evolutionary pressure for single-strands of DNA to have approximate equivalences of the Watson-Crick pairing bases. Thus, Chargaff's second parity rule is consistent with single strands of DNA having considerable potential for forming secondary structure.

However, loops with some numerically unpaired bases make up a large part of most structures, so that the second parity rule for single-stranded nucleic acid would be expected to be much less precise than the first parity rule for nucleic acid in duplex form. Accordingly, it must be noted that in some organisms the second parity rule shows remarkable precision (e.g. see the above base composition of Vaccinia virus). Explanations for this include the possibility of a requirement for some base-pairing between loops, which may not necessarily be closely colocalized in a single DNA strand. Thus, we begin to recognize the possibility of long-range base pairing interactions between distant loops [8].

Much of the above discussion might appear laborious to someone with a mathematical turn of mind. For example, some observations (Fig. 4-2) appear as a simple application of the binomial theorem (i.e. of complementary trinucleotides there are 4 pairs in the $\mathbf{S}_3$ (**SSS**) group, 12 pairs in the $\mathbf{S}_2\mathbf{W}$ group, 12 pairs in the $\mathbf{W}_2\mathbf{S}$ group, and 4 pairs in the $\mathbf{W}_3$ group). However, non-mathematical members of the species *Homo bioinformaticus* should usually be able to get there, albeit slowly, by the careful use of pencil and paper as indicated here (Table 4-3). And there are now powerful computers and programs (e.g. principle component analysis) that can take us into realms that previously only the mathematically gifted could contemplate.

Summary

Prose, poetry and palindromes can be seen as informational devices that trade-off increasing degrees of redundancy in order to increase error-detecting power. The most extreme of these is the palindrome which, if a general restriction, would severely compromise the normal transfer of information between humans, while greatly decreasing the chances of error. That hereditary information in the form of DNA sequences is palindrome-like, suggests that evolutionary pressures for error-detection may be at least as powerful as those for the encoding of primary messages. Given limits on genome space, there are potential conflicts between different forms and levels of information. Classical Darwinian selective forces in the environment acting on an organism's form and function provide *extrinsic* constraint, but genomes are also under *intrinsic* constraint. In palindromes there is a one-to-one pairing relationship between symbols (letters, bases). In DNA this finds expression as Chargaff's second parity rule, namely Chargaff's first parity rule for duplex DNA also applies, to a close approximation, to single-stranded DNA. Selective forces generating the second parity rule equivalences may have acted at higher oligonucleotide levels than the level of single bases. Diminished second parity rule equivalences in coding regions suggests diminished potential for secondary structure (stem-loops) in these regions.

Chapter 5

Stems and Loops

> "No system consisting of these elements could possibly have the properties that atoms were known to have. ... In Bohr's paper of 1913 this paradox was met by introducing the notions of stable orbits and jumps between these orbits. ... These were very irrational assumptions, which shocked ... many physicists ... . The crucial point ... is the appearance of a conflict between separate areas of experience, which gradually sharpens into a paradox and must then be resolved by a radically new approach."
>
> Max Delbrück (1949) [1]

The sequence of a single-stranded nucleic acid is referred to as its primary structure. When it forms a duplex by pairing with a complementary single strand it adopts a secondary structure. Chargaff's first parity rule for duplex DNA was consistent with the Watson-Crick idea of a base in one strand of the duplex pairing with a complementary base in the other strand of the duplex (*inter*strand base pairing), thus stabilizing the secondary structure (Fig. 2-1). By the same token, the existence of a parity rule for single strands of nucleic acid (see Chapter 4), suggested *intra*strand base pairing. At least by virtue of the composition of the stems in stem-loop secondary structures, there should be approximately equivalent quantities of the classical pairing bases. Do single-stranded nucleic acids have the potential to form such intrastrand secondary structures? If so, is this a chance event, or are adaptive forces involved? These questions began to be addressed when the sequences of various tRNAs and bacterial viruses became available in the 1970s. It became evident that nucleic acid structure ("flaps") is a form of information that has the potential to conflict with other forms.

Transfer RNA

Having adopted a stem-loop secondary structure, a single-stranded nucleic acid can participate in more elaborate intrastrand bonding to generate higher-ordered structures (tertiary, quarternary, etc.). However, for many purposes nucleic acid structure can be adequately discussed in terms of secondary structure. This can be simply drawn in two dimensions without showing the pairing strands as double helices (Fig. 5-1).

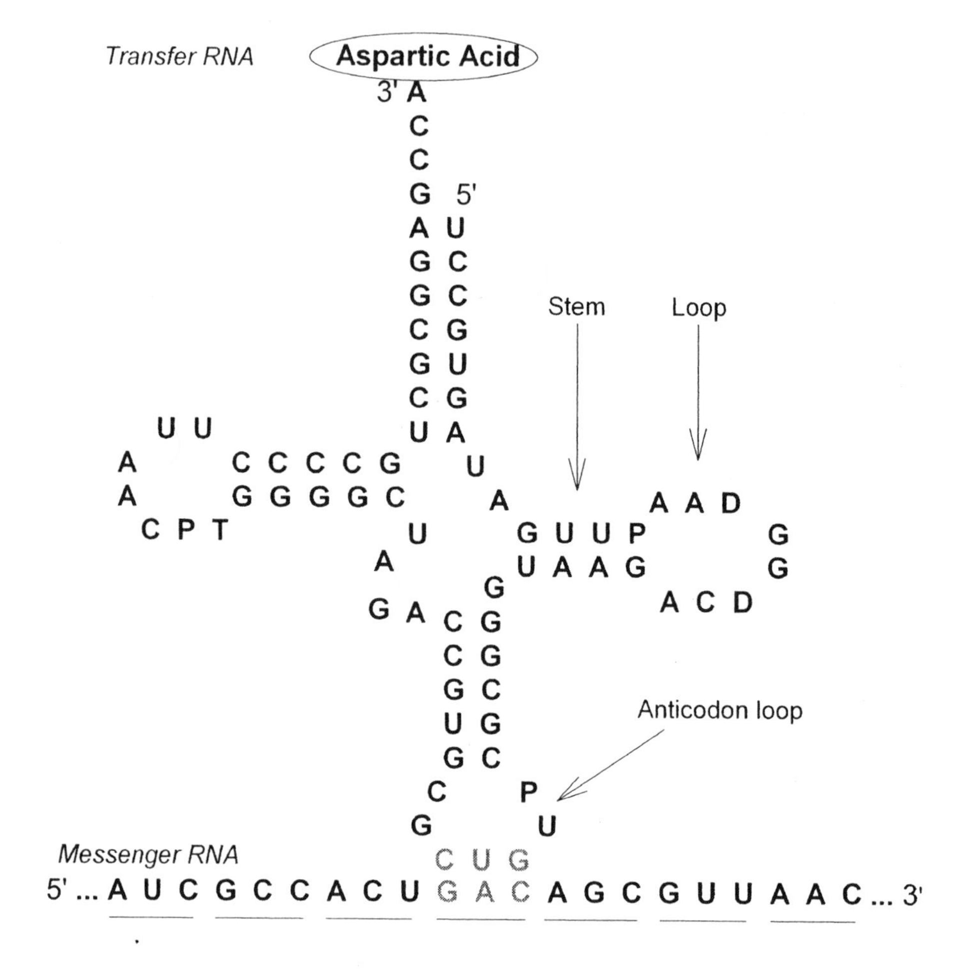

Fig. 5-1. Stem-loop secondary structure of a transfer RNA (tRNA) molecule, in simple two-dimensional form. Successive triplet codons in the messenger RNA are underlined (bottom). The mRNA codon for aspartic acid (**GAC**, in grey) interacts with ("kisses") the complementary sequence at the tip of the anticodon loop of the tRNA (**GUC**, in grey). Note that in RNA there is some pairing between **G** and **U** (as well as the more usual **A–U** pairing). Furthermore, tRNA molecules often have a few unusual bases (here designated P, D and T) that are not normally found in mRNA molecules. Some bases may be methylated (not shown here; see Chapter 15)

When the first transfer RNA (tRNA) sequences became available it was found they could be folded so that stem-loop-containing structures were generated. In any cell engaged in protein synthesis there must be at least twenty tRNA types, corresponding to the twenty amino acids. The 3' end of each tRNA type is attached to a particular amino acid that will be transferred to a growing chain of amino acids (hence "*transfer* RNA"). The chain of amino acids (peptide chain) will give rise to a protein.

At the tip of one of the loops (the anticodon loop) there is a three base sequence (5' **GUC** 3' in Fig. 5-1) that is able to base pair with the complementary sequence (5' **GAC** 3') of a codon in mRNA. This is a brief, reversible interaction, which can be referred to as "kissing" (see Chapter 6). The base pairing follows Watson and Crick in that the pairing strands are anti-parallel – an arrangement applying both to the intrastrand pairing within the tRNA molecule, and to its transient interstrand pairing with the codon sequence in the mRNA. We should note that in RNA, unlike DNA, there is usually the base **U** instead of **T**. Like **T**, **U** pairs best with **A**, but a weaker pairing with **G** can also occur. We should also note that the base in tRNA before (on the 5' side of) the anticodon triplet is almost invariably **U**, and the base after (on the 3' side of) the anticodons triplet is usually a purine (**R**). The significance of this will be discussed in Chapter 7.

The finding of distinctive secondary structures in tRNAs made sense, since a particular tRNA functions cooperatively with a specific enzyme that adds a specific amino acid to the 3' end of the tRNA. The specificity of interaction with the enzyme (a protein) is largely determined by the structure of the tRNA, which allows the enzyme to distinguish it from other tRNA types. Apart from tRNAs there are other RNAs whose roles are critically linked to their structure. These include RNAs that, like tRNAs, take part in protein synthesis (e.g. ribosomal RNAs; rRNAs). Much of the cytoplasm is concerned with protein synthesis (e.g. it contains many rRNA-containing particles known as "ribosomes" that act as protein-assembly "work-benches"). Accordingly, like other components of the "machinery" for protein synthesis, tRNAs and rRNAs are abundant, collectively making up about 96% of total cell RNA.

From the above one might infer that the RNA species concerned with the programming (templating) of ribosomes, namely mRNA, would not demonstrate much in the way of secondary structure. Indeed, such structure might even interfere with the translation process. Yet, every RNA is a transcript derived by copying (transcribing) the sequence of another (usually genomic) nucleic acid (usually DNA) following the Watson-Crick base-pairing rule (see Chapter 6). If it were advantageous for genomic nucleic acid to have secondary structure then mRNAs might have a structure by default. Genomes of viruses should be informative in this respect, since viruses employ their host cell's machinery for protein synthesis and often do not have information for tRNAs and rRNAs in their genomes.

The Bacteriophage Paradox

In some viruses that infect bacteria the genome is RNA (e.g. bacteriophages R17 and MS2), and in others the genome is DNA (e.g. bacteriophage T4). Genomes of bacteriophage (Greek: *phagein* = to eat) consist of tightly packed genes from which mRNAs are transcribed for translation into proteins by the host's translation machinery. These proteins of bacteriophage origin assist it to control the host and, late in the cycle of infection, include coat proteins into which new phage genomes are packaged prior to release of virus particles from the infected cell.

In classical Darwinian terms it was assumed that bacteriophages that encoded optimal proteins were best adapted to their environment. Thus, the environment acting on virus proteins would have selected for survival the "fittest" viruses whose genes encoded optimal proteins. Yet, paradoxically, it was observed that the base sequence seemed to serve the needs of nucleic acid *structure* just as much as the protein-encoding function. Indeed, the needs of nucleic acid structure sometimes seemed to be served *better* than those of the protein-encoding function. Although not widely perceived at the time, this paradox shattered the Darwinian orthodoxy and pointed to the need for "a radically new approach."

At what level might structure be of most selective importance? This could be at the level of mRNAs themselves, or of the genes that encoded them, or of the entire genome. Winston Salser in 1970 opted for importance at the mRNA level [2]:

> "RNA phage R17 has very extensive regions of highly ordered base pairing. It has seemed likely that this might be necessary to allow phage packaging. ... [We] were therefore somewhat surprised to find that T4 messengers [mRNAs], which do not have to be packaged, also have a very large amount of secondary structure. ... Our results suggest that a high degree of secondary structure may be important in the functioning of most mRNA molecules. ... The possible functions of such extensive regions of base-pairing are unknown."

In 1972 Andrew Ball drew further implications [3]:

> "The selection pressure for specific base pairing in a messenger RNA severely limits its coding potential" ... [so that] ... "there is a pressure for some amino acid sequences to be selected ac-

cording to criteria which are distinct from the structure and function of the proteins they constitute."

A solution to the paradox, which we will elaborate in Chapters 8 and 10, was arrived at when better methods for calculating RNA secondary structure became available. It will be shown in this chapter that, for many mRNA sequences, the energetics of the folding of natural sequences are better than those of sequences derived from the natural sequences by shuffling the order of the bases. Natural sequences must be shuffled and then folded many times to arrive, by chance, at a structure approaching the stability of the folded natural sequence [4, 5]. It appears that "the hand of evolution" has arranged the *order* of bases to support structure, sometimes at the expense of the protein-coding function.

Genome-Level Selection

Although abundantly present in cells, tRNAs and rRNAs are encoded by a relatively small part of genomes. In microorganisms (e.g. bacteria) the sequences of mRNAs are more representative of the sequences of the corresponding genomes. If many mRNAs have highly significant secondary structure, then the corresponding genomic regions (i.e. the genes from which the mRNAs are transcribed) should also have this potential. Indeed, as indicated above, the primary evolutionary pressure for the elaboration of mRNA secondary structure might have been at the *genomic* level rather that at the mRNA level. If so, regions of a genome that are *not* transcribed into mRNAs might also demonstrate potential for secondary structure.

When folding programs were applied to the sequences of individual DNA strands, it was found that there is indeed considerable potential for secondary structure *throughout* genomes. Stem-loop potential in DNA is not restricted to regions encoding mRNA (or rRNA or tRNA), but is also present in intergenic DNA and, in the case of intron-containing organisms, also in introns (indeed, it is greater in introns than exons; see Chapter 10). Stem-loop potential, greater than that of the corresponding shuffled sequences, is widely and abundantly distributed throughout the genomes of all species examined (Fig. 5-2).

Calculation from Single Base Pairs

With a knowledge of the Watson-Crick base-pairing rules, given the sequence of a tRNA one should be able to arrive at the structure shows in Figure 5-1 with pencil and paper by a process of trail and error. This would be quite laborious and with longer sequences it becomes impracticable. So computers are used. For present purposes it is not necessary fully to understand the programs (algorithms) that allow computers to calculate elaborate secondary structures, but the principles are relatively straightforward. Even so, it is conceptually quite tricky. Don't worry if it doesn't click the first time through. It took one of the masters of folding, Michael Zuker, several decades of exclusive focus to find the still far-from-perfect algorithm.

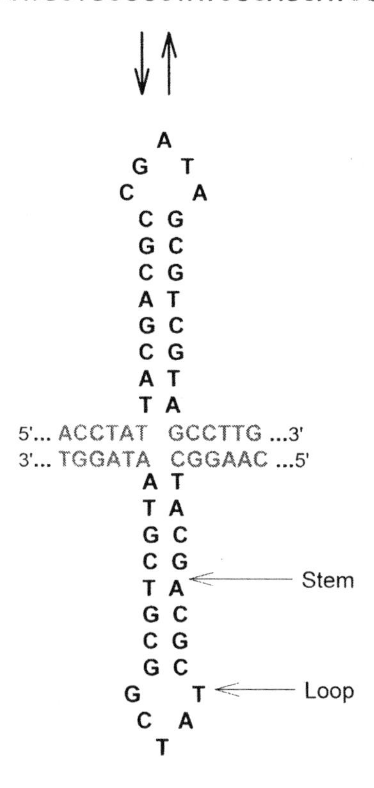

Fig. 5-2. Reversible extrusion of stem-loop secondary structures from duplex DNA. Here there is classical Watson-Crick pairing in the stems. However, just as **G** and **U** can pair weakly in RNA (see Fig. 5-1), so some extruded DNA structures have pairing between **G** and **T** . Since the process is reversible, then, unless otherwise restrained (e.g. by interactions with proteins) a segment of DNA should *vibrate* between the two conformations. The frequency should be lower for **GC**-rich DNA where the structures are more stable, than for **AT**-rich DNA where the structures are less stable. It follows that sequences of a given (**G+C**)% should have close vibration frequencies and, according to Muller's vibration model (see Chapter 2), should therefore have the potential to seek each other out. Since secondary structure stabilities are also affected by base order (see later), then when two sequences have similar base orders they should have even closer vibration frequencies

The energetics of nucleic acid folding can be considered simply in terms of *base composition* and *base order*. In other words, the total folding energy can be decomposed into two components, a base composition-dependent component and a base order-dependent component. The latter is determined by subtracting the base composition-dependent component from the total folding energy.

First a computer gives you a value for the total folding energy of a natural sequence. The tricky part is then getting the base composition-dependent component. To do this you have to destroy the base order-dependent component by shuffling to get a randomized sequence with the same base composition. Then you get the computer to fold this sequence and calculate the folding energy. The trouble with this is that, on a chance basis, the randomized version of the sequence that the computer has first generated might just happen to have a particular base order that generates an idiosyncratic energy value. To allow for this, you get the computer to shuffle and refold the original sequence, and then recalculate the folding energy, several more times. Each successive randomized version has a particular base order different from the original base order of the natural sequence, but *retains* the original base composition. Thus, the average folding energy of the set of different randomized versions of the natural sequence reflects this commonality – their base composition. In short, the base composition composition-dependent component is determined by shuffling and refolding a sequence several times, thus destroying the base order-dependent component, and then taking the average folding energy of the resulting structures.

To begin, we can, following Ignacio Tinoco and his coworkers [6], assign "stability numbers" of 1 and 2 to **A-T** and **G-C** base pairs, respectively. This acknowledges in crude form the greater strength of the pairing (H-bonding) interaction between **G** and **C**, compared with that between **A** and **T** (Table 2-1). A series of hypothetical stem-loop structures that might be extruded from duplex DNA are shown in Figure 5-3. To simplify, we allow only a single stem, and disallow sliding of strands relative to each other. Furthermore, the role of the loop is ignored.

The first "stem" consists of two sets of 8 consecutive **A** residues, which do not complement each other. Thus, if this were part of a natural DNA sequence, either the sequence would remain unextruded from duplex DNA (i.e. the two sets of consecutive **A** residues would remain paired with two sets of consecutive **T** residues on the complementary strand), or a large loop (rather than a stem) would form.

In the second stem a quarter of the bases are **T** residues. With the value 1 assigned to an **A-T** base pair, the total stability of the stem can be scored as 4, with a high score meaning high stability. Since **T** residues are infrequent, it is likely that **T**s will be opposite **A**s, rather than opposite **T**s (i.e. if **A**s and **T**s in the proportions 3:1 were allowed to randomly combine, there would be more **A-T**s than **T-T**s). Thus base *composition*, rather than base *order*, can be considered to make the major contribution to the score. This is a fundamental point to which we will return.

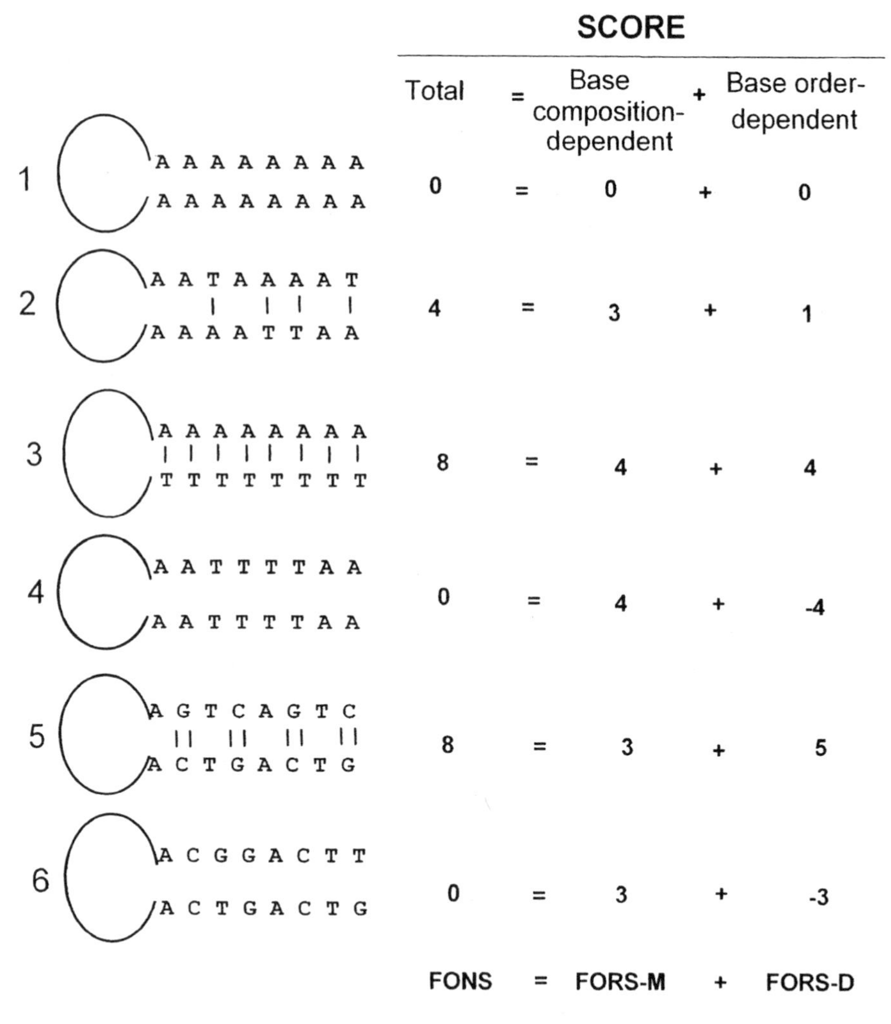

Fig. 5-3. Relative contributions of base *order* and base *composition* to the stability of stems in DNA stem-loop structures. The weakly bonding **A-T** base-pairs (the **W** bases) are assigned a score of one. The strongly bonding **G-C** base-pairs (the **S** bases) are assigned a score of two. Absence of base-pairing scores as zero. The stability of a stem is quantitated as the total stability score, which may be contributed to both by base order and by base composition. The terms FONS, FORS-M and FORS-D are described in the text

The third stem consists of equal proportions of **A**s and **T**s. The total possible number of **A-T** base-pairs (eight) is formed. The average number of **A-T** pairs that would be formed if 8 **A**s and 8 **T**s were randomly mixed would be 4 (try it by shaking the letters in a bag and picking out pairs). Thus, the contribution to the score attributable to base composition alone is 4. By subtraction we can determine that the base order-dependent component of the score is 4.

The fourth stem has the *same* base composition as the third stem, so that the potential base composition-dependent contribution to the stability remains at 4. However, the *distribution* of **A**s and **T**s is such that no **A-T** base-pairs form. Thus, the total score is zero, and by subtraction we determine that the contribution of base order is minus 4. The bases are ordered so as to *oppose* the random tendency for a sequence of 8 **A**s and 8 **T**s to form 4 **A-T** intrastrand base-pairs. Stem-loop extrusion from duplex DNA would be opposed. Alternatively, if associated with an extruded structure, the order of bases would favour loop, rather than stem, formation.

The fifth and sixth stems contain equal proportions of all four bases (i.e. their base compositions are identical). In the fifth stem only four complementary pairs are present. These are **G-C** pairs, to each of which we assign the score of 2, for a total score of 8. The relative contributions of base composition and base order are 3 and 5, respectively (as will be shown below). The sixth stem has no base pairing, so the total score is zero. The potential contribution of base composition remains at 3 so that, by subtraction, the contribution of base order is minus 3.

In summary, Figure 5-3 shows that the correct bases may be present in the correct proportions, but if base order is inappropriate there may be no stem (zero total score). Stem stability depends on base composition in two ways. Complementary bases must be present in equal proportions, and the more **G-C** (rather than **A-T**) pairs there are, the higher will be the stability.

Role of Base Composition

If the fifth stem in Figure 5-3 were a natural sequence, then we could call the total score (which sums to 8) the "folding of natural sequence" (FONS) value. How do we calculate the relative contributions of base composition and base order to that score? As indicated above, the stem has a unique characteristic, its base order, and two other characteristics that it shares with a large set of other possible DNA sequences, its length and base composition. The natural sequence is but one member of a hypothetical set of sequences that share length and base composition. Any *average* characteristic of this hypothetical set must reflect their shared lengths and base compositions. This is another, quite fundamental, point.

By keeping length constant, we can focus on the role of base composition. If the order of bases is randomized (shuffled), keeping the length constant, then members of the set which differ only in base order, are obtained. The sixth stem in Figure 5-3 is one member of the set. Figure 5-4 shows 10 other members of the set. These happened, by chance, to be those I *first* generated by consecutively shuffling base order in the fifth stem in Figure 5-3. Associated with each member is a total score. Each score is a "*folding of randomized sequence*" (FORS) value for the fifth stem in Figure 5-3.

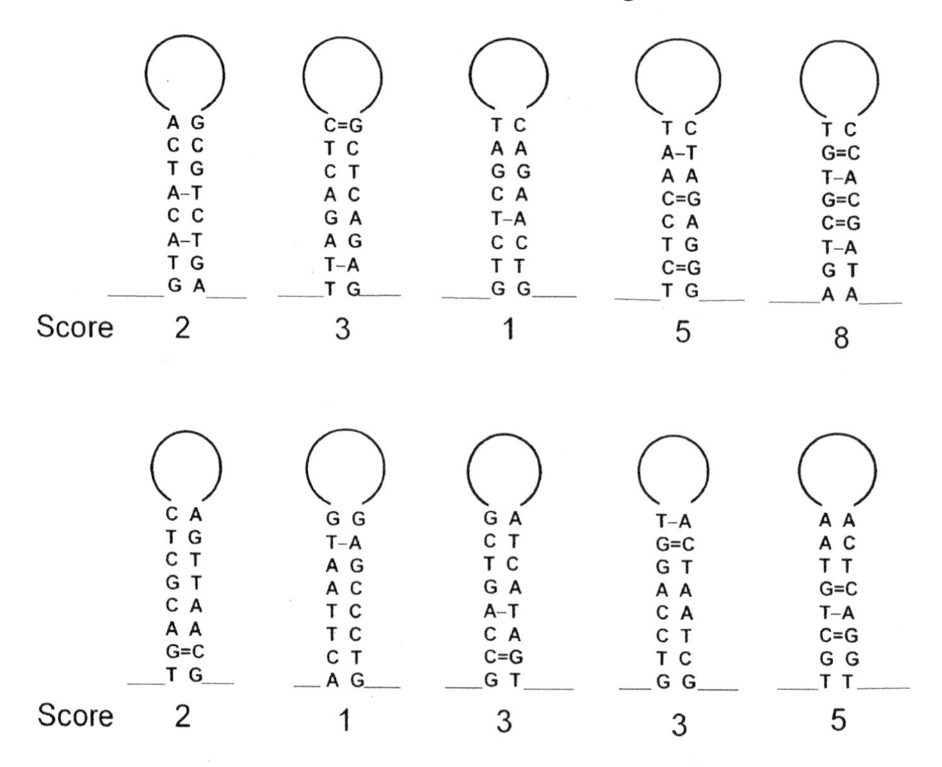

Fig. 5-4. A set of stem-loop structures that happened to be generated *first* by successively randomizing (shuffling) the order of bases in the fifth stem in Figure 5-3, and then refolding

It is seen that most members score less than the natural sequence. The mean score of the set of 10 is 3.3, which approximates to 3. This is the "*folding of randomized sequence mean*" (FORS-M) value, which is an *average* characteristic of the set, and thus should be *base order-independent*. The value provides a measure of the contribution of base composition to the FONS value for the 5th stem of Figure 5-3. By subtraction from the FONS value, the contribution of base order is found to be 5. This is the "*folding of randomized sequence difference*" (FORS-D) value, which provides a measure of the contribution of the primary sequence (base order) to the stability of stem-loop structures in a natural sequence of given length. Thus, FONS = FORS-M + FORS-D. In the present case, the FORS-D value is positive, and is significantly different from zero. This makes it likely that the sequence of the fifth stem in Figure 5-3 had accepted mutations to increase the formation of complementary base pairs, thus enhancing its stem potential.

It will be noted that four of the stems derived by randomization (Fig. 5-4), have lower total scores than 3. An extreme example of this is the sixth stem in Figure 5-3. To generate *by chance* a stem with no base-pairs would usually required more randomizations than the ten used to generate Figure 5-4. Thus, the sixth stem of Figure 5-3 is highly improbable. If the sixth stem were part of a real natural sequence, then the low FORS-D value would imply that base order had been working *against* base composition in determining the total score [4].

Six complementary dinucleotide pairs	AA	TT	-0.9
	AC	GT	-2.1
	AG	CT	-1.7
	CA	TG	-1.8
	CC	GG	-2.9
	GA	TC	-2.3
Four self-complementary dinucleotide pairs	TA	TA	-1.1
	GC	GC	-3.4
	CG	CG	-2.0
	AT	AT	-0.9

Table 5-1. The ten complementary base pairs in DNA (see Table 4-1) with approximate values for pairing energies (in negative kilocalories/mol.). Values for DNA and RNA are different, but of the same order. Note that the **W** base-rich pairs have less negative values (indicating weak pairing) than the **S** base-rich pairs. Since **G** can pair very weakly with **T** (or **U**), precise analysis of stem-loop potential also requires values for complementary base pairs that include this non-classical pairing [7]

Calculation from Dinucleotide Pairs

If the bases of the complementary Watson-Crick base-pairs just paired with each other, and neighbouring bases had no influence on this, then refinement of computer folding programs would largely require finding precise values to replace Tinoco's two crude "stability numbers." However, when double helices are formed, the flat bases *stack* above each other (like piles of coins, or "rouleaux"). This displaces water molecules. The increased freedom given to the water molecules (see Chapters 12, 13) makes a major contribution to the energetics of helix formation. The actual value depends on the nature of the two flat bases that form the "sandwich" from which water molecules are liberated. Thus, sequence (i.e. base-order) is of much importance in determining the stability of a nucleic acid duplex.

It is observed that this base order-dependent component can be accounted for by moving from Tinoco's two stability numbers for single base pairs, to ten numbers for dinucleotide pairs (Table 5-1). Each of these ten numbers has been determined using chemical methods [7]. For most purposes there is no need to go to higher sequence levels (e.g. use numbers corresponding to the 32 complementary trinucleotide pairs). A sequence can be decomposed into a set of overlapping dinucleotides (Table 5-2) and the corresponding numbers summed to determine the total folding energy. A computer is able repeatedly to fold a single-stranded nucleic acid, each time calculating the score and discarding folding patterns corresponding to poor scores, until it arrives at a folding pattern the score of which cannot be improved upon.

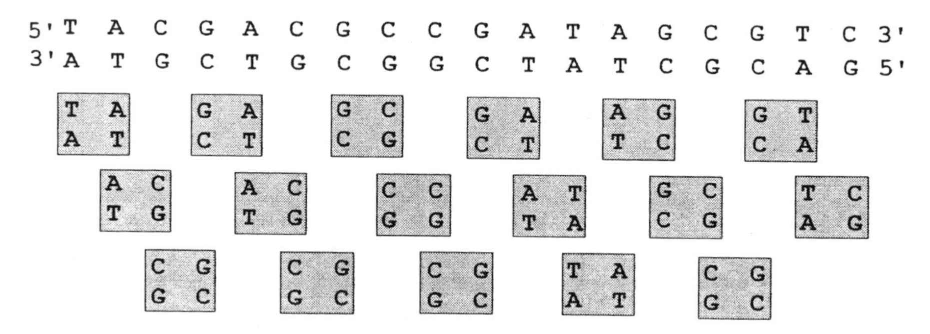

Table 5-2. Decomposition of a duplex, (which might be part of a long stem in a stem-loop structure), into individual members of the set of 10 complementary dinucleotides. Each of these can be scored using values from Table 5-1. The sum of the scores provides an overall energy value for the stem (in negative kilocalories/mol)

How do we determine that a calculated structure (Fig. 5-5) corresponds to the actual structure the nucleic acid adopts within cells? Support for calculated structures has been obtained in cases where nucleic acids have been crystallized and structures determined by X-ray crystallography. Also there are enzymes (nucleases) that recognize specific features of nucleic acid structure. A good calculated structure allows prediction as to which nucleases a nucleic acid will be susceptible.

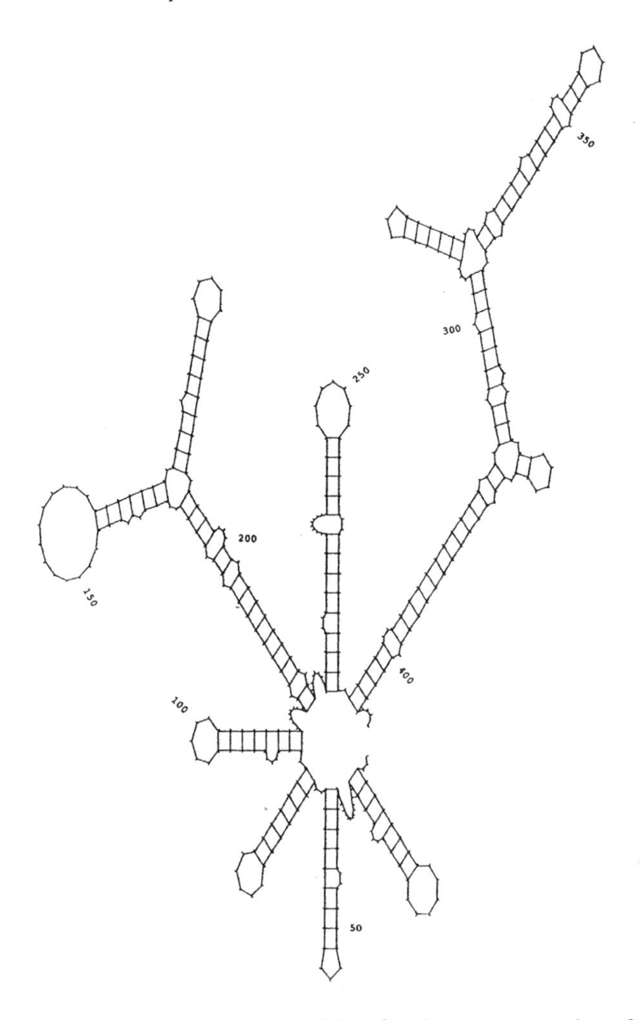

Fig. 5-5. Calculated structure of the mRNA for the oncoprotein c-fes. The laddering indicates individual base pairings. For present purposes all structures are "stem-loop," but a few would be considered "bulge loops." The structure includes some weak pairings between **G** and **U** (not shown). Extruded DNA segments corresponding to this sequence (exons) have some pairings between **G** and **T**. Similar, highly significant, structures are generated when non-genic and intronic sequences are folded similarly. Stem-loop potential is widely and abundantly distributed throughout genomes

RNA Structure and Conflict

All mRNAs have a coding region, namely a region containing a sequence of non-overlapping codons that must be "read," one at a time, by corresponding tRNAs, to allow the synthesis of a specific sequence of amino acids. As a linear sequence, mRNAs may begin with a 5' non-coding region that is followed by the coding region, and may then terminate with a 3' non-coding region.

Figure 5-5 shows the calculated structure of the 408 base sequence of the human mRNA encoding the cancer-related protein (oncoprotein) cFes [8]. This structure has a calculated stability of about –300 kilocalories/mole. Do not worry what the energy units mean, but note that the energy value is negative, as are the energy values for complementary dinucleotide pairs shown in Table 5-1. Greater negativity corresponds to greater stability. This is because, by convention, the stability of a structure is assessed by the energy made available when the structure forms. The structure itself loses energy in this process, and it is this net energy loss (loss being negative, gain being positive) that is taken as a measure of stability. This contrasts with the positive scoring scheme used earlier in the chapter when calculating structures using the Tinoco numbers.

If one were to shuffle the natural c-fes mRNA sequence and then calculate the energetics of the resulting structure, the chances would be great that the stability *value* would be considerably "higher" (i.e. would be much less negative) than that of the natural sequence (i.e. the stability itself would be *less*). One would have to test shuffle and calculate folding values many times in order to obtain, by chance, a structure of energy value similar to that of the natural sequence. Thus, it is extremely improbable that the structure shown in Figure 5-5 would be arrived at by chance. The structure of the natural sequence is of high statistical significance [4, 5].

It should be noted that the structure encompasses all parts of the mRNA, including the coding region. Thus, there is a potential conflict between the "need" of the nucleic acid sequence for secondary structure and the "need" for it to encode an optimally functioning protein (Fig. 5-6). By the same token, it is possible that there will be a conflict between the "needs" of the *genes* encoding tRNAs and rRNA for secondary structures appropriate for their functions in the nucleus, and the "needs" of the tRNAs and rRNAs themselves to have structures appropriate to their functions in the cytoplasm.

The genes (DNA) have essentially the same sequences as their RNA products. Yet, the structures of tRNAs and rRNAs, arrived at in response to selective pressures operating during protein synthesis in the cytoplasm, may not optimally served the "needs" for structure of their corresponding genes. These genes may be presumed to have arrived at their structures in response to selective pressures operating during their function in the nucleus (or in the cytoplasm in organisms, or cell cycle stages, when there is no defined nucleus). Thus, the sequences of these nucleic acid molecules may be a compromise between competing demands operating at genomic (i.e. DNA) and cytoplasmic (i.e. RNA) levels. One possible solution is to edit mRNAs en route from the site of transcription to the site of translation, so that their sequences end up different from the sequences of the corresponding genes (see also Chapter 10) [9].

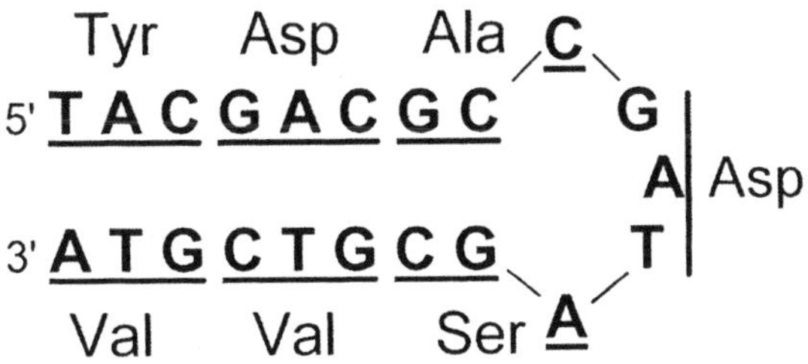

Tyr Asp Ala C
5' TAC GAC GC G
A Asp
3' ATG CTG CG T
Val Val Ser A

Fig. 5-6. Potential conflict between the "need" of a nucleic acid for a particular structure and its need to encode a particular amino acid sequence. Sequence 3.1 of Chapter 3 is here folded into a stem-loop structure. If it were part of the protein-coding region of an mRNA, it could encode the amino acid sequence **TyrAspAlaAspSerValVal**. If stem-formation is essential, then there is less flexibility in the encoding of amino acids. For example, the codon **TAC** encoding tyrosine, **Tyr**, *must* exist opposite the codon **GTA** encoding valine, **Val**. If the protein "wants" the tyrosine, then it *must* accept the valine, and vice versa. Otherwise, the encoding nucleic acid must weaken, or dispense with, the stem

DNA as Substrate

Proteins are macromolecules that often act as enzymes catalyzing changes in other molecules, their specific substrates, which may be micromolecules such a glucose or fatty acids, or macromolecules (e.g. DNA). In the course of evolution there are changes in proteins and in the DNAs that encode those proteins (Fig. 1-1). Substrates do not usually participate in this directly. For example, glucose is a preferred energy source in many organisms. As organisms evolve, specific enzymes may change and so improve the utilization of glucose. But glucose itself has no say in this. There is no way it can change to lighten, or impede, the task of the enzymes.

For some enzymes, however, DNA is a substrate. DNA both encodes the enzymes (locally in the regions of their genes), and is either their local or general substrate. We have seen here that, by virtue of changes in the order or composition of its bases, DNA can change a character such as stem-loop potential. Thus, it has a means of lightening or impeding the tasks of the enzymes that act on it. More than this, the *constancy* of glucose means that the enzymes of glucose metabolism can count of glucose being the same from generation to generation. The *limited constancy* of DNA means that many of its enzymes are on a treadmill. They must change as their substrate changes (see Chapter 9).

Summary

Transcription of DNA generates single-stranded RNAs that may operate by virtue of their structures (tRNAs, rRNAs), or by virtue of their encoding of proteins (mRNAs). The former consist of a small number of types which, while quantitatively abundant, are collectively encoded by a small part of genomes. The latter consist of a large number of types which, while each quantitatively sparse, are collectively encoded by a much larger part of genomes. All types of RNA fold into highly significant (non-random) secondary "stem-loop" structures. While the selective forces affecting tRNA and rRNA structures largely relate to their roles in the cytoplasm, those affecting mRNA structures largely relate to the need for structure (stem-loop potential) of the genes from which they were transcribed – a need that may conflict with the need to optimize the sequence of an encoded protein. Since non-genic regions of genomes also reveal the potential for highly significant stem-loop structure, the potential is likely to be a genome-wide response to some selective force. Higher ordered structures of single-stranded nucleic acids may be calculated from the base-pairing energies of overlapping dinucleotides, which are fundamental units of nucleic acid structure. Contributions to the energetics of such structures decompose into base composition-dependent and base order-dependent components. The latter is determined by subtracting the base composition-dependent component from the total folding energy. The base composition composition-dependent component is itself determined by shuffling and refolding a sequence several times, thus destroying the base order-dependent component, and then taking the average folding energy of the resulting structures.

Chapter 6

Chargaff's Cluster Rule

> "Another consequence of our studies on deoxyribonucleic acids of animal and plant origin is the conclusion that at least 60% of the pyrimidines occur as oligonucleotide tracts [runs] containing three or more pyrimidines in a row; and a corresponding statement must, owing to the equality relationship [between the two strands], apply also to the purines."
>
> Erwin Chargaff (1963) [1]

A nucleic acid sequence has three fundamental characteristics – its length, its base composition, and its sequence. If you know the sequence then you can calculate length and base composition with great precision. However, before the emergence of sequencing technologies in the 1970s there were biochemical methods that could provide values for length and base composition, albeit less precisely. Furthermore, Chargaff and his coworkers developed a method for evaluating a particular sequence characteristic – base clustering – that could distinguish DNA samples on the basis of sequence differences.

Base Clusters

It is a chemical property of DNA that under acid conditions it loses its purines (**R**), but retains it pyrimidines (**Y**). The loss of purines makes it easier to break the molecule chemically at the sites from which purines have been stripped. This leaves free segments (oligonucleotides) containing clusters of pyrimidines. The following two lines show a sequence with the pyrimidine clusters that would result if the sequence were broken in this way:

```
YYRYYYYRRRRYRYRRYRRRRYYYYYYYRYY
YY YYYY    Y Y  Y    YYYYYYY YY      (6.1)
```

Pyrimidine clusters can be separated on the basis of size differences to generate a distinct profile for any particular DNA molecule. From sequence

6.1, for example, we obtain 3 **Y**s, 2 **YY**s, 0 **YYY**s, 1 **YYYY**s, 0 **YYYYY**s, 0 **YYYYYY**s and 1 **YYYYYYY**s.

It is easy to dismiss such clusters as trivial. After all, in an average audience you may see several members of the same sex together in a row. Males and females are usually not evenly distributed. There are statistical techniques for telling whether clustering is likely to be a chance phenomenon. Chargaff found that the clustering of pyrimidines was more than expected on a random basis. Furthermore, with the knowledge that DNA was usually in duplex form with pyrimidines in one strand pairing with purines in the other, it was easy to infer that purine clusters must complement pyrimidine clusters. Thus the above single-strand would appear in duplex form as:

5'**YYRYYYYRRRRYRYRRYRRRRYYYYYYYRYY**3'
3'**RRYRRRRYYYYRYRYYRYYYYRRRRRRRYRR** 5' (6.2)

Chargaff's observation has been found, like his first and second parity rules, to apply to many genomes, and so can be considered a species-invariant feature of DNA.

Although the clusters themselves are not random, are they randomly distributed in the genome, or are they more prevalent in certain regions? Assuming that, like many things on this planet, random distribution (chaos) is the default state (see Chapter 12), what selective forces might have allowed clustered bases to persist? A cluster, by definition, locally violates Chargaff's second parity rule; but a cluster of pyrimidines might be locally matched by purines (clustered or relatively unclustered), so the violation might usually not extend beyond several bases. On the other hand, if clusters were themselves clustered (i.e. pyrimidine clusters clustering with pyrimidine clusters, and purine clusters clustering with purine clusters), violations might be more extreme.

Clusters of Clusters

The cluster observation was extended by work from Waclaw Szybalski's laboratory in the 1960s, which showed that clustering of clusters in microorganisms is most evident in transcriptionally active regions, and that the nature of the clustering of clusters (purine or pyrimidine) relates to transcription direction [2]. The "top" strand of part of a DNA duplex that is transcribed contains pyrimidine clusters if transcription is to the left of the promoter (where the enzyme RNA polymerase initiates transcription), and purine clusters if transcription is to the right of the promoter (Fig. 6-1). Base clustering does not necessarily imply an extensive local conflict with Chargaff's second parity rule. For example, a run of **T** residues, might be accompanied by a corresponding number of dispersed **A** residues, so that **A**% $\cong$ **T**%.

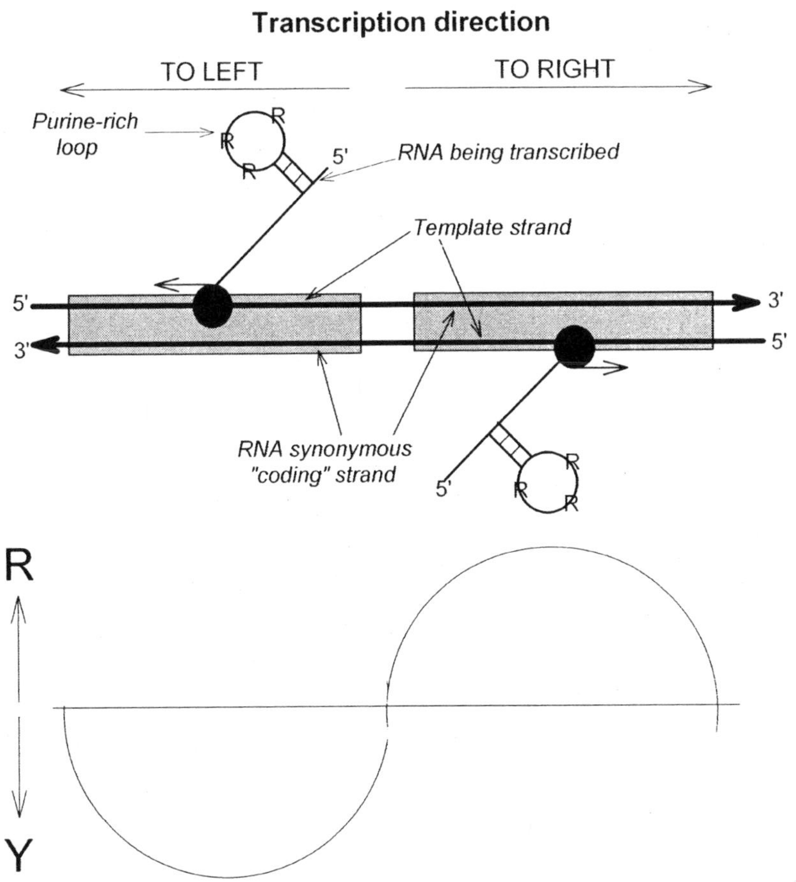

Transcription direction

TO LEFT

TO RIGHT

Purine-rich loop

RNA being transcribed

Template strand

RNA synonymous "coding" strand

R

Y

Fig. 6-1. Purine-loading of loops in RNAs is reflected in deviations from Chargaff's second parity rule in the corresponding DNA regions from which the RNAs are transcribed. Heavy black horizontal arrows refer to the "top" and "bottom" strands of duplex DNA. Two genes (boxes) are transcribed from a central promoter (site of binding of RNA polymerases), one to the left, and one to the right. Black balls with thin horizontal arrows (indicating the direction of transcription) are RNA polymerases that have moved either left or right from the promoter transcribing RNA molecules, which are shown projecting at an angle from the horizontal axis. As the RNA molecules grow they assume stem-loop structures with **R**-rich loops (purine-loading). The graphical representation at the bottom shows that in a region of leftward transcription the "Chargaff difference" for the top (template) strand favours pyrimidines ("**Y**-skew"). In a region of rightward transcription the "Chargaff difference" for the top (RNA-synonymous) strand favours purines ("**R**-skew"). If the RNA in question is mRNA, then RNA-synonymous strands are also "coding strands"

However, Oliver Smithies showed that there are distinct local deviations from the second parity rule, which again correlate with transcription direction [3]. Thus, clustering of clusters can result in local deviations from the second parity rule in favour of the clustered bases. When transcription of mRNA is to the left, the top strand is the "mRNA template" strand (pyrimidine-rich), and the bottom strand is the "mRNA synonymous," or "coding," strand (purine-rich). When transcription of mRNA is to the right, the top strand is the coding strand (purine-rich), and the bottom strand is the template strand (pyrimidine-rich). It follows that, whether arising from a gene transcribed to the left or to the right, the transcribed RNAs themselves tend to be purine-rich ("purine-loaded"). As will be considered further in Chapters 8 and 9, purine-loading is extreme in thermophiles (Fig. 6-2).

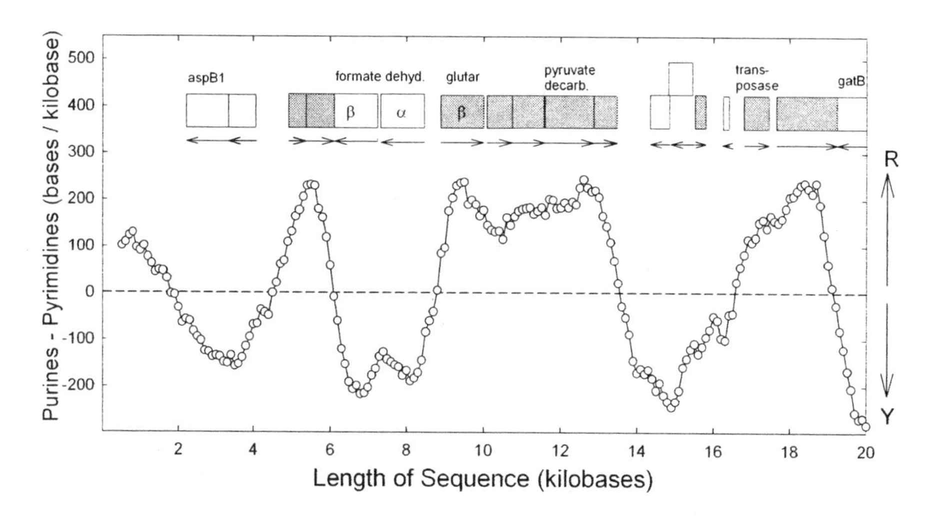

Fig. 6-2. Purine or pyrimidine excess in the top strand of the first 20 kb of the genomic sequence of the thermophile *Methanococcus jannaschii*, a bacterium-like microorganism classified as a species of archaea. A 1 kilobase sequence window was moved in steps of 0.1 kilobase and base compositions were determined in the window at each step. Total pyrimidines were subtracted from the total purines to give the recorded values (in bases/kilobase). When there are equal quantities of purines and pyrimidines, the difference ("Chargaff difference") between purines and pyrimidines is zero. Thus, points above the dashed horizontal line correspond to purine excess, and points below the dashed horizontal line correspond to pyrimidine excess. The locations of genes (open reading frames) are shown as boxes (white when transcribed to the left and grey when transcribed to the right). Transcription directions are also marked by horizontal arrows below each box. While most genes correspond to proteins that are currently only hypothetical, some have been tentatively identified, and their abbreviated names are shown above the boxes

Smithies used non-overlapping "windows" of a constant size (approx. 0.1 kilobase) to examine base composition. Bases were counted in a window and then the window was moved along the sequence, and the count repeated. Thus, a base composition profile for a genomic region was constructed. When sequences of much larger genomes became available, it was found that window sizes between 0.5 kilobase and 1 kilobase (overlapping or non-overlapping) were optimal for determining the locations of genes (open reading frames) and their transcriptional orientation in previously uncharted DNA [4–6]. In most cases, the transcripts were mRNAs that could be translated into the corresponding proteins. However, the purine-loading phenomenon is a characteristic of all transcripts, whether they are templates for a protein product (mRNAs), or exist in their own right (e.g. tRNA, rRNAs and other RNAs with various functions)[7].

Polarity

The text you are now reading was written in the direction you are reading it – most helpful if you had wanted to read it as it was being written. Looking over my shoulder, you could have begun reading before the sentence was complete. But the nature of the present medium is such that there is a considerable delay between the time of composition and the time of reading. So the sentence could have been written backwards, for all you could care. Or the middle words could have been written first, and then words at the ends added later. All you want is that the final product be in the left-to-right order with which you are familiar.

As discussed in Chapter 4, we write nucleic acid sequences from left to right beginning with the 5' end. Because we have, by convention, chosen to write them this way, it does not necessarily follow that nucleic acid sequences are assembled in biological systems in the same way. Yet, it is a chemical property of the nucleotide "building blocks" from which nucleic acids are composed (Fig. 2-2), that the 5' end of one is joined to the 3' end of the nucleotide that has already been added to the growing chain (Figs. 2-3, 2-4). In biological systems nucleic acid synthesis has polarity, beginning at the 5' end and proceeding to the 3' end. This applies not only for replication – the copying of DNA information into DNA information, but also for transcription – the copying of DNA information into RNA information.

Furthermore, a nucleic acid sequence is "read" *biologically* in the same direction as it has been "written" *biologically*. Translation – the decoding of mRNA information to form a sequence of amino acids – proceeds from 5' to 3'. Proteins have their own polarity, beginning and ending at what, for chemical reasons, are known as the amino-terminal, and carboxyterminal ends, respectively. Proteins are "written" with this polarity. But proteins fold into unique structures, so are not "read" by the molecules, with which they interact three-dimensionally to generate the phenotype (Fig. 1-4), in the same way as they are synthesized. Similarly, nucleic acid transactions that require secondary and higher ordered structures (see Chapter 5), are usually uninfluenced by the way the nucleic acid sequence was assembled.

However, in some cases there are alternative, sometimes energetically equivalent, folding patterns. Folding can begin at the end of a protein or nucleic acid that is synthesized first, before the synthesis of the other end is complete. This early folding can influence the folding path adopted by a later synthesized part of a molecule. To this extent, the order of synthesis of the primary sequence can influence the folding pattern, and hence the subsequent three-dimensional interactions, of a macromolecule. Furthermore, a macromolecule that is synthesized rapidly may not be able to follow slow folding paths. In this circumstance there may be selection for codons that are translated slowly (codon bias; see Chapter 9), perhaps because of a deficiency (quantitative or qualitative) in the corresponding transfer RNAs. This slows the synthesis rate and allows slow folding processes to occur. Even so, the selection of a particular folding pathway may not be able to proceed without the aid of other molecules (molecular chaperones; see Chapters 12 and 13).

Origin of Replication

The enzyme RNA polymerase initiates transcription at distinct sites in the genome (promoters) that have certain general features (e.g. often there is the base sequence **TATA**, known as the "TATA box"). The RNA transcript is then synthesized sequentially by the addition of ribonucleotides (nucleotides with ribose as the pentose sugar), one at a time, until appropriate termination signals are encountered (again, often there is a distinctive base sequence). Because there is usually a need to transcribe a particular RNA at a particular time, the enzyme has to begin and end transcription at distinct sites where transcription can be regulated.

However, since an entire chromosome (often circular in microorganisms) has to be replicated, in theory the enzyme DNA polymerase could initiate DNA replication at any site by joining together deoxyribonucleotides (nucleotides with deoxyribose as the pentose sugar) according to the dictates of the parental DNA template. The enzyme could then proceed from that starting point round the circular genome until returning to the starting point.

Yet, presumably to facilitate regulation, there is usually one distinct origin of replication in a circular chromosome (say 12 o'clock), from which DNA replication proceeds leftwards and rightwards, bidirectionally. Thus, when the chromosome is half-replicated there are at least two DNA polymerases tracking in different directions (say at 9 o'clock and 3 o'clock; Fig. 6-3). They complete replication by meeting at a point in the circle approximately opposite the starting point (say, at 6 o'clock). This is the point of termination.

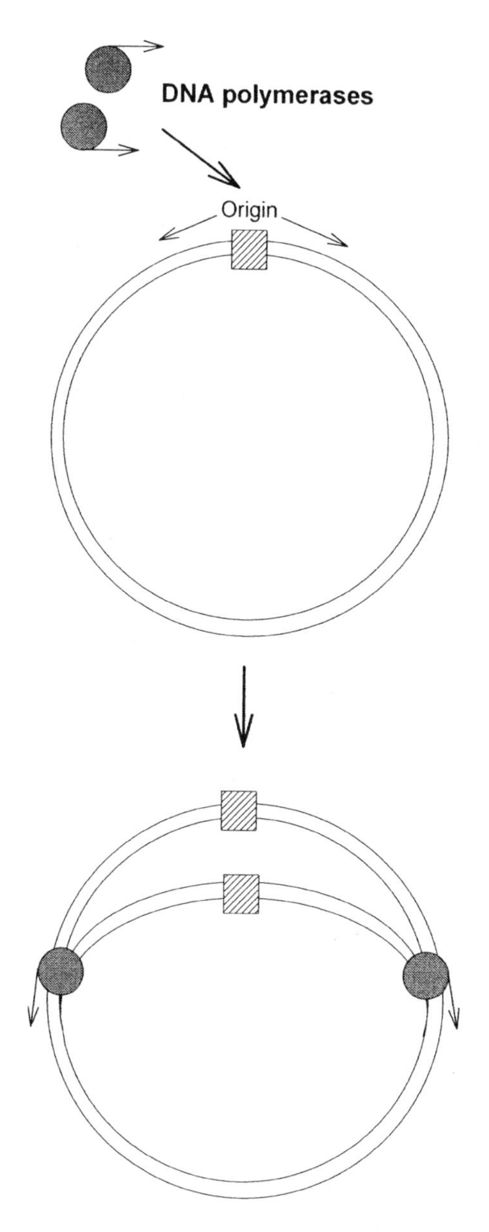

Fig. 6-3. Bi-directional replication of a circular chromosome from a single origin (box) by DNA polymerases (grey balls) that attach at the origin of replication. One DNA polymerase travels anticlockwise and the other DNA polymerase travels clockwise, so that at any moment there are two Y-shaped replication forks as shown in Figure 2-3. It is possible that there is more than one polymerase at each replication fork – at least one for the "leading" strand and at least one for the "lagging" strand (see Fig. 6-6.)

In a short linear chromosome, the origin is usually at, or near, one end. In a long linear chromosome there may be multiple origins so that segments of the chromosome may be replicated simultaneously by independent DNA polymerases, and then the segments joined together by other enzymes ("ligases").

Origins and terminations of replication often have distinctive sequence characteristics, but these alone are usually insufficient for us to determine origins and terminations by inspection of the DNA sequence. However, Szybalski and his colleagues found for the circular lambda phage genome, that genes to the left of the origin of replication were transcribed to the left, and genes to the right of this origin were transcribed to the right [8]. Since leftward-transcribing and rightward transcribing genes are distinguishable by top strand pyrimidine-loading and purine-loading respectively, it follows that the switch from pyrimidines to purines, and vice-versa, provides a method of determining origins of replication and sites of termination (Fig. 6-4).

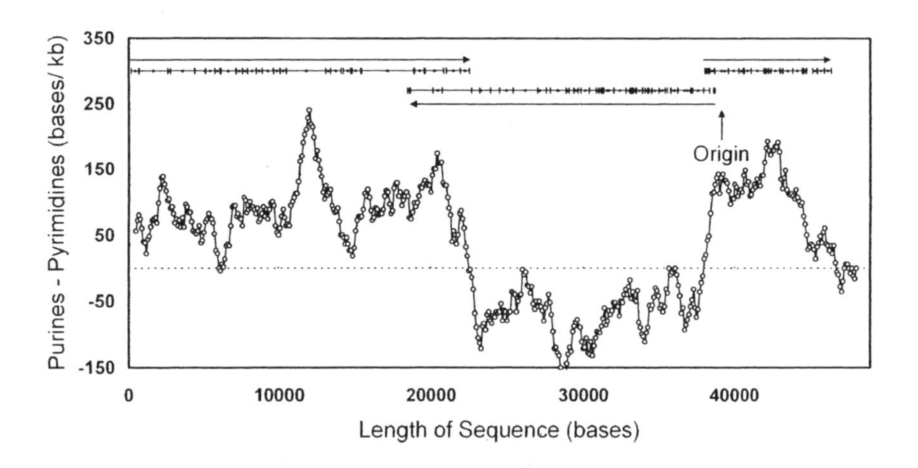

Fig. 6-4. Purine or pyrimidine excess in the top strand of the 48 kilobase genomic sequence of lambda phage. Base frequencies were determined in 1 kilobase windows, which were moved in steps of 0.1 kilobase. Since numbers of pyrimidines are subtracted from numbers of purines, positive values indicate top-strand purine-loading, and negative values indicate top-strand pyrimidine-loading. Horizontal lines with vertical cross-hatching show the location of genes. Long continuous horizontal arrows show the collective transcriptional orientation of groups of genes. The vertical arrow refers to the origin of replication within the interval 38686-39585 bases, from which replication proceeds bidirectionally to terminate around base 21000. Thus, the horizontal rightward arrow at top left, corresponding to rightward movement of a DNA polymerase around the circular chromosome, is an extension of the horizontal rightward arrow at top right

The possible general nature of this became evident from studies in organisms considered higher in the evolutionary scale [3]. In the circular genome of SV40 virus, which infects monkeys, genes are transcribed to the left, to the left of the origin of replication, and here **C% > G%**; genes are transcribed to the right, to the right of the origin of replication, and here **G% > C%**. When further sequences became available the relationship to the origin of replication was found to be a feature of many microbial genomes [9]. Instead of determining the origin of replication by differences between purines and pyrimidines in moving sequence windows ("skew analysis"), an interesting variation is to allow the differences to summate ("cumulative skew"; Fig. 6-5).

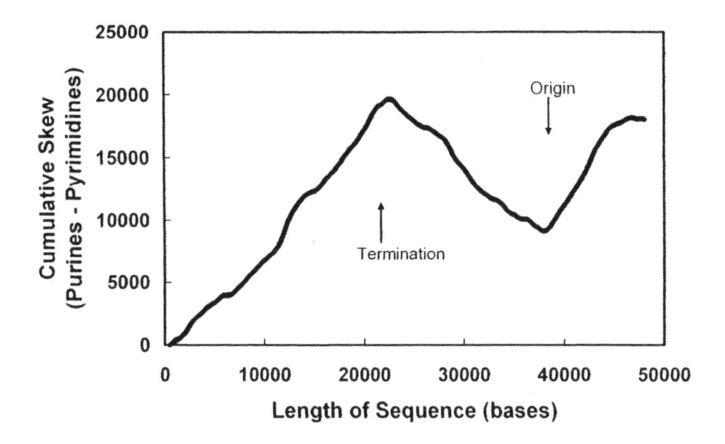

Fig. 6-5. "Cumulative skew" presentation of the lambda phage data shown in Figure 6-4. Instead of plotting directly, values for consecutive windows from left to right are progressively summed so that the curve slopes upwards in regions of rightward transcription and downwards in regions of leftward transcription. Origin and termination regions for DNA synthesis are indicated by inflections around 21000 bases (termination) and 39000 bases (origin)

Yet, although often highly informative, for some genome sequences skew plots (Chargaff difference plots) or cumulative skew plots (cumulative Chargaff difference plots) give ambiguous results. While reasons for this ambiguity remain uncertain, one factor causing ambiguity is the movement of genes, or groups of genes, from one part of a genome to another (transposition). This movement provided an opportunity to address the chicken-and-egg question as to whether replication direction can determine transcription direction, or vice-versa. If a rightward-transcribing gene to the right of the origin of replication were transposed to the left of the origin of replication, but kept its orientation so it was still rightward-transcribing, would it retain its purine-loading (i.e. would transcription overrule replication)? The answer seems to be that it begins to accept mutations from purines to pyrimidines (i.e. replication overrules transcription). The mutations can sometimes change amino acids, so that a gene transposed in this way has increased opportunities for functional change [10].

Why should there be a relationship between replication direction and transcription direction? One suggestion is that it is necessary not to interrupt transcription when replication is occurring in the same part of the genome. Perhaps there is less disruption if RNA polymerase and DNA polymerase move in the same direction on a strand of DNA that is acting as a template for synthesis of complementary anti-parallel strands of RNA or DNA, respectively [11]. One possibility is that if DNA polymerase and RNA polymerase were not moving in the same direction, they might collide "head-on." A result of such a collision might be that DNA in the region could end up irreversibly knotted [12].

Leading and Lagging Strands

The facts that nucleic acid synthesis is directional, from 5' to 3', and that the two strands of the DNA duplex are antiparallel (Fig. 2-4), has the interesting implication that one of the two child strands at the Y-shaped replication fork (Fig. 2-3) has to be synthesized in the opposite direction to the direction in which the replication point itself is moving. This is shown in Figure 6-6 in which the two strands are labeled as "leading" and "lagging," depending on whether DNA synthesis is continuous ("leading strand"), or discontinuous, being synthesized in fragments that are later joined up ("lagging strand"). As drawn, this does not seem very elegant. Surely Nature could do better? Nature probably does. A three dimensional model with a multiplicity of mobile molecular actors might do her more justice.

In passing we should recall that the two parental strands in the stem of the Y are actually coiled around each other as a double helix (Fig. 2-1). Furthermore, when DNA synthesis is complete the two new duplexes go their separate ways. To separate the parental strands for child strand DNA synthesis, the helix has to be uncoiled (unwound). If the entire chromosome is circular, this means that for every uncoiling step there has to be, first a transient strand breakage, then an uncoiling, and then a rejoining of the broken ends. Long linear DNA molecules are anchored in such a way that they usually also require this. The uncoiling is carried out by enzymes (topoisomerases) which, when a DNA duplex is not being replicated, tend to uncoil it slightly without strand breakage. This "negative supercoiling" places the helix under stress, a stress that can be partly relieved by the extrusion of stem-loops (see Fig. 5-2). Thus, stem-loop extrusion is normally favored energetically by the negatively supercoiled state of DNA duplexes.

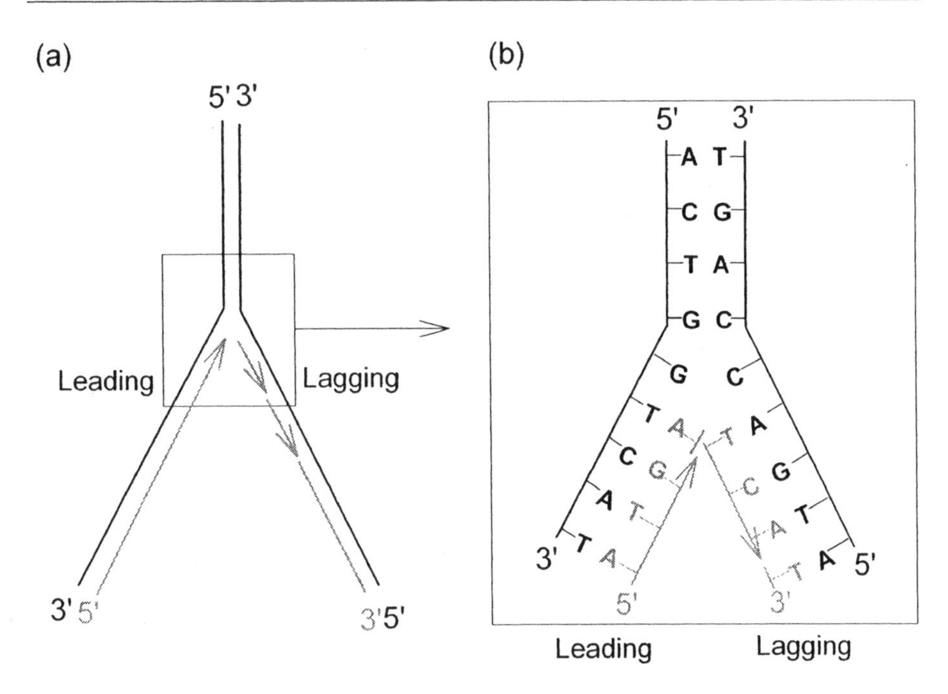

Fig. 6-6. Leading and lagging strand synthesis of DNA at the inverted Y-shaped point of DNA replication. Since parent (black) and child (grey) strands are antiparallel, on the left limb of the replication fork the "leading" child strand can be synthesized in the same direction as overall replication is occurring (with the separation of the two parental strands; see Fig. 2-3). On the right limb of the replication fork, the "lagging" strand is synthesized in a direction opposite to that in which overall replication is occurring. Discontinuous child strand DNA synthesis generates fragments that are joined up to make a continuous strand, so that overall replication occurs in the 3' to 5' direction. The detailed figure with bases (boxed in *(b)*) is part of a larger scale figure without bases *(a)*, which indicates the initial fragmentary nature of the lagging strand

Purine loading

Chargaff's main interest in the base cluster phenomenon was that, prior to nucleic acid sequencing technology, it provided some measure of the uniqueness of the base order of a nucleic acid. Szybalski's main interest was that the clustering might played a role in the control of transcription. This implied an evolutionary selection pressure for clustering so that organisms with clusters would better control transcription than organisms that did not have clusters. However, following a better understanding of Chargaff's second parity rule as a reflection of nucleic acid secondary structure (see Chapters 4 and 5), a case could be made that a selection pressure for clustering had arisen at the level of complete RNA transcripts (i.e. after transcription had occurred), and was related to the secondary structure of individual RNA molecules.

An important implication of what may be called "Szybalski's transcription direction rule" is that RNAs, in general, tend to be purine-loaded (Fig. 6-7). This was suggested by Chargaff's work on the base composition of RNA extracted from various species, but his data would then have mainly reflected the compositions of the most abundant RNA type, the rRNAs [13, 14].

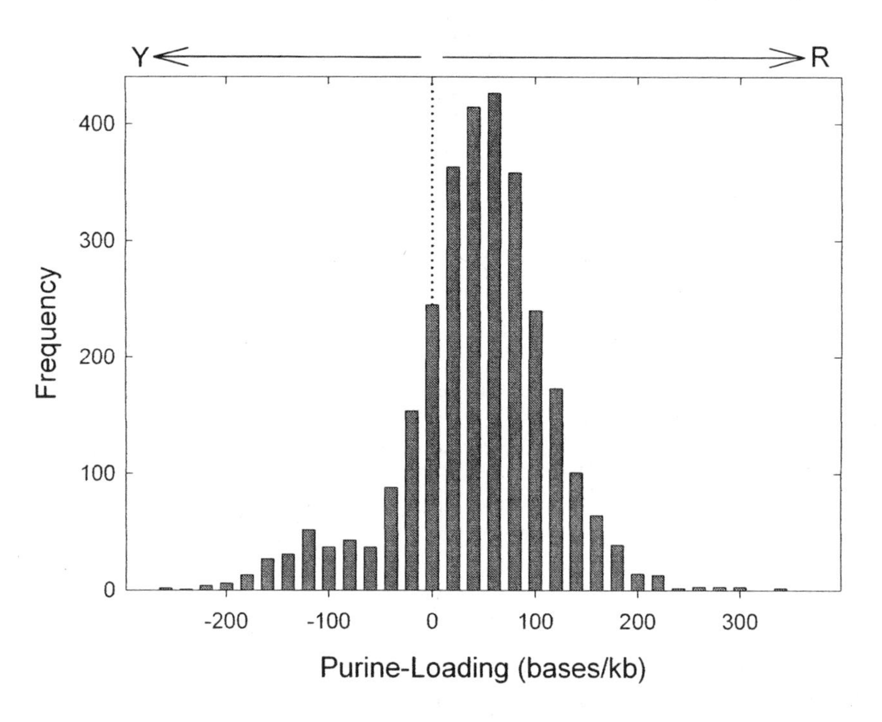

Fig. 6-7. Distribution of purine-loading among 3000 species. Positive X-axis values indicate purine-loading (**R** > **Y**). Negative values indicate pyrimidine-loading (**Y** > **R**). The purine-loading value for all human genes (excluding mitochondria) is 42 bases/kilobase, so that, on average, there are 42 more purines than pyrimidines for every kilobase of coding sequence. The shoulder with negative purine-loading values (i.e. pyrimidine-loading) corresponds mainly to mitochondria. Purine-loading of coding regions was calculated from codon usage tables for all species represented in the GenBank database (in 1999) by four or more genes. The purine-loading value (bases/kilobase) for a species is calculated as 1000[(**G**-**C**)/**N**] + 1000[(**A**-**T**)/**N**], where **G**, **C**, **A**, and **T** correspond to the numbers of individual bases, and **N** corresponds to the total number of bases in the codon usage table for that species. This measure of the purine-loading of RNAs disregards 5' and 3' non-coding sequences in mRNAs. When purine-loading is zero, (**A**+**G**)% is 50. This may be calculated from the formula: Purine-loading = 20[(**A**+**G**)%] – 1000

Given that RNAs in general tend (i) to be loaded with runs of purines, and (ii) to have an elaborate secondary structure, where in the structures would purine clusters be found? Since, for base-pairing in duplex stems, purine clusters must be matched with complementary pyrimidine clusters, and since pyrimidine clusters are scarce in RNAs, purine clusters should occupy the unpaired regions of RNA secondary structures, namely the loops. Indeed, this is where they are found in calculated mRNA and ribosomal RNA (rRNA) structures [5, 15, 16]. Why should it be advantageous for an RNA to "load" its loops with purines?

A possible answer derives from the studies of Jun-ichi Tomizawa on the way a "sense" RNA molecule and its complementary "antisense" RNA molecule interact, prior to forming a double-strand duplex molecule (dsRNA) [17, 18]. In some organisms such sense-antisense interactions are important in regulation. In some experimental situations, antisense RNAs and DNAs are employed to bind to complementary sense mRNAs to impair their translation [19]. Tomizawa found that sense sequences in RNA search out complementary antisense sequences through "kissing" interactions between the tips of the loops of stem-loop structures. If Watson-Crick base-pairing between the loops is achieved, then the formation of a length of dsRNA becomes feasible (Fig. 6-8). Thus, if RNA loops were purine-loaded, "kissing" interactions would decrease (because purines pair poorly with purines), and hence the probability of forming dsRNA would decrease.

Why would *failure* to form dsRNA be of selective advantage? Or, in other words, why would formation of dsRNA be disadvantageous? One explanation arises from the physico-chemical state of the fluid part of the cytoplasm (the cytosol), where macromolecules are highly concentrated. The "crowded" cytosol, is an environment that besides encouraging interactions between different parts of a molecule (*intra*molecular interactions), also encourages interactions between molecules (*inter*molecular interactions; see Chapter 13).

That RNA-RNA interactions would be affected can be deduced from first principles. Cytosolic characteristics such as salt concentration and hydrogen ion concentration (i.e. degree of acidity or alkalinity) should have been fine-tuned to support the rapid and accurate interaction between codons in mRNA sequences and the anticodons of tRNA molecules that are so vital for protein synthesis. As a by-product of this, interactions between RNA molecules in general would also have been facilitated. However, when so tied up in RNA-RNA interactions, mRNAs might be less available for association with ribosomes where protein synthesis occurs. Purine-loading should militate against mRNA-mRNA interactions, while not compromising mRNA-tRNA interactions.

Whatever the merits of this argument, which at the time of this writing remain to be investigated [20], it would follow that RNAs might have achieved the same result by adopting pyrimidine-loading rather than purine-loading (since pyrimidines pair poorly with pyrimidines). All the RNAs in a given cytosol should adopt the same strategy (purine-loading or pyrimidine-loading), so this can be compared with the decision made within a country for all vehicles to drive either on the left side or on the right side of the road. Yet, within the RNAs of a species there is variation in the extent of purine-loading. Indeed, some "maverick" RNAs showing pyrimidine-loading.

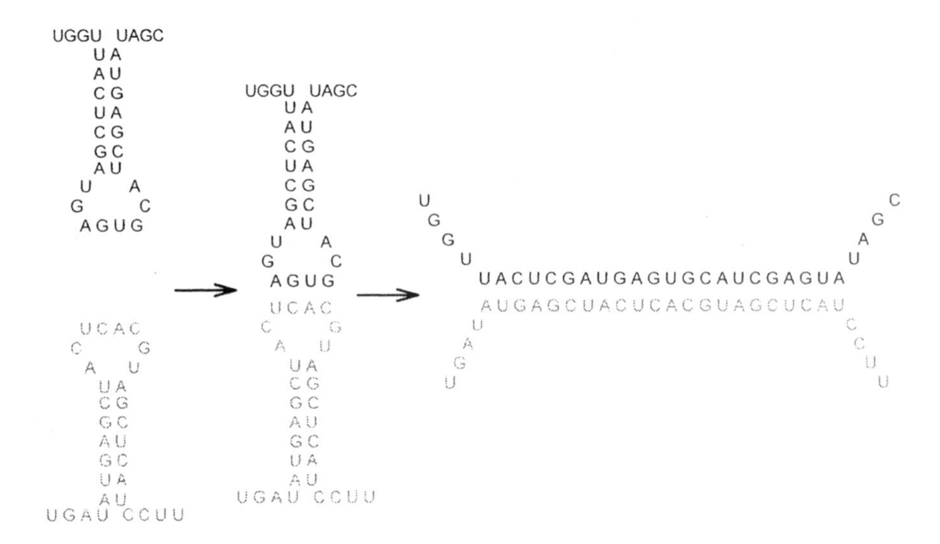

Fig. 6-8. A stem-loop "kissing" model for the initiation of hybridization between two RNA species (black and grey base letters). The physico-chemical state of the crowded cytosol (see Chapter 13) is conducive to intramolecular interactions, so that single-stranded RNAs in solution rapidly adopt energetically stable stem-loop structures. These structures must be disrupted if two free RNA species are to form a mutual duplex. Accordingly, there are two steps in duplex formation. The two RNAs first interact transiently at the tips of stem-loops ("kissing"), mainly by conventional **G-C** and **A-U** base-pairing. If this pairing is sufficiently stable, it propagates progressively, and the stem-loop structures unfold to generate a duplex. In the present case, the pairing cannot proceed beyond 22 base-pairs (about two helical turns), since beyond this limit base-complementarity is lost. Here, as in Figure 5-2, all duplex structures are dynamic, undergoing vibratory partial disruptions ("breathing"), which, as the critical temperature is approached, result in increasing degrees of strand separation. Thus, the reverse process probably occurs in one step (i.e. there might not be a distinct multistep path). For this, the duplex would have to be heated to a critical temperature when the two strands would suddenly separate and each would immediately adopt the stable stem-loop configurations shown at the left

Pyrimidine-Loading

From the average distribution of loading (purine versus pyrimidine) among species (Figure 6-7) it can be seen that, although most species purine-load, some pyrimidine-load. Mitochondria, the intracellular organelles concerned with energy generation, generally pyrimidine-load their RNAs. A cell may have its cytosolic RNAs "driving" on the purine side of the road, but its mitochondria, which exist in the same cytosol (but have their own genomes, cytosols, and RNA types), "drive" their RNAs on the pyrimidine side. Thus, some aspect of the intra-mitochondrial environment may have been particularly conducive to pyrimidine-loading of mitochondrial RNAs. On the other hand, during evolution the first mitochondrial "Eve" may just have happened to have its RNAs pyrimidine-loaded, and this feature was retained in all subsequent mitochondrial lineages.

Since transcription in mitochondria tends to be unidirectional (i.e. most genes are orientated together in one strand), there is usually a strong purine/pyrimidine asymmetry between the two strands of mitochondrial DNA (i.e. Chargaff's second parity rule is strongly violated). The "top," pyrimidine-loaded, mRNA-synonymous, strand is likely to be less vulnerable to mutation than the "bottom," purine-loaded, strand, which, in its role as template, is likely to be more exposed to the alkaline intra-mitochondrial environment. Could this explain the evolution of pyrimidine-loading? Alkali favors deamination of cytosine (see Chapter 15), while militating against depurination. Thus, it would appear adaptatively advantageous to assign pyrimidines to the less exposed top strand. A purine-rich template strand should be less vulnerable to mutation (or perhaps easier to repair), than a pyrimidine-rich template strand.

Mitochondria generate a highly mutagenic witch's brew of toxic oxygen derivatives (superoxides and hydrogen peroxide). It is perhaps for this reason that there are multiple "back-up" copies of the genome (at least ten in each human mitochondrion), that may facilitate DNA repair (see Chapter 14). The maintenance of mitochondrial function is of fundamental importance, not only because of their role in energy generation (ATP synthesis), but also because mitochondria can release proteins that instigate "programmed cell death" (apoptosis; see Chapter 13). A failure to maintain the integrity of mitochondrial DNA would imperil mitochondrial function and hence the life of their host cell. It is perhaps for this reason (among others) that there are multiple "back-up" mitochondria per cell, and even suggestive evidence that mitochondrial DNA, or perhaps intact mitochondria, can transfer between cells [21].

Mitochondrial genomes are attractive subjects for bioinformatic investigation. Generally they are quite small (16.6 kilobases in humans), and many sequences are available. Furthermore, wherever their host cells go, there must mitochondria go. Thus, if a host cell has adapted to grow in an extreme environment, it is likely that its mitochondria have similarly adapted.

In some species groups there is a reciprocal relationship between the degree of purine-loading and **(G+C)%** (see Chapter 9). When purine-loading is high, **(G+C)%** is low. Conversely, when purine-loading is low (i.e. pyrimidine-loading is high), **(G+C)%** is high. This does not apply to mitochondria. Across all species examined, mitochondria have pyrimidine-loading (top strand) *and* low **(G+C)%** (both likely to decrease vulnerability to deamination of cytosine). We should also note that, while most viruses are purine-loaded like their host cells, some are strongly pyrimidine-loaded, and these have high **(G+C)%** values (e.g. HTLV1 see Chapter 8; EBV see Chapter 11). This will be further considered in Chapter 12 where the role of double-stranded RNA (dsRNA) as an intracellular alarm signal is reviewed.

Summary.

The two DNA strands of a gene are the mRNA-synonymous "coding" (codon-containing) strand and the mRNA-template (anticodon-containing) strand. Clusters of clusters of purines are general features of coding strands of DNA, which usually contain an excess of purines ("purine-loading"). Accordingly, since purines pair with pyrimidines, complementary clusters of clusters of pyrimidines are general features of the corresponding template strands, which usually contain an excess of pyrimidines ("pyrimidine-loading"). The clustering of clusters within a gene locally violates Chargaff's second parity rule. This permits prediction of transcriptionally active regions in uncharted DNA and, in some cases, the origins and termination sites of DNA replication. RNA transcription and DNA replication appear to proceed optimally when the enzymes performing these functions (polymerases) are moving in the same direction along the DNA template. Since the stems of nucleic acid stem-loop structures require equal numbers of complementary bases, clusters tend to occupy loop regions. Here they would militate against the loop-loop "kissing" interactions that precede the formation of nucleic acid duplexes. Thus, the loading of loops with bases that do not strongly base-pair with each other should decrease unproductive interactions between nucleic acids, so leaving them freer to engage in productive interactions, such as those between mRNAs and tRNAs that are critical for protein synthesis.

Part 3 Mutation and Speciation

Chapter 7

Species Survival and Arrival

> "We have all grown up in the greatest confidence that we all knew ... what Darwin meant. I am very tired of having some excessively loosely expressed truism, such as that 'all defective deer must be devoured by tigers', put forward as 'the ordinary Darwinian argument'."
>
> Ronald Fisher (1930) [1]

If it had inherited a mutated gene that impaired its ability to run, then, relative to its companions, a deer would be defective and thus subject to selective devourment by predators. This is Tennyson's "Nature, red in tooth and claw," and Darwin's "natural selection." In 1862, shortly after reading Darwin's *Origin of Species by Means of Natural Selection* while breeding sheep in New Zealand, the 27 year old Butler appeared to accept "the ordinary Darwinian argument," albeit with some important caveats [2]:

> "That the immense differences between the camel and the pig should have come about in six thousand years is not believable; but in six million years it is not incredible... . Once grant the principles, once grant that competition is a great power in Nature, and that changes in circumstances and habits produce a tendency to variation in the offspring (no matter how slight that variation may be), and unless you can define the possible limit of such variation during an infinite series of generations, unless you can show that there is a limit, and that Darwin's theory oversteps it, you have no right to object to his conclusions."

Yet in 1930 the great statistician Ronald Fisher began the preface of his book, *The Genetical Theory of Natural Selection*, with the statement: "Natural Selection is not Evolution" (Fisher's capitalization) [3]. In so doing he drew attention to a question that had much concerned Darwin. To what extent can evolution be due to natural selection? Seeming to look to causes beyond natural selection, Fisher placed two quotations in a prominent position at the opening of his first chapter. The first was from an 1856 letter by Darwin that downplayed the evolutionary role of "external conditions" (e.g. tigers) [4]:

> "At present ... my conclusion is that external conditions do *extremely* little, except in causing mere variability. This mere variability (causing the child *not* closely to resemble its parent) I look at as *very* different from the formation of a marked variety or new species." [Darwin's italics]

The second was from a paper by William Bateson, who by 1909 had become familiar with Butler's writings [5]:

> "As Samuel Butler so truly said: 'To me it seems that the "Origin of Variation",whatever it is, is the only true "Origin of Species"'."

However, despite the caution implied by these quotations, Fisher set out to show, using mathematical procedures referred to by biologist Ernst Mayr as "bean bag genetics," that natural selection was, in essence, evolution. Natural selection was responsible not only for species *survival*, but also for species *arrival*. Thus, Fisher fanned the classical Darwinist flame long tended by Alfred Wallace and August Weismann, which had flickered but never faded under the attacks of Bateson and others in the early decades of the twentieth century. We will here first review the mutational processes that give rise to variant organisms that may be differentially selected. We will then considers a special type of mutation, which causes a chromosomal "repatterning" that may be conducive to a different kind of selection, so leading to an "arrival" of species.

Separating Process from Result

Members of a biological species vary. We acknowledge as a "variation" an observed character in an organism that is different from the character seen in most other members of its species. Hence the organism would be called a "variant." The difference may be in *number*, or in *arrangement*, or in an actual *character itself.* Long before the nature of mutations at the DNA level was understood, Bateson suggested that these distinctions were fundamental. Variations in the number or arrangement of body parts, he deemed "meristic" and "homeotic" variations, respectively. Variations in the substances of which the parts were composed (i.e. their characteristics), he deemed "substantive" variations. Thus, a variant organism with an extra, but otherwise normal, finger (Fig. 7-1) would be a meristic variant (Greek: *meros* = part). A variant organism with rearrangements of body parts (e.g. an insect's leg where its antenna should be, or a flower's stamen where a petal should be) would be a homeotic variant (Greek: *homeo* = same). In Bateson's words:

"The essential phenomenon is not that there has merely been a change, but that something has been changed into the *likeness* of something else." A variant organism with loss of normal colouration (e.g. an albino) would be a substantive variant [6].

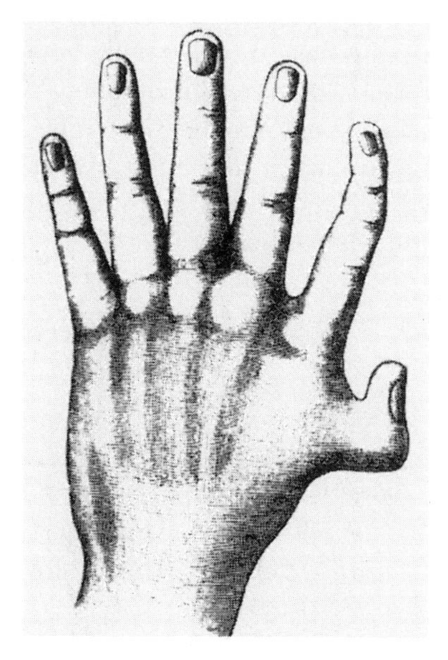

Fig. 7-1. A meristic variation (from William Bateson's 1894 text)

It was difficult for Bateson to imagine that variations in number or arrangement could have originated in the same way as variations in characters themselves. He could not separate the cause of a variation (now regarded as a "spontaneous" or induced mutation in DNA) from the actual variation observed in the organism itself. He could not separate the mutational (variant-generating) *process* from the *result*.

We now know that mutations at the DNA level may sometimes produce no observable effect. Among mutations producing an observable effect, some happen to result in meristic variations, some happen to result in homeotic variations, and some happen to result in substantive variations. It is true that there are different types of mutation (changes in DNA), but a given type of DNA mutation does not correspond to a particular type of observed variation. A single base change from, say, **T** to **G** ("micromutation") can produce extreme changes in structure or physiology (phenotype). A deletion of a large segment of DNA ("macromutation") can result in no phenotypic change.

Note that, in the scientific literature, a mutation at the DNA level, and the observed result of that mutation (a "variant" or "mutant" organism) may both be called "a mutation." So whether the word refers to cause or effect must be determined by context. By the same token, "micromutation" and "macromutation" may refer, as above, to an actual mutation at the DNA level, or to the extent of a change in phenotype. Again, context is important since, Goldschmidt used these terms to distinguish fundamentally different mutational processes (Table 3-1). A mutation at the DNA level may be repaired, sometimes within seconds. In this case it will usually be both chemically and biologically undetectable (i.e. with current technology we can only hypothesize that such mutations occur). If not repaired, it will be an "accepted mutation." This acceptance initially occurs within an individual and then may spread, sometimes by positive selection (see below), sometimes randomly (e.g. "drift" in small populations), to other members of a species.

First Position	Second Position: U		C		A		G	
U	UUU	Phe	UCU	Ser	UAU	Tyr	UGU	Cys
	UUC	Phe	UCC	Ser	UAC	Tyr	UGC	Cys
	UUA	Leu	UCA	Ser	UAA	Stop	UGA	Stop
	UUG	Leu	UCG	Ser	UAG	Stop	UGG	Trp
C	CUU	Leu	CCU	Pro	CAU	His	CGU	Arg
	CUC	Leu	CCC	Pro	CAC	His	CGC	Arg
	CUA	Leu	CCA	Pro	CAA	Gln	CGA	Arg
	CUG	Leu	CCG	Pro	CAG	Gln	CGG	Arg
A	AUU	Ile	ACU	Thr	AAU	Asn	AGU	Ser
	AUC	Ile	ACC	Thr	AAC	Asn	AGC	Ser
	AUA	Ile	ACA	Thr	AAA	Lys	AGA	Arg
	AUG	Met	ACG	Thr	AAG	Lys	AGG	Arg
G	GUU	Val	GCU	Ala	GAU	Asp	GGU	Gly
	GUC	Val	GCC	Ala	GAC	Asp	GGC	Gly
	GUA	Val	GCA	Ala	GAA	Glu	GGA	Gly
	GUG	Val	GCG	Ala	GAG	Glu	GGG	Gly

Table 7-1. The genetic code as it would appear in mRNA. At the DNA level **T** would replace **U**. Amino acids are represented as three letter abbreviations. **UAA**, **UAG** and **UGA** are stop codons that terminate a growing chain of amino acids. The amino acids also have single-letter abbreviations, usually corresponding to their first letter. Exceptions are **Phe** (F), **Tyr** (Y), **Trp** (W), **Gln** (Q), **Arg** (R), **Asn** (N), **Lys** (K), **Asp** (D), and **Glu** (E)

The Genetic Code

We refer to *the* genetic code because, with few exceptions, all species use the same code. The code defines a relationship between amino acids and groups of nucleic acid bases (Table 7-1). Each position in a triplet codon can be occupied by one of four bases. There are four alternatives for the first position, four alternatives for the second position, and four alternatives for the third position. Thus, there are 4 x 4 x 4 = 4^3 = 64 possible triplets combinations (see Table 4-2). Sixty-one code for the twenty amino acids normally found in proteins, and the remaining three provide alternative signals for stopping the synthesis of a protein.

Since there are more triplet codons than amino acids, and all codons are potentially employable, then the code is described as "degenerate." There is not a reversible, one-to-one, relationship between codon and amino acid. If you know the codon you know the amino acid, but if you know the amino acid you do not know the codon. The only exceptions are **UGG**, which usually encodes only the amino acid tryptophan, and **AUG**, which usually encodes only the amino acid methionine. The latter codon is special since, in the appropriate context, **AUG** provides a signal for methionine to begin the sequence of a protein. Thus, **AUG** and the three chain-termination triplets (**UAA**, **UAG**, **UGA**) provide punctuation signals to a cell's protein synthesizing machinery, so that it "knows" when to start and stop translating the sequence of an mRNA template.

The degeneracy of the code mainly relates to the third position of codons. Thus, **GGN** is a generic codon for glycine. The third base (**N** = any base) can be any of the four bases and the encoded amino acid will still be glycine. A mutation from **GGU** to **GGC** would be a non-amino-acid-changing mutation, otherwise known as a "synonymous" mutation (i.e. the mutation might occur, but the encoded amino acid would be the same, and so would have the same name; Greek: *syn* = same, *onyma* = name). Mutations in the first and second codon positions are usually amino-acid-changing mutations, otherwise referred to as "non-synonymous" mutations (i.e. the encoded amino acid would be different, and so would have a different name).

A nucleic acid sequence can display some flexibility in that it can change considerably with respect to third codon positions while still encoding the same protein (see sample sequences 3.1 and 3.2 in Chapter 3). However, this flexibility has its limits. Extreme shifts in base composition can affect the amino acid composition of a protein. Amino acids are divided into various groups on the basis of their physical and chemical properties (e.g. acidic, basic, hydrophilic, hydrophobic). For a position in a protein sequence that must have a basic amino acid, but is otherwise flexible, a genome rich in **A** and **T** (i.e. low (**G**+**C**)%) will tend to have the codons **AAA** or **AAG**, which encode the basic amino acid lysine; for the same position, a genome rich in **G** and **C** (i.e. high (**G**+**C**)%) will tend to have one of the **GC**-rich codons for the basic amino acid arginine (**CGN**).

Mutation is usually a random process and, within a short time frame, two mutational events are unlikely to occur in close proximity in DNA (see later). Thus, one base in a triplet codon is likely to change at a time. This one-step mutational process means that a mutation in the first codon position will tend to change an amino acid to one vertically related to it in Table 7-1 (e.g. **CUU** to **GUU** shifts coding from leucine to valine). A mutation in the second codon position will tend to change an amino acid to one horizontally related to it (e.g. **CUU** to **CAU** shifts coding from leucine to histidine). A mutation in the third codon position may change the amino acid only if there is more than one type of amino acid in the same box (e.g. **CAU** to **CAA** shifts coding from histidine to glutamine).

This implies that a particular amino acid is related, through mutation, to a certain set of other amino acids; in the case of mutations affecting first and third codon positions the amino acid replacements often show some similarity to the original amino acid in their physical and chemical properties. Thus, leucine and valine (related through first position mutation) are both classed as "water-fearing" (hydrophobic) amino acids, likely to locate to the interior of a protein. Histidine and glutamine (related through third position mutation) are both "water-loving" (hydrophilic) amino acids, which may locate to the surface of a protein. To some extent, the genetic code would appear to buffer the immediate effects of a mutation in that an original amino acid would be replaced with an amino acid with somewhat similar properties. By this token, mutations involving second codon positions are the most likely to be lethal (e.g. leucine-histidine interchange).

When the actual mutational changes undergone by proteins are examined, it is seen that many of them fit the one-step pattern, but there are also many amino acid substitutions (e.g. leucine to aspartic acid) that would require more than one step (e.g. **CUU** to **GUU** or **CAU**, and then either **GUU** to **GAU**, or **CAU** to **GAU**). Thus, the mutational process may take more than one path, and this may require that an organism with an intermediate mutant form (**GUU** or **CAU** in this example) be viable for a number of generations to provide time for a second mutation to occur. Each of the twenty amino acids can mutate to one of the nineteen others, for a total of 190 possible interchanges. Of these, 75 require only single base substitutions. The remaining 115 require substitutions in two or three bases.

Since there are three bases in a codon and each base can be substituted by one of three other bases, a given codon is related by one-step mutation to nine other codons. Given 61 codons for amino acids, there are 549 (i.e. 61 x 9) possible codon base substitutions (183 for each codon position). Of these 549, it can be determined from the genetic code (Table. 7-1) that 134 would be non-amino acid-changing (synonymous) and 415 would be amino acid-changing (non-synonymous). The relationships of different types of mutations to codon positions are shown in Table 7-2. Mutations may be from purine to purine, or from pyrimidine to pyrimidine ("transition mutations"), or may be between purine and pyrimidine ("transversion mutations").

Mutations	Codon Position			
	1	2	3	All
Total	**183**	**183**	**183**	**549**
Synonymous	**8**	**0**	**126**	**134**
Amino acid-changing	**175**	**183**	**57**	**415**

Table 7-2. Distribution of non-amino acid-changing (synonymous) and amino acid-changing (non-synonymous) mutations among the three codon positions, as calculated from the genetic code. There are many technical problems in arriving at accurate values for the numbers of synonymous and amino acid-changing differences between two similar nucleic acid sequences. Mathematically-inclined biologists spend much time on this. However, for many purposes it suffices to assume that amino acid-changing mutations involve first and second codon positions, and synonymous mutations involve third codon positions (see boxes)

RNY Rule

A codon in mRNA, and its anticodon at the tip of a tRNA loop, transiently "kiss" to form a short helical duplex (see Fig. 5-1). The specificity of this interaction appears to depend on complementary base pairing of three or less of the bases in a triplet codon. However, codon-anticodon interactions are more than triplet-triplet interactions. Neighboring bases are drawn in so that there is a quintuplet-quintuplet interaction. This is because a major component of the interaction energy between two complementary bases derives from the stacking interactions with their neighbors (see Chapter 5). Thus, the bases in mRNA on the 5' side of the first base of a codon and on the 3' side of the third base of a codon influence the recognition of the codon by the corresponding anticodon.

Whereas the triplet anticodon sequences in tRNAs vary according to the amino acid specificity of the tRNA, the flanking bases that can affect tRNA-mRNA interactions show regularities [7, 8]. On one side of tRNA anticodons there is a relatively invariant pyrimidine and on the other side there is a relatively invariant purine. Accordingly, it would seem advantageous for a codon in mRNA when interacting with its anticodon to be *preceded* in the mRNA by an **NNY** codon (so that the **Y** in the codon can pair with the invariant purine in the tRNA), and to be *followed* in the mRNA by an **RNN** codon (so that the **R** in the codon can pair with the invariant pyrimidine in the tRNA). Thus, if we express our central pairing codon as **NNN** (where **N** is any base), the sequence of codons should read ...,**NNY**, **NNN**, **RNN**,... . Since, codons are translated successively, it follows that codons that, while encoding amino acids correctly, can *also* obey an "**RNY**-rule," will be preferred. Thus, all central codons and their two flanking codons, should ideally read ...,**RNY**, **RNY**, **RNY**,... [9]. The tendency of codons to adopt a generic **RNY** form is so evident that it can assist detection of protein-coding regions in uncharted sequences [10]. The mRNA codons in Figure 5-1 have been arranged to be consistent with the rule.

The sequence of a codon in mRNA also exists at the level of the corresponding gene in duplex DNA. Here, in the mRNA-synonymous ("coding") DNA strand, **RNY** codons base-pair with their complements (**RNY** anticodons) in the other DNA (mRNA "template") strand. Thus, the **RNY** pattern is present in both strands (Fig. 7-2).

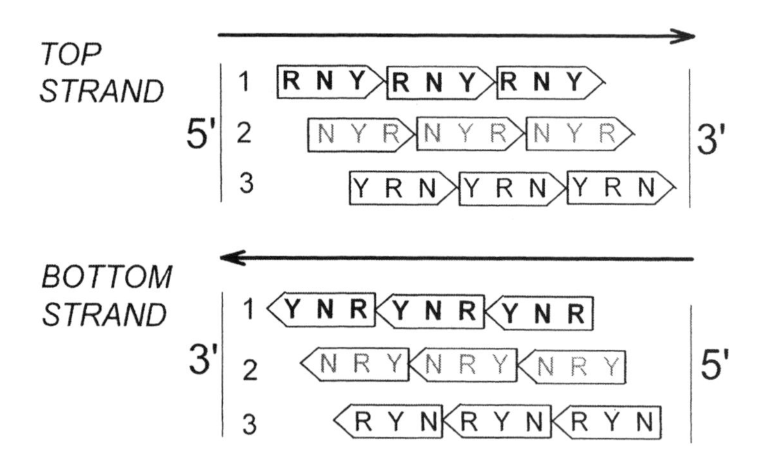

Fig. 7-2. The **RNY** pattern is in both DNA strands. Since the genetic code is a non-overlapping triplet code, there are three potential reading frames in the top strand of DNA, and three potential reading frames in the bottom strand of DNA. The pattern **RNY** can only correspond to one of the three frames and, since **R** always pairs with **Y** in duplex DNA, whichever frame **RNY** occupies in the top strand, it must also occupy in the bottom strand (here frame 1). Arrowed boxes indicate the polarity of coding triplets

In the case of amino acids whose codons contain **Y** in the first position or **R** in the third position, "**RNY**-pressure" on codons is in potential conflict with protein pressure (the pressure to insert a particular amino acid in a particular position in a protein). Also to be accommodated are **GC**-pressure (see Chapter 8) and **AG**-pressure (purine-loading pressure; see Chapter 6). Indeed, in species under high **AG**-pressure third codon positions tend to acquire purines and the **RNY** rule may no longer apply (see Chapter 9).

Negative and Positive Selection

If reproductive success is impeded by a mutation, then selection of organisms with the mutation is negative. If reproductive success is promoted then the selection is positive. These are two, mutually exclusive, consequences of a mutation in a nucleic acid sequence. The extreme imperative of negative selection is: if you mutate, you die. Thus, the broad population of non-mutators remains and the few mutators die. The extreme imperative of positive selection is: if you do *not* mutate, you die. Thus, the broad population of non-mutators dies, and the few mutators flourish (i.e. there is a population "bottleneck" from which only the mutators emerge). Occupying the middle-ground between these extremes are "neutral" mutations, and mutations that may lead to either weak positive or weak negative selection. In the latter cases there will, by definition, be effects on the number of descendents, but only in the long-term.

Whether a base mutation will lead to negative or positive selection depends on the part of a gene-product (usually a protein) that it affects. A mutation affecting the active site of an enzyme will usually disturb enzyme function and this may impair the function of an organism to the extent that it will produce fewer descendents than will organisms without the mutation (i.e. its fitness to reproduce its kind is impeded). In the extreme, this is ensured by the death of the organism. On the other hand, a mutation affecting an antigen at the surface of a pathogen may allow it to evade the immune defenses of its host. A pathogen mutating in this way may produce more descendents than will pathogens without the mutation (i.e. its fitness to reproduce its kind is enhanced). In the extreme this is ensured by the death of pathogens that do not have the mutation.

Nucleic acid bases that are evolving slowly (i.e. they are conserved among related organisms) are likely to affect functions subject to negative selection (i.e. organisms with mutations in the bases are functionally impaired). Nucleic acid bases that are evolving very rapidly (i.e. they are not conserved among related organisms) are likely to affect functions subject to positive selection (i.e. organisms with mutations in the bases are functionally improved). So a determination of evolutionary rate can assist the distinction between bases under positive selection, and bases under negative selection, and hence can provide information on the function of a nucleic acid segment. For this, base differences between sequences can be calibrated against some temporal scale (e.g. the period from the present to the time of evolutionary divergence of sequences from a common ancestral sequence). Accumulation of a large number of differences (base substitutions) in a short time can be taken to indicate positive selection. However, accurate temporal calibration is difficult.

A more direct way to evaluate the mutational rate of a nucleic acid segment can be understood by considering a nucleic acid in the same way as we consider a language with different levels of information (see Chapter 3). A public speaker conveys both a message (primary information) and an accent (secondary information). Imagine requiring each member of a group of competing speakers, one at a time, to read a given text to a large audience. The speakers are informed that they will be timed to determine the slowest, and that the audience will be polled to determine the most incomprehensible. The slowest and the most incomprehensible speakers will then be eliminated. The performance will be repeated, after which the slowest and most incomprehensible speakers will again be eliminated. This process will continue until a winner emerges.

Initially, each speaker relays both the text (primary information) and an individual accent (secondary information). However, under pressures both to speak rapidly and to be understood, speakers with more deviant accents are soon eliminated. Speakers are under strong pressure to eliminate personal idiosyncrasies of accent (i.e. to mutate their secondary information). The pressures for fast and coherent speech will progressively decrease the diversity of the secondary information among surviving group members. The final sound of the text will probably be the same for any large group of speakers exposed to the same large audience. Thus, the divergent accents of the initial multiplicity of speakers converge on a single accent to which the hearing of the average member of the audience is best attuned.

In a competition where there is no pressure for speed, idiosyncrasies of accent are less likely to interfere with comprehension (i.e. the diversity of secondary information is tolerated). The speakers can then compete on the basis of emphasis, charm, body-language, or what you will (other forms of information), but not on the basis of accent.

Viewed from this perspective, it can be seen that a nucleic acid segment which is evolving rapidly with respect to its primary information (e.g. the sequence of a protein) may not be able to accommodate some of the other forms of information ("accents") that it might be carrying. These other forms, perhaps evolving leisurely under negative selection, include the ordering of bases to support stem-loop potential, as will be shown in Chapter 10. Thus, sequences under positive selection are also likely to be sequences where one or more forms of secondary information are impaired. On this basis the type of selection can be evaluated in a sequence of DNA without temporal calibration and without a need to compare with sequences in other genomes.

There is another way to evaluate the type of selection. The further apart two genes, the more likely they are to be separated by recombination. The frequency of recombination may then be sufficient to allow separation of a nucleic acid segment containing a gene that is evolving *slowly* from neighboring segments in the region. There can be gene shuffling (see Chapters 8 and 14). Thus, in one individual within a population a slowly evolving conserved gene may occur in association with a particular neighboring gene. In another individual the conserved gene may occur in association with a variant form of the neighboring gene (especially likely if the neighboring gene belongs to a polymorphic gene family). Among members of the population there is sequence diversity in the region.

However, in the case of a nucleic acid segment containing a gene that is evolving *rapidly* (i.e. individuals containing it are being positively selected), there may not be time for separation from neighboring segments through recombination. Thus, neighboring genes of particular types will tend to remain attached (linked). They will "hitchhike" through the generations with a positively selected gene. Variant forms of neighboring genes will be lost from the population in the course of this "selective sweep." Consequently, among members of a population, variation in the region of a positively selected gene is decreased. Sequence diversity is lost locally due to the imposition of a "bottle-neck" on the population by the positively selected gene. This way of evaluating the type of selection requires that sequences from different individuals be compared.

Neutral Mutations

The idea of neutral mutations was attractive to those seeking an internal frame of reference for evaluating mutation rates, and for determining whether a mutation would impede or promote reproductive success. Mutations in third positions of codons often do not change the nature of the encoded amino acid, and hence do not change the corresponding protein (Tables 7-1, 7-2). Accordingly, any anatomical or physiological characteristics of an organism that depend on the protein do not change. It was tempting to consider such synonymous mutations as "neutral."

The most obvious advantage of the use of a particular codon, rather than a synonymous one, is that certain codons might be translated more rapidly, or more accurately. This is indeed of evolutionary significance in the case of certain unicellular organisms where the rate of protein synthesis is critical (bacteria, yeast) [11]. Here synonymous mutations may not be neutral. But in many organisms the rate of protein synthesis is not critical. In these cases arguments that mutations in third codon positions might not be neutral tended to be overlooked.

To provide a relatively time-independent, internal, frame of reference for determining the form of selection (negative or positive), it was found convenient to compare the ratio of amino acid-changing (non-synonymous) base substitution mutations (d_a) to non-amino acid-changing (synonymous) base substitution mutations (d_s) in genes. For this, it was usual to align similar (orthologous) sequences from members of different species. At points where the bases were not the same, the differences were scored as either amino acid-changing or synonymous. The procedure assumed that synonymous mutations were adaptively neutral, and hence reflected a "background" rate of accepted mutation in a genomic segment of interest. Thus, the ratio (d_a/d_s) within a nucleic acid segment seemed capable of providing an index of the rate at which that segment was evolving. A high ratio suggested that the segment was under positive selection. A low ratio suggested that the segment was under negative selection.

In many cases values for synonymous mutations (d_s) are significantly above zero, and determinations of ratios agree with biological expectations. This suggests that third codon position mutations can indeed be neutral. Yet it is not unusual to find values for synonymous mutations at or close to zero. This is particularly apparent with certain genes of the malaria parasite, *Plasmodium falciparum*. Some interpreted this as revealing a recent population bottle-neck (i.e. a shrinking of the population, the surviving members of which become founders for a subsequent population expansion) [12, 13]. Following the bottle-neck (i.e. a loss of population diversity) there would have been insufficient time for new synonymous mutations, to accumulate. Thus, at the extreme, existing species members could have been derived from one "Eve" of recent origin.

However, others proposed that zero values for synonymous mutations could result from high conservation of bases that, while not determining the nature of an encoded amino acid, *do determine something else* (unspecified secondary information). This would violate the neutral assumption, so that both the recent origin argument, and calculations based on ratios (d_a/d_s), would be invalid. Favouring this view, it was shown that, more than most genomes, that of the malaria parasite is sensitive to some of several non-classical selective factors, which affect third codon positions and collectively constitute the "genome phenotype." Some examples of this will be given in Chapter 11.

Genome Phenotype

Sometimes an amino acid can be encoded from among as many as six possible synonymous codons (Table 7-1). In 1974 this prompted US geneticist

Walter Fitch to remark that "the degeneracy of the genetic code provides an enormous plasticity to achieve [nucleic acid] secondary structure without sacrificing specificity of the message" [14]. Yet, sometimes this "plasticity" (flexibility) is insufficient, so that, with the exception of regions of genes under strong positive Darwinian selection pressure (see Chapter 10), genomic pressures can "call the tune".

When this happens, genomic pressures modify the amino acid sequence (non-synonymous codon changes), sometimes at the *expense* of protein structure and function. A protein has to adapt to the demands of the environment, but it also has to adapt to pressures that have derived, not from the conventional environment acting upon the conventional ("classical") phenotype, but from other environments, including what may be called the "reproductive environment" acting on the "genome phenotype." Thus, Italian biochemist Giorgio Bernardi noted in 1986 [15] that:

> "The organismal phenotype comprises two components, the classical phenotype, corresponding to the 'gene products,' and a 'genome phenotype' which is defined by [base] compositional constraints."

To further illustrate this, let's consider a set of homologous (orthologous) genes that carry out essentially the same function in mice and rats – two species that diverged from a common ancestral species some ten million years ago. Although they have many similar characters, mice are anatomically and physiologically different from rats, and we would expect to find these phenotypic differences reflected in differences in some of their proteins (i.e. differences in the classical phenotype). For the protein-coding regions of individual genes of the set, geneticists Kenneth Wolfe and Paul Sharp in Ireland compared base substitutions (DNA divergence) with amino acid substitutions (protein divergence) [16].

In Figure 7-3 each point corresponds to an individual gene. Plotted against the percentage DNA divergence between mouse and rat for each gene are the corresponding protein divergences (Fig. 7-3a), and the corresponding amino acid-changing (d_a) and synonymous (d_s) divergences (Fig. 7-3b). Genes differ dramatically in the extent of divergence. While some proteins (bottom left of Fig. 7-3a) have not diverged at all, the corresponding DNA sequences have diverged a little. Some proteins (top right of Fig. 7-3a) have diverged more than 20% and the corresponding DNA sequences are also highly diverged. The two divergences are linearly related with an intercept on the X-axis corresponding to a DNA divergence of 5.4%.

So it appears that for ten million years some proteins (bottom left of Fig. 7-3a) have remained unchanged. Presumably any organisms with mutations in these proteins have died before they could reproduce their kind (i.e. they have been negatively selected), and so no mutations are found in modern organisms. On the other hand, organisms with synonymous mutations that do not change the nature of an encoded amino acid would not have been counter-selected so severely. Thus, for a gene whose protein is unchanged there is a DNA divergence of about 5.4% (intercept on X-axis), implying that some, albeit only a few, synonymous mutations have been accepted. This is shown in Figure 7-3b where, as expected, the plot for amino acid-changing mutations (d_a) resembles that in Figure 7-3a, and the plot for synonymous mutations (d_s) extrapolates back close to zero.

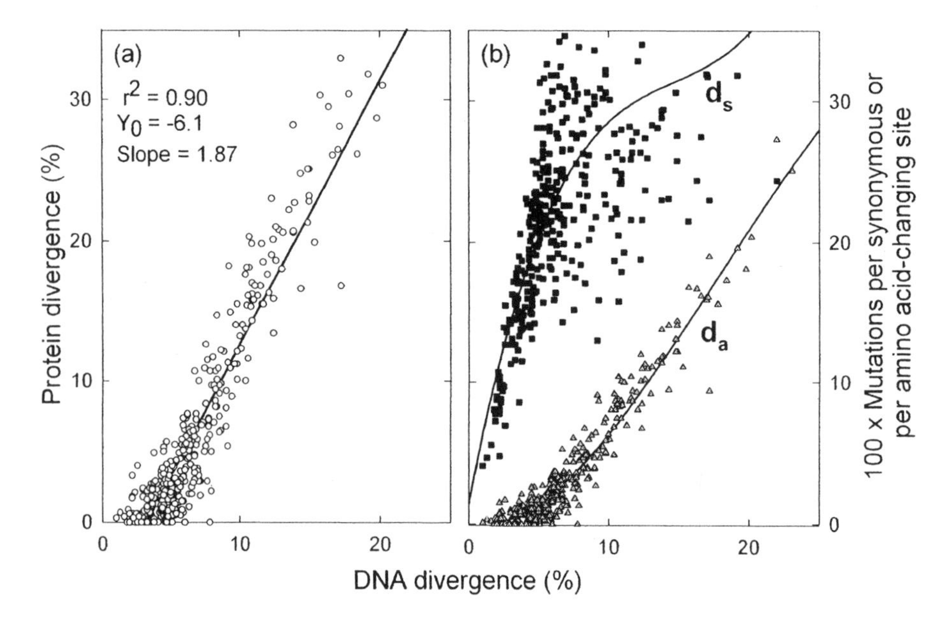

Fig. 7-3. Divergence between 363 genes of mouse and rat species. For each gene is shown the relationship between DNA divergence (bases %) and either *(a)* protein divergence (amino acids %), or *(b)* mutations per synonymous site (black squares) and per amino acid-changing site (grey triangles). For convenience, in *(b)* each value is multiplied by 100 and points are fitted to third order regression lines. Note from Table 7-2 that more bases in the DNA of coding regions are available for amino acid-changing mutation than for synonymous mutation. Thus, in *(b)* values are expressed as mutations relative to the number of sites (bases) available for that class of mutation. Since there are fewer sites available for synonymous mutation, when mutation rates are high (or over long time scales), synonymous sites are more likely to become saturated, and so are unable to accept more forward mutations. This might be responsible for the curve of d_s to the right in *(b)*. Data for preparing this figure were kindly provided by Kenneth Wolfe

The linear relationship shows that amino acid-changing mutations, and synonymous mutations are *correlated*. If a gene has zero or a few amino acid-changing mutations (i.e. it is likely to have been under negative selection), then it will have few synonymous mutations, and will display a low overall DNA divergence. If a gene has many amino acid-changing mutations (i.e. it is likely to have been under some degree of positive selection), then it will also have many synonymous mutations and will display a high overall DNA divergence. This is a general observation and is not confined to the mouse-rat divergence. Why should there be this correlation? Why is the synonymous DNA divergence so low in genes with low protein (i.e. low amino acid) divergences?

It seems likely that, within the group of codons that encode a particular protein, the demands of the conventional phenotype and of the genome phenotype interrelate. An accepted amino acid-changing mutation often cannot help change one or more aspects of the genome phenotype. By the same token, an accepted mutation that changes the genome phenotype may happen to be amino acid-changing and so may also change the conventional phenotype. For example, a mutation from the codon **ACA** to **AAA** may be to the selective advantage of an organism because it causes a lysine to be substituted for a threonine; but the mutation might also marginally affect DNA conformation (stem-loop potential; Chapter 5), purine-loading (Chapter 6) and (**G**+**C**)% (Chapter 8). A mutation from the codon **AGC** to **AGG**, while not affecting (**G**+**C**)%, causes an arginine to be substituted for serine, which may be of selective advantage to an organism; but the mutation might also marginally affect DNA conformation, purine-loading, and **RNY**-pressure. A pressure to purine-load might change **ACA** to **AAA**, or **AGC** to **AGG,** thus conferring adaptive advantages; but the mutations could also cause proteins to encode lysine or arginine "against their wills" (i.e. purine-loading would "call the tune").

Most importantly, it should be noted that *genes with similar degrees of divergence do not necessarily colocalize in a genome*. A gene encoding a protein with zero divergence may be the neighbor of a gene encoding a protein with 20% divergence. Thus, *the borders of a region within which amino acid-changing mutations and non-amino acid-changing mutations can coadapt may approximate to the borders of an individual gene*. In this respect each gene is a "selfish" autonomous entity in the sense implied by George Williams (see Chapters 8–10).

Compensatory Mutations

We should now qualify the above assertion that, mutation being essentially a random process, within a short time frame two mutational events are unlikely to occur in close proximity in DNA. If by "in close proximity" we included codons within a common gene, then the chances of a second mutational event being favored by a prior mutational event in the same gene are quite high. For example, the above amino acid-changing mutation from **ACA** to **AAA** may have been accepted *primarily* because it changed a threonine to a lysine. As a by-product the purine content of the gene increased. This may have made it propitious for the subsequent acceptance of a synonymous purine-to-pyrimidine substitution mutation elsewhere in the gene, thus restoring the original degree of purine-loading without further changing amino acids. Currently, it is not clear how short the time frame between the two mutations would usually be, or how close within a gene the compensatory second mutation would usually be to the first.

Sometimes the second mutation will be a back-mutation at the same site. But a back-mutation is unlikely if the prior forward-mutation is lethal. Thus, the glycine codon **GGA** can forward-mutate to **TGA**, which is a stop codon and so is likely to be immediately lethal. When a **GGA** codon exits a genome in this way (because organisms with the mutation are no longer around to reproduce), it is not readily restored. For this reason the frequency in the population of **GGA** relative to the other three glycine codons should diminish. Although glycine in a certain position might be optimal for the function of a protein, the amino acids substituted for glycine when the latter three glycine codons mutate might be consistent with viability for a time sufficient for a back-mutation to glycine to occur, so restoring optimal protein function.

It can be seen in Figure 7-3b that as overall DNA divergence increases, the plot for protein divergence (d_a) increases linearly, whereas the plot for synonymous divergence (d_s) curves to the right. Thus, at high degrees of divergence, in many, but not all, genes the d_a/d_s ratio increases. This might be interpreted as revealing that many widely diverged genes have been under positive selection pressure (acting on the conventional phenotype). However, it should be noted that the ratio increase is due more to a decline in d_s than to an increase in d_a. So the plots show that, at high degrees of divergence, synonymous mutations tend to be constrained in some genes, but not others. Bases at synonymous sites may be conserved, or the sites may have become mutationally "saturated," so that no more forward mutations can be accepted. Thus, there are pressures on the genome phenotype that resist change (decrease d_s) despite concurrent pressures on the conventional phenotype that promote change (increase d_a).

RNA viruses, because they evolve rapidly (see Chapter 8), are very suitable for such studies. Isabelle Novella and her colleagues demonstrated the *parallel evolution* of various virus strains as they adapt to new conditions (presumed to be primarily due to the acceptance of amino acid-changing mutations). However, concurrent synonymous mutations do not occur randomly as classical neutral theory would predict. The *same* synonymous mutations are accepted independently in *different* evolving strains [17]. Thus, either synonymous mutations independently contribute to strain adaptations (e.g. affecting RNA structure), or they are secondary to (compensatory for) primary adaptations at amino acid-changing sites (or vice-versa).

From all this it can be seen that the synonymous mutation rate (d_s), although often correlated with the amino acid-changing mutation rate (d_a), has a life of its own. Its utility as a frame of reference for evaluating changes in d_a (i.e. the d_a/d_s ratio) may be somewhat limited [12]. We will return to this later.

Limits of Natural Selection

Despite Darwin's belief that "external conditions ... cause mere variability" [4], the generation of DNA mutations is a random process that is essentially independent of any Darwinian environmental factor. These mutations, in turn, can generate organisms ("mutants," "variants") with variant characters. Favorable or disfavorable variant characters favor or disfavor the organisms bearing them, so that their descendents accordingly increase or decrease in the population by natural selection. Whatever the selection, be it positive or negative, it is the classical *natural* selection of Darwin, who used the term to contrast with the *artificial* selection carried out by man in his role as breeder or horticulturalist. Natural selection is generally acknowledged as a major driving force of linear, *within-species*, evolution.

However, within the constraints of a species, the ability of Nature to "invent" new phenotypes is limited. As discussed in Chapter 3, Shakespeare could explore the theme of personal ambition in a castle in Scotland in one play, but other themes required that new plays be written. For real novelty, just as Shakespeare's works diverge, so must those of Nature. The only figure in Darwin's great book, The Origin of *Species by Means of Natural Selection*, depicted this divergence. In even simpler form than Darwin's original, evolution can be represented by an inverted letter Y, with a linear stem and two linear arms (Fig. 7-4). Thus, at least two evolutionary processes are distinguished: – linear, within-species, evolution (represented by the linear stem and the two linear arms), and divergent, between-species, evolution (represented by the point of divergence from stem to arms). The title of Darwin's book suggested that both were driven by natural selection, although he cautioned [18]: "I am convinced that natural selection has been the most important, but not the exclusive, means of modification."

Natural selection favors or disfavors organisms by virtue of the characters they possess, which we now know to be encoded by genes in DNA sequences. Thus, natural selection in the classical Darwinian sense is selection acting on gene-products – a process that can be referred to as "genic" selection. There is little disagreement that natural genic selection plays an important role in the process of linear, non-branching, within-species, evolution ("species survival"). However, throughout the nineteenth and twentieth centuries its role in the branching that creates two linear growing arms from one stem ("species arrival") was more contentious. Was species arrival due to business-as-usual natural selection, or was there a need for an underlying "genetic revolution"? Overcoming constraints in human thought sometimes requires powerful "paradigm shifts." Perhaps, in similar fashion, constrained species need to be periodically jolted out of their complacency? If so, what could be the ultimate molecular basis of this genetic kick-in-the-pants paradigm shift? Before considering this, we should first review some basic genetics as set out by the man who, in the early 1860s, while Samuel Butler was breeding sheep in New Zealand, was breeding peas in Moravia – Gregor Mendel.

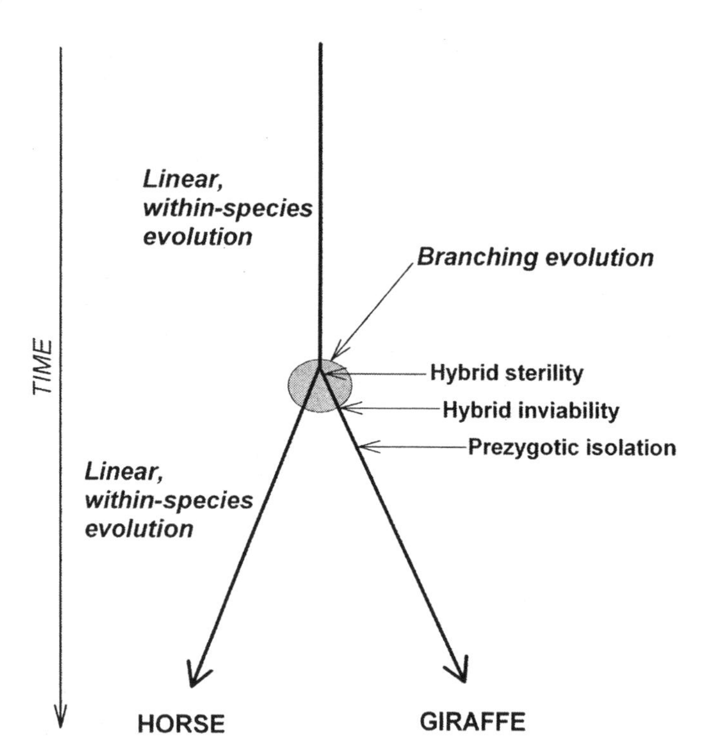

Fig. 7-4. Divergence into two species, which is preceded and followed by phases of linear, within-species, evolution. Successive barriers to reproduction supplement or replace earlier ones. In the *general* case, at the point of divergence there is the chromosomal pairing barrier (a postzygotic barrier producing hybrid sterility). This may be followed by a developmental barrier (a post-zygotic barrier producing hybrid inviability), and then by a gamete transfer barrier (a prezygotic barrier)

Blending Inheritance

Do the characters of two parents blend in their children, or do children inherit certain discrete characters from one parent and other discrete characters from the other (see Chapter 3)? If blending were the rule, then variation in a population would be constantly decreasing unless opposed by some differentiating force (i.e. fresh mutations). An analogy is drawn with the mixing of black and white paints to produce a uniform grey paint from which the originals would not be recoverable. On the other hand, if characters were inherited discretely, although within a line of inheritance half the parental characters would, on average, be lost in each generation, those that were transmitted would be undiluted. Variation within a population would tend to remain constant depending to the relative rates of gain and loss of mutations. In 1857, two years before the publication of *The Origin of Species*, Darwin inclined to the latter view [19]:

> "I have lately been inclined to speculate, ... that propagation by true fertilization will turn out to be a sort of mixture, and not true fusion, of two distinct individuals, or rather of innumerable individuals, as each parent has its parents and ancestors. I can understand in no other view the way in which crossed forms go back to so large an extent to ancestral forms."

Consistent with this, his friend the botanist Joseph Hooker noted in 1860 (20):

> "A very able and careful experimenter, M. Naudin, performed a series of experiments at the Jardin des Plantes at Paris, in order to discover the duration of the progeny of fertile hybrids. He concludes that the fertile posterity of hybrids disappears, to give place to the pure typical form of one or other parent."

Shortly thereafter the suspicions of Darwin, Naudin and Hooker were confirmed by Mendel's breeding studies with peas [21]. When he crossed pure tall plants with pure small plants all the resulting hybrid offspring were tall. The small character had disappeared. When these offspring were crossed among themselves, the small character re-emerged in the next generation. On average, there were three tall plants to one small plant. The tall character was considered to be "dominant" to the small character ("recessive"). However, Mendel's studies remained largely unnoticed until 1900. Someone who may have paid attention was Romanes. He cited Mendel in 1881 and in 1894 wrote [22]:

> "I have found, after several years experimenting with rats [and] rabbits ... , that one may breed scores and hundreds of first

crosses between different varieties, and never get a single mongrel throwing intermediate characters – or indeed any resemblance to one side of the house. Yet, if the younger are subsequently crossed *inter se* (i.e. brothers and sisters, or first crossings) the crossed parentage at once repeats itself. *Ergo*, even if the pups which are to be born appear to give a negative result, keep them to breed from with one another."

With hindsight the ability of characters to retain their identities through generations of breeding is rather obvious. Your father was male and your mother was female, yet you have inherited this characteristic – their sex – quite discretely. You are either male or female. You are not a mixture (a hermaphrodite). Despite their mothers being female, boys usually do, in time, become as masculine as their fathers. Despite their fathers being male, girls usually do, in time, become as female as their mothers. Darwin reiterated this point in later correspondence with Wallace [23]:

> "My dear Wallace ... I do not think you understand what I mean by the non-blending of certain varieties. It does not refer to fertility; an instance will explain. I crossed the Painted Lady and Purple sweet peas, which are very differently coloured varieties, and got, even out of the same pod, both varieties perfect but none intermediate. Something of this kind I should think must occur at least with your butterflies and the three forms of Lythrum; tho' these cases are in appearance so wonderful, I do not know that they are really more so than every female in the world producing distinct male and female offspring."

In his usual colorful way, Butler made the same point in 1877 [24]:

> "The memory [stored information in the embryo] being a fusion of its recollections of what it did, both when it was its father and also when it was its mother, the offspring should have a very common tendency to resemble both parents, the one in some respects, the other in others; but it might also hardly less commonly show a more marked recollection of the one history than of the other, thus more distinctly resembling one parent than the other. And this is what we observe to be the case ... so far as that the offspring is almost invariably either male or female, and generally resembles rather the one parent than the other."

Characters Determined by Single or Multiple Genes

If a particular character is determined by one gene for which two types (alleles) exist among members of a species, say T and t, then individual diploid organisms can either be pure *TT* (homozygote), Tt (heterozygote), or pure *tt* (homozygote). There should normally be similar numbers of diploid *Tt* males and diploid *Tt* females in the species. Each *Tt* individual will have inherited a *T* gene (allele) from one parent and a *t* gene (allele) from the other. A particular chromosome will contain either *T* or *t* at a particular "slot" or "address" (locus), thus defining *T* and *t* as allelic genes, or alleles, that are mutually exclusive with respect to location on a single chromosome (Fig. 7-5)

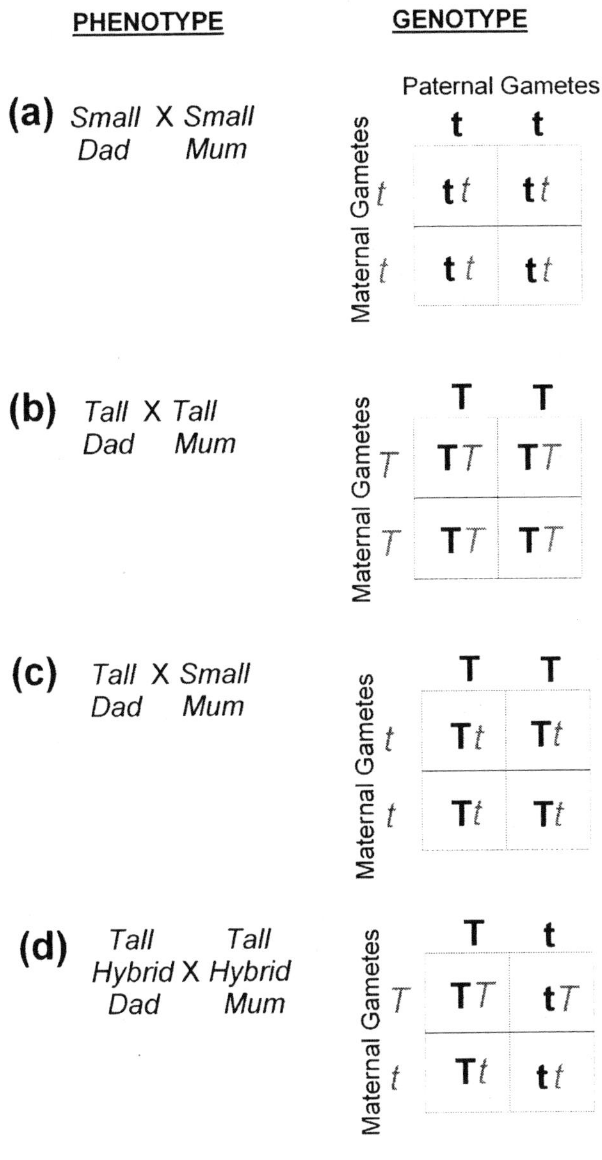

PHENOTYPE | **GENOTYPE**

(a) *Small* X *Small*
Dad *Mum*

Paternal Gametes

Maternal Gametes	**t**	**t**
t	**t***t*	**t***t*
t	**t***t*	**t***t*

(b) *Tall* X *Tall*
Dad *Mum*

Maternal Gametes	**T**	**T**
T	**T***T*	**T***T*
T	**T***T*	**T***T*

(c) *Tall* X *Small*
Dad *Mum*

Maternal Gametes	**T**	**T**
t	**T***t*	**T***t*
t	**T***t*	**T***t*

(d) *Tall* *Tall*
Hybrid X *Hybrid*
Dad *Mum*

Maternal Gametes	**T**	**t**
T	**T***T*	**t***T*
t	**T***t*	**t***t*

Fig. 7-5. Random distribution of allelic genes for tallness (T) and smallness (t) according to Mendel. A diploid organism inherits one allelic gene for a particular character (e.g. height) from its father, and one allelic gene for the same character from its mother. The paternal and maternal genes are designated "allelic" because they are potential alternative versions ("alleles") and occupy corresponding positions ("slots") on homologous chromosomes. The diploid state can be written as "TT" for a tall organism and "tt" for a small organism. Since the alleles are the same in each organism, then the organisms are individually homozygous ("pure") for the height character. From these organisms two types of haploid gametes (with either the paternally- or maternally-derived character) would be produced. Thus, the homozygous parents from which gametes are derived might be designated T_PT_M, and t_Pt_M (with P and M referring to the corresponding parents of the parents; i.e. the four grandparents). However, for simplicity the subscripts are omitted.

Each box shows the genotype of four children that would be produced, *on average*, if four maternal gametes and four paternal gametes met. For crosses when male and female homozygotes are of the same kind, all children are of the parental type [see *(a)* and *(b)*]. For crosses when male and female homozygotes are of different kinds [see *(c)*], all children are heterozygotes with respect to the height character (i.e. Tt). If T is dominant to t, then all heterozygotes appear tall. If a male heterozygote is crossed with a female heterozygote [see *(d)*], gametes of type T and t are produced in equal quantities by both parents and, of four children, *on average* three appear tall (TT, Tt, tT) and one appears small (tt). These are the famous Mendelian ratios. If T had not been dominant, some interaction between the T and t genomes might have produced a new phenotypic character, such as intermediate height. Note that Tt and tT ("reciprocal crosses") are equal heterozygotes; the order simply relates to the convention of placing the genome of paternal origin first, so that these might also be designated as T_Pt_M, and t_PT_M

If two *Tt* parents mate, then, with random mixing of chromosomes during meiotic production of gametes, the probability of a *TT* child is 25%, of a *Tt* child is 50%, and of a *tt* child is 25% (Fig. 7-5d). However, on a chance basis ("random drift"), whenever fertilization occurs a *t* spermatozoon from one parent might happen to always meet a *t* ovum from the other parent, so that all their children would be *tt*, and the *T* genes the parents had been carrying through the generations would be lost. Of course, fresh *T* genes could be recreated by fresh mutations. If the difference between the *T* and *t* alleles were a single base in DNA causing a difference in an amino acid at a particular position in an encoded protein, then *t* could retain the potential to mutate to *T*, so allowing generation of a new *Tt* individual. Such mutations being rare, only one allele would mutate at a time. Thus, pure *tt* parents, already most unlikely to produce *Tt* offspring by mutation, would be even more unlikely to produce *TT* offspring by mutation.

All this implies a one-to-one relationship between an allelic gene and a character (unigenic character trait). In such circumstances there is true non-blending inheritance. However, many characters are the product of the interactions of several non-allelic genes (multigenic character traits). In this circumstance there may, indeed, be blending inheritance (i.e. parental characters appear to blend in their children). For example, genes *A*, *B*, *C* and *D* that collectively determine tallness may be contributed by one parent, and the corresponding allelic genes *A'*, *B'*, *C'* and *D'* that collectively determine smallness may be contributed by the other parent. If tallness and smallness are otherwise neutral in their phenotypic effects, then members of the population will have all possible combinations, and their children will inherit these alleles in all possible combinations. Depending on the various dominance-recessive relationships, among a large number of children there could be a smooth (binomial) distribution of heights between the rare extremes of "pure" tall (*ABCD*) and of "pure" small (*A'B'C'D'*). In 1894, six years before Mendel's work emerged, Bateson had some feeling for this [6]:

> "An error more far-reaching and mischievous is the doctrine that a new variation must immediately be swamped, … . This doctrine would come with more force were it the fact that as a matter of experience the offspring of two varieties, or of variety and normal, does usually represent a mean between the characteristics of the parent. Such a simple result is, I believe, rarely found among the facts of inheritance. …Though it is obvious that there are certain classes of characters that are often evenly blended in the offspring, it is equally certain that there are others that are not. In all this we are still able only to quote case against case. No one has [yet] found general expressions differentiating the two classes of characters, nor is it easy to point to any one character that uniformly follows either rule. Perhaps we are justified in the impression that among characters which blend or may blend evenly, are especially certain quantitative characters, such as stature; while characters depending on differences in number, or upon qualitative differences, as for example colour, are more often alternative in their inheritance."

Following the discovery of Mendel's work, Bateson and Saunders noted [25]:

> "It must be recognized that in, for example, the stature of a … race of man, a typically continuous character, there must certainly be on any hypothesis more than one pair of possible allelomorphs [alleles]. There may be many such pairs, but we have no certainty that the number of such pairs, and conse-

> quently of the different kinds of gametes, are altogether *unlimited* even in regard to stature. If there were even so few as, say, four or five pairs of possible allelomorphs, the various homo- and hetero-zygous combinations might, on seriation, give so near an approach to a continuous curve [i.e. blending], that the purity of the elements would be unsuspected" [Bateson's italics].

Mendel himself had reached this conclusion when studying beans (*Phaseolus*), rather than his favourite plant, the pea (*Pisum*) [21]:

> "Apart from the fact that from the union of a white and a purple-red (crimson) colouring a whole series of colours results, from purple to pale violet and white, it is also striking that among 31 flowering plants only one received the recessive character of the white colour, while in Pisum this occurs on the average in every fourth plant. ... Even these enigmatic results, however, might probably be explained by the law governing Pisum if we might assume that the colour of the flowers and seeds of *Ph. multiflorus* is a combination of two or more entirely independent colours, which individually act like any other constant character in the plant."

The above "pure" combinations *ABCD* and *A'B'C'D'* could correspond to the "pure lines" of Johannsen (see Chapter 3). If bred with their own kind (e.g. an *ABCD* male gamete meets an *ABCD* female gamete), they would not be diluted or "swamped" (i.e. they would retain their positions near the surface of Jenkin's sphere; see Fig. 3-2). However, if crossed (e.g. an *ABCD* male gamete meets an *A'B'C'D'* female gamete), then their rare distinctive characters (extremes of tallness and smallness) might not be seen among the children (i.e. there could be a regression towards the mean).

Pattern Change

As set out in my book, *The Origin of Species, Revisited* [26], in his 1886 theory of "physiological selection,"Romanes distinguished the mechanisms of the two evolutionary processes that we now call genic (associated with linear evolution), and non-genic (associated with branching evolution). Unlike most other evolutionists, Bateson, then only 25, immediately took to the idea [27]:

> "The scheme thus put will at least work logically, while the other, as left by Darwin, would not. I did not suppose Romanes would even write as good a paper, ... it is a straight forward, common-sense, suggestion."

After various false leads, Bateson went on to infer that an inherited "residue" or "base" (like the base of a statue), distinct from genes, would played an important role in the branching process [28]. Perhaps Goldschmidt in Germany had this in mind when in 1917 he depicted genes ("particulate hereditary factors") as moveable elements reversibly anchored in specific order to a fixed chromosome base (Fig. 7-6) [29].

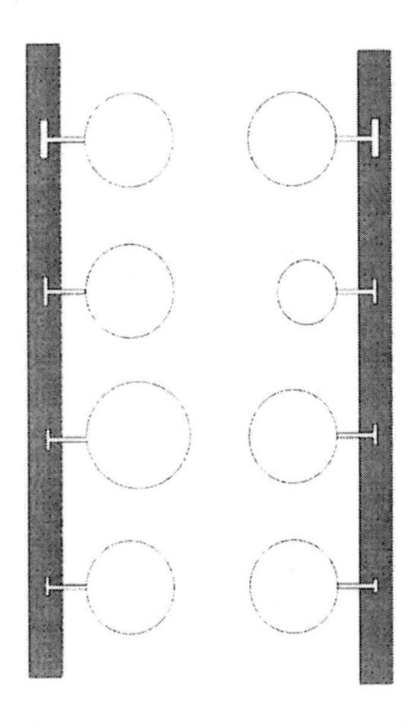

Fig. 7-6. Model of a chromosome as seen by Goldschmidt in 1917. Like Bateson, he distinguished the genic factors (white balls) from a "base," or "residue" (long grey rectangles), upon which the factors rested: "The force that anchors a particulate hereditary factor to its chromosome is denoted by a right-angled anchor, whose size corresponds to the quantity of the force." We now know that genes consist of a collection of purines and pyrimidines that are "anchored" to a phosphate-pentose sugar chain. Since purines are liberated from DNA under acid conditions more readily than pyrimidines (see Chapter 6), it follows that purine-rich genes can be considered as anchored more loosely than pyrimidine-rich genes. However, it will be proposed (see Chapter 8) that the best modern depiction of Bateson's abstract "residue" is the (**G** +**C**)%. Goldschmidt probably arrived at his model in an attempt to explain why the chromosomes, seen so clearly by light microscopy during cell division, seemed to disappear in non-dividing cells. De Vries had proposed that working gene copies would pass from nucleus to cytoplasm where they would program cell function (i.e. mRNA to the modern mind), but Goldschmidt seems to have thought that the genes themselves would move to-and-fro between the nucleus (where they would be visible but not functional), and the cytoplasm (where they would functional but not visible)

In 1940, Goldschmidt, now in the USA, postulated a mutational underpinning of the duality, referring to the process of genic, linear, non-branching, evolution as due to "micromutations," and to the process of non-genic, branching, evolution, as due to "systemic mutations." Changes in genes, which could be followed through the generations by examining the assortment of the characters they represent among the children, he considered to be examples of "microevolution" occurring within a single "reaction system." Systemic mutations were changes in the overall "pattern" of chromosomes that were responsible for divergent, between-species, "macroevolution." Chromosomal repatterning was an on-going process that could be "neutral" to the extent that there would be no necessary change in genic characters [30]:

> "So-called gene mutation and recombination within an interbreeding population may lead to a kaleidoscopic diversification within the species, which may find expression in the production of subspecific categories [races, varieties], if selection, adaptation, isolation, migration, etc., work to separate some of the recombination groups... . But all this happens within an identical general genetic pattern which may also be called a single reaction system... . The change from species to species is not a change involving more and more additional atomistic changes, but a complete change of the primary pattern or reaction system into a new one, which afterwards may again produce intraspecific variation by micromutation. One might call this different type of genetic change a systemic mutation, though this does not have to occur in one step."

Goldschmidt distinguished his postulated repatterning due to "systemic mutations" from chromosomal "macromutations" that were sometimes visible by light microscopy (segment deletions, inversions, transpositions, or duplications). He proposed that chromosomal "repatterning" might proceed slowly and progressively, without necessarily producing any change in the structure or function of organisms, until a new species emerged that was reproductively isolated from the old one by virtue of the new pattern being "incompatible" with that of the old:

> "A systemic mutation (or series of such), then, consists of a change of intrachromosomal pattern... . Whatever genes or gene mutations might be, they do not enter this picture at all. Only the arrangement of the serial chemical constituents of the chromosomes into a new, spatially different order; i.e. a new chromosomal pattern, is involved. The new pattern seems to emerge slowly in a series of consecutive steps ... . These steps may be

> without a visible effect until the repatterning of the chromosome ... leads to a new stable pattern, that is, a new chemical system. This may have attained a threshold of action beyond which the physiological reaction system of development, controlled by the new genetic pattern, is so basically changed that a new phenotype emerges, the new species, separated from the old one by a bridgeless gap and an incompatible intrachromosomal pattern."

By "incompatible" Goldschmidt was referring to differences between the chromosomes of two potential parents. These chromosomes would consequently not be able to cooperate functionally and/or to pair properly at meiosis within their child. Thus, the two parents would be reproductively isolated in that, even if they could produce children, those children would be inviable or infertile:

> "An unlimited number of patterns is available without a single qualitative chemical change in the chromosomal material, not to speak of a further unlimited number after qualitative changes (model: addition of a new amino acid into the pattern of a protein molecule). ... These pattern changes may be an accident, without any significance except for creating new conditions of genetic isolation by chromosomal incompatibility... ."

In striving to give some meaning to his concept of pattern, Goldschmidt wrote in bioinformatic terms:

> "Let us compare the chromosome with its serial order to a long printed sentence made up of hundreds of letters of which only twenty-five different ones exist. In reading the sentence a misprint of one letter here and there will not change the sense of the sentence; even the misprint of a whole word (rose for sore) will hardly impress the reader. But the compositor must arrange the same set of type into a completely different sentence with a completely new meaning, and this in a great many different ways, depending upon the number of permutating letters and the complexity of the language (the latter acting as a 'selection'). To elevate such a model to the level of a biological theory we have, or course, to restate it in chemical terms."

Four years before Oswald Avery showed that DNA was the form in which hereditary information was transferred through the generations [31], and thirteen years before Watson and Crick presented their model for DNA, it was not unreasonable to think of chromosomal patterns in terms of amino acid, rather than nucleotide, sequences. Thus Goldschmidt wrote in 1940:

> "I do not think that an actual chemical model can yet be found. But we might indicate the type of such a model which fulfills at least some, though not all, of the requirements. It is not meant as a hypothesis of chemical chromosome structure, but only as a chemical model for visualizing the actual meaning of a repatterning process … . Let us compare the chromosome to a very long chain molecule of a protein. The linear pattern of the chromosome is then the typical pattern of the different amino acid residues."

A need for a change in reaction pattern for divergent evolution to occur finds an easy metaphor in human political systems. "Every boy and gal born into the world alive, is either a little Liberal, or else a little Conservative," went the Gilbert and Sullivan ditty as one Gladstone administration operating within a linear liberal reaction pattern was succeeded by a Disraeli administration operating within a linear conservative reaction pattern. Typically, dictatorships allow only linear progress. At the time of this writing (2003) to change reaction pattern in the Middle East has required outside military intervention.

Reproductive Isolation

To ask about the "arrival" of a new species is to ask about the nature of the process by which one line diverges into two lines. Members of one line are unable to reproduce with members of the other and, on this basis, are defined as members of independent species. For reproductive purposes, members of a species assort with their own kind. In other words, for reproductive purposes members of a species positively assort with their own kind (members of the same species), and negatively assort with other kinds (members of other species). The origin of species is the origin of reproductive isolation. Members of two species, being reproductively isolated, are unable to swap genes, or any other segments of their DNAs. So a character encoded by a gene tends to remain discrete as it moves "vertically" through the generations while remaining within species bounds (Fig. 1-1).

Until a species becomes extinct, reproduction within that species is a continuous cyclic process proceeding from gametes, to zygotes, to adults with gonads, where new gametes are produced to continue the cycle. It is a characteristic of cyclic processes that they may be arrested by an interruption at *any* stage. Thus, reproductive isolation will be achieved when the cycle is interrupted at some stage. The problem of the origin of species is essentially that of determining what form that interruption *first* takes, and at what stage in the cycle it *usually* occurs.

There are three fundamental barriers to cycle operation, one operating to prevent gametes meeting (the transfer or prezygotic barrier), and two operating after the gametes have merged to form the single-celled zygote, which then divides repeatedly to produce a multicellular embryo and eventually a mature, gonad-bearing, adult (postzygotic barriers). The Montagues and the Capulets had only the gamete transfer barrier to keep apart Romeo and Juliet. Nature also has the two post-zygotic barriers. Nature may permit Romeo to fertilize Juliet, but then, cruelly, may not allow the two parental sets of chromosomes to cooperate effectively, either for development of the embryo or child ("hybrid inviability"), or for the pairing process (meiosis) within the adult gonad that is required for the formation of gametes ("hybrid sterility").

At the bottom of the inverted Y in Figure 7-4 it is obvious that today's horses and giraffes are prezygotically isolated. Anatomical, physiological and psychological differences conspire to make this absolute. As we move back through time, up along the limbs of the inverted Y, one by one, these differences diminish and disappear. Eventually a stage is reached where copulation and fertilization can occur. However, gene products (e.g. proteins) do not cooperate in the embryo and development fails. We are at the hybrid inviability barrier. Moving closer to the fork of the Y, gene products become increasingly compatible. The hybrid inviability barrier eventually falls and development proceeds normally. A child is born and it grows into an adult. Only one barrier, the hybrid sterility barrier, remains.

Hybrid sterility is an intrinsic isolating barrier. But, as in the case of the Montagues and the Capulets, external (extrinsic) barriers also have the potential to reproductively isolate. A group within a species can become geographically isolated from the main species group. Under the protection of this "allopatric" reproductive isolation, the groups can independently evolve to such an extent that, should the geographical barrier be removed, members of one group are no longer reproductively compatible with members of the other. For example, there might have been mutations in genes determining the structures of the organs of copulation. Biologists have long disagreed on whether such genic changes can account for origins of species as they have *generally* occurred over evolutionary time [26]. Darwin found hybrid sterility particularly puzzling [18]:

> "On the case of natural selection the case is especially important, inasmuch of the sterility of hybrids could not possibly be of any advantage to them, and therefore could not have been acquired by the continued preservation of successive profitable degrees of sterility."

He here wrote about a hybrid that, by virtue of its sterility, would be dead in an evolutionary sense since it could no more continue the line than a slow deer, which had been eaten by a tiger. But what mattered were the *parents* of the sterile hybrid, which itself was merely the expression of *their* incompatibility. Darwin seemed not to see that at least one of the parents might have been on a "journey" (by accumulating mutations) towards a new species, thus renouncing membership of the old one. As long as it was traveling in the right direction, the further that parent "traveled" from the old species the closer it *must* have got to the new one. Sterility of the offspring produced by crossing with a member of the old species would be considered disadvantageous to a parent when it was still perceived as close to most members of that old species (i.e. when it had not "traveled" far by mutation). But when perceived as a potential member of the new species, increasing degrees of sterility of the offspring produced by crossing with members of the old species (as the member of the new species continued "traveling" and accumulating further mutations) could be considered extremely "profitable," since it would eventually lead to "species arrival" (i.e. complete reproductive isolation from the old species, but complete reproductive compatibility with other members of the new species).

First Mutations Were Synonymous?

To conclude, we consider again the data for mouse and rat, two species deemed to have diverged from a common ancestor (Fig. 7-3). Can we infer from the sequences of two modern genomes the likely nature of the first mutations that distinguished the mouse and rat lineages ten million years ago? More simply, can we infer whether those mutations were likely to be amino acid-changing (non-synonymous), so implying that changes in the conventional (genic) phenotype) had been instrumental in initiating the divergence process (i.e. classical Darwinian natural selection)? Alternatively, were the first mutations synonymous, so implying that changes in the genome phenotype had been instrumental in initiating the divergence process?

Given that each gene has a specific mutation rate, with a proportionate number of synonymous and amino acid-changing mutations (Fig. 7-3), the problem is set out in simple terms in Figure 7-7. Given a *sustained and uniform* proportionality within a gene, the plots should extrapolate back to zero. If *at the outset* one type of mutation made a disproportionate contribution, then there should not be extrapolation to zero. Hence, the nature of the mutations, which, by their accumulation contributed most to the *initial* divergence from a common ancestor is hinted at in Figure 7-3. Indeed, since DNA divergence increases with time, the X-axis can be viewed as a time axis [13]. Extrapolating back to zero divergence (zero time) it can be seen that synonymous mutations (i.e. third codon position mutations) are the most likely candidates (Fig. 7-3b). Synonymous differences begin to appear at zero time, whereas amino acid-changing differences (involving first and second codon positions) emerge later. In other words, the divergence of proteins (Fig. 7-3a) is likely to be *secondary* to whatever initiated the divergence process.

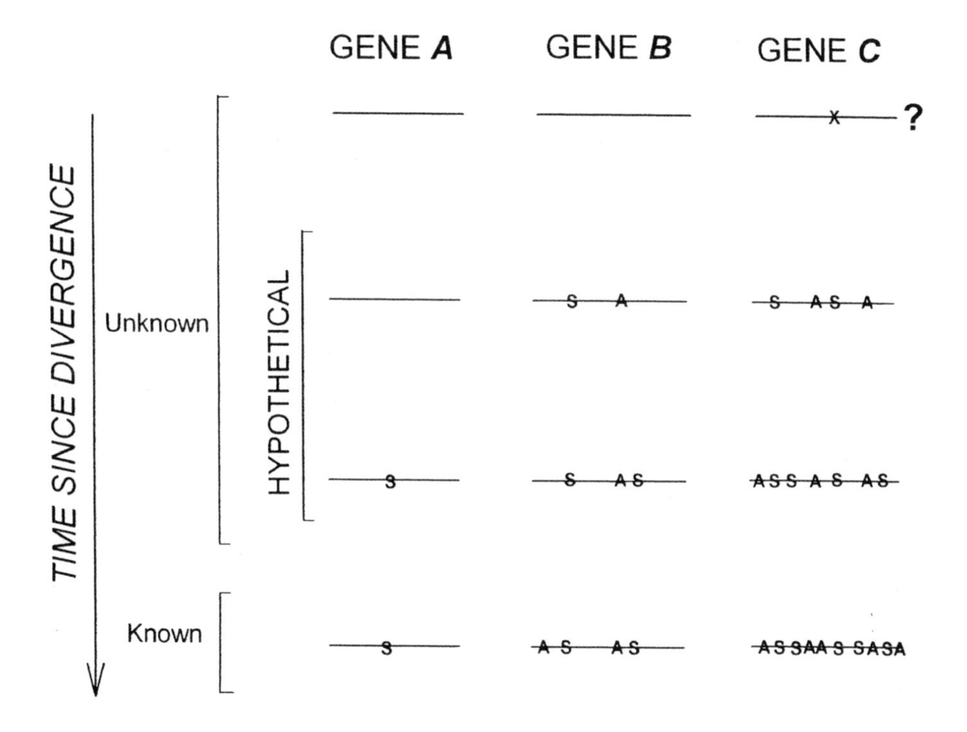

Fig. 7-7. Were the first mutations (X) associated with the mouse-rat divergence, synonymous (S) or amino acid-changing (A; non-synonymous)? Genes *A, B* and *C* have low, intermediate, and high mutation rates, respectively. This is inferred from the number of known substitutions that distinguish the modern genes (bottom three lines). Under the assumption that rates of synonymous and amino acid changing mutations are positively correlated (i.e. proportionate), hypothetical intermediate substitution levels at progressive intervals following the divergence are shown, with proportionate numbers of synonymous (S) and amino acid-changing (A) mutations in each gene

From Figure 7-7 it might be argued from the bottom line ("known") that in gene *A* we see merely high conservation relating to some important phenotypic function (e.g. histone proteins). But, as indicated by the lack of synonymous mutations in *A* (relative to *B* and *C*), there has been a proportionate high conservation of functions related to the genome phenotype. Thus, the above inference from the extrapolation in Figure 7-3 seems reasonable.

For those who, like me, are addicted to metaphors, the following may help. At the time of the divergence ("acceleration" of a new species "to cruising speed"), synonymous mutations ("gasoline consumption"), mainly in genes of class *C*, would have been *differentially* accepted ("high gasoline consumption relative to distance traveled"). After the divergence ("attainment of cruising speed"), both synonymous mutations ("gasoline consumption"), and amino acid-changing mutations ("distance traveled"), would have been accepted *proportionately* in genes of all classes. Synonymous mutations, being non-amino acid changing, are a useful indicator of mutations occurring concomitantly both extragenically, and within introns. The latter mutations, in concert with intragenic synonymous mutations, would have driven the divergence process. Indeed, synonymous, intronic, and extragenic mutation rates (all three) are often found to be closely correlated.

However, the case that species divergence is mainly driven by synonymous mutations, does not rest on Figure 7-3 alone. In the next chapter we consider a possible intrinsic, essentially non-genic, biochemical basis for Goldschmidt's postulated pattern change, which can lead to hybrid sterility, and hence to a sympatric origin of species. In this, we shall be taking advantage of the special properties of third codon positions that are generally free from protein-encoding constraints. While being closely located to first and second codon positions, which thus serve as "controls," third codon positions provide a "window" for evaluating some important non-genic pressures that operate, free from protein-encoding constraints, in introns and in non-genic regions of a genome.

Summary

Anatomical or physiological variations that are inherited are due to inherited changes (mutations) in base sequences of DNA. Mutations that change genes can affect the *conventional phenotype* resulting in linear within-species evolution, often under the influence of natural selection (species survival). Changes in the conventional phenotype are usually associated with amino acid-changing (non-synonymous) mutations in the first or second bases of triplet codons. DNA mutations can also result in changes in the *genome phenotype*. These changes include synonymous (non-amino acid-changing) mutations, usually in the third bases of codons. Each gene in a genome has distinctive rates of acceptance of amino acid-changing and synonymous mutations, which are positively correlated. A gene with few amino acid-changing mutations also has few synonymous mutations. A gene with many amino acid-changing mutations also has many synonymous mutations. Two genes may be closely located but differ greatly in their mutation acceptance rates. Thus, each gene is an independent mutational entity. Synonymous changes may be of particular importance in changing the "pattern" of a genome, so facilitating branching evolution (species arrival).

Chapter 8

Chargaff's GC rule

> "DNA is in its composition characteristic of the species from which it is derived. This can ... be demonstrated by determining the ratios in which the individual purines and pyrimidines occur .... There appear to exist two main groups of DNA, namely the 'AT type,' in which adenine and thymine predominate, and the 'GC type,' in which guanine and cytosine are the major constituents."
>
> Erwin Chargaff (1951) [1]

Evolutionary selective pressures sometimes act to preserve nucleic acid features at the expense of encoded proteins. That this might occur in the case of nucleic acid secondary structure was noted in Chapter 5. That this might also apply to the species-dependent component of the base composition, **(G+C)**%, was shown by Sueoka in 1961 [2]. The amino acid composition of the proteins of bacteria is influenced, not only by the demands of the environment on the proteins, but also by the **(G+C)**% of the genome encoding those proteins.

From the genetic code (Table 7-1) it can be inferred that **AT**-rich genomes tend to have more codons for the amino acids **Phe**, **Tyr**, **Met**, **Ile**, **Asn**, and **Lys**. These amino acids have single-letter abbreviations that collectively spell "FYMINK." **GC**-rich genomes tend to have more codons for the amino acids **Gly**, **Ala**, **Arg** and **Pro**. These amino acids have single-letter abbreviations that collectively spell "GARP." If you know the FYMINK/GARP ratio of an organism's proteins, then you can make a good guess as to its **(G+C)**%. Sueoka's observation has since been shown to apply to a wide variety of animal and plant species. Could an organism's **(G+C)**% have arisen randomly and yet be so powerful that it could force a gene to encode an amino acid "against its will" (i.e. to encode an amino acid that might not best serve the needs of the encoded protein, and hence, of the organism)? Can a selective force be recognized?

Uniformity of (G+C)%

Chargaff's "**GC** rule" is that the ratio of (**G**+**C**) to the total bases (**A**+**G**+**C**+**T**) tends to be constant in a particular species, but varies between species. Sueoka further pointed out that for individual "strains" of *Tetrahymena* (ciliated *protozoa*ns) the (**G**+**C**)% (referred to as "GC") tends to be uniform throughout the genome:

> "If one compares the distribution of DNA molecules of *Tetrahymena* strains of different mean GC contents, it is clear that the difference in mean values is due to a rather uniform difference of GC content in individual molecules. In other words, assuming that strains of *Tetrahymena* have a common phylogenetic origin, when the GC content of DNA of a particular strain changes, all the molecules undergo increases or decreases of GC pairs in similar amounts. This result is consistent with the idea that the base composition is rather uniform not only among DNA molecules of an organism, but also with respect to different parts of a given molecule."

Again, this observation has since been shown to apply to a wide variety of species, although many organisms have their genomes finely sectored into regions ("homostability regions" or "isochores") of low or high (**G**+**C**)% (see later). Sueoka also noted a link between (**G**+**C**)% and reproductive isolation for strains of *Tetrahymena*:

> "DNA base composition is a reflection of phylogenetic relationship. Furthermore, it is evident that those strains which mate with one another (i.e. strains within the same 'variety') have similar base compositions. Thus strains of variety 1 ..., which are freely intercrossed, have similar mean GC content."

The Holy Grail

It seems that, in identifying (**G**+**C**)% as the component of the base composition that varies between species, Chargaff had uncovered what can now be recognized as the "holy grail" of speciation postulated by the Victorian physiologist George Romanes [3]. Romanes had drawn attention to the possibility of what we would now call non-genic variations (germ-line mutations that usually do not affect gene products). As manifest in the phenomenon of hybrid sterility, these would tend to isolate an individual reproductively from most members of the species to which its close ancestors had belonged, but not from individuals that had undergone the same non-genic variation. Romanes held that, in the general case, this isolation was an essential *precondi-*

tion for the preservation of the anatomical and physiological characteristics (genic characteristics) that were distinctive of a new species.

In the early decades of the twentieth century William Bateson also postulated non-genic inherited variations that tend to remain relatively constant (vary only within narrow limits) *within* a species, but would vary *between* species (i.e. a species member would not differ from its fellow species members, but would differ from members of allied species). The non-genic variations, in whatever was responsible for carrying hereditary information from generation to generation (not known at that time), would have the potential to lead to species differentiation, so that variant individuals (constituting a potential "not-self" incipient species) would end up not being able to reproduce with members of the main species ("self" species).

Reproduction being unsuccessful, the main species can be viewed as constituting a "reproductive environment" that *moulds* the genome phenotype ("reprotype") by negatively selecting (by denying reproductive success to) variant organisms that attempt (by mating and producing healthy, fertile, offspring) to *recross* the emerging interspecies boundary. Thus, the main species positively selects *itself* by negatively selecting variants. Should these variants find compatible mates, then they might accumulate as a new species that, in turn, would positively select itself by negatively selecting further variants. This is "species selection," a form of group selection that many biologists have found hard to imagine. Indeed, Richard Dawkins, having scorned the "argument from personal incredulity," was obliged to resort to it when confronted with the possibility of species selection: "It is hard to think of reasons why species survivability should be decoupled from the sum of the survivabilities of the individual members of the species" [4].

When the latter sentence is parsed its logic seems impeccable. Hold tight, and we will see if we can work it out. "*The* species" is the established main species, members of which imperil themselves only marginally, if at all, by mating with (denying reproductive success to) members of a small potentially incipient species. Thus, in reproductive interactions between a main and an incipient species, survivability of the main species is coupled *negatively* to the sum of the survivabilities of individual members of the incipient species (i.e. it survives when they do not survive), much more than it is coupled *positively* to the sum of the survivabilities of its own individual members (i.e. it survives when they survive). In this sense, main species survivability is *coupled* to the sum of the survivabilities of individual members of the incipient species, and *decoupled* from the sum of the survivabilities of its own individual members.

Of course, by individual survivabilities is meant, not just mere survival, but survival permitting unimpeded production of fertile offspring. Survival of members of an incipient species occurs, not only when classical Darwinian phenotypic interactions are favourable (e.g. escape from a tiger), but also when reprotypic interactions are favourable (e.g. no attempted reproduction with members of the main species). Tigers are a *phenotypic* threat. Members of the main species are a *reprotypic* threat [3].

Individual members of a main species that are involved (when there is attempted crossing) in the denial of reproductive success to individual members of an incipient species, are like individual stones in the walls of a species fortress against which the reproductive arrows of an incipient species become blunted and fall to the ground. Alternatively, the main species can be viewed as a Gulliver who barely notices the individual Lilliputian incipients brushed off or trampled in his evolutionary path. Just as individual cells acting in collective phenotypic harmony constitute a Gulliver, so individual members of a species acting in collective reprotypic harmony constitute a species. That harmony is threatened, not by its own members, but by deviants that, by definition, are no longer members of the main species (since a species is defined as consisting of individuals between which there is no reproductive isolation). These deviants constitute a potential incipient species that might one day pose a phenotypic threat to the main species (i.e. they will become part of the environment of the latter).

It is true that a member of a main species that becomes irretrievably pair-bonded with a member of an incipient species (e.g. pigeons) will leave fewer offspring, so that both members will suffer the same fate (have decreased survivability in terms of number of fertile offspring). But, in the general case, one such infertile reproductive encounter with a member of an incipient species will be followed by many fertile reproductive encounters with fellow members of the main species. Members of the main species are most likely to encounter *other members of the main species* – hence, there will be fertile offspring. Members of an incipient species, being a minority, are also most likely to encounter members of the main species – hence, there will be infertile (sterile) offspring. Much more rarely, a member of an incipient species will encounter a fellow incipient species member with which it can successfully reproduce – an essential precondition for species divergence.

Once *branching* (reproductive isolation) is initiated (Fig. 7-4), the natural selection of Darwin should help the branches sprout (extend in length). Natural selection would favour *linear* species differentiation by allowing the survival of organisms with advantageous genic variations, and disallowing the survival of organisms with disadvantageous genic variations. These genic variations would affect an organism's form and function (the classical phenotype).

Darwin thought that natural selection might itself suffice to bring about branching. Indeed, it *appears* to do so in certain circumstances, as when segments of a species have become geographically isolated from each other.

However, here the branching agency is whatever caused the geographical isolation, *not* natural selection. Speciation requires isolation in some shape or form. The problem of *the* origin of species is that of determining what form isolation takes *in the general case*. In his faith in the power of natural selection, Darwin was like the early chemists who were satisfied with atoms as the ultimate basis of matter. But for some chemists phenomena such as swinging compass needles (magnetism), falling apples (gravity), and (later) radioactivity, were manifestations of something more fundamental in chemistry than atoms. Likewise, for some biologists the phenomenon of hybrid sterility seemed to manifest something more fundamental in biology than natural selection [3].

Romanes referred to his holy grail (speciating factor) as an abstract "intrinsic peculiarity" of the reproductive system. Bateson described his as an abstract "residue" with which genes were independently associated. Goldschmidt's was an abstract chromosomal "pattern" caused by "systemic mutations" that would not necessarily affect genic functions (see Chapter 7). These are just what we might expect of (**G**+**C**)%. Indeed, in bacteria, which when so inclined intermittently transfer DNA in a sexual fashion [5], differences in (**G**+**C**)% appear early in the speciation process [6], in keeping with Sueoka's above observations in ciliates.

As shown in Chapter 3, where different *levels* of genetic information were considered, a metaphor for the role (**G**+**C**)% might play in keeping individuals reproductively isolated from each other, is their accent [7]. A common language brings people together, and in this way is conducive to sexual reproduction. But languages can vary, first into dialects and then into independent sub-languages. Linguistic differences keep people apart, and this difference in the reproductive environment can militate against sexual reproduction.

At the molecular level, we see similar forces acting at the level of meiosis – the dance of the chromosomes. In the gonad similar paternal and maternal chromosomes (homologues) align. The early microscopists referred to this as "conjugation." If there is sufficient sequence identity (i.e. the DNA "accents" match), then the band plays on. The chromosomes continue their minuet, progressing through various check-points [8], and gametes are formed. If there is insufficient identity (i.e. the DNA "accents" do not match) then the music stops. Meiosis fails, gametes are not formed, and the child is sterile – a "mule." Thus, the parents of the child (their "hybrid") are reproductively isolated *from each other* (i.e. unable to generate a line of descendents due to hybrid sterility), but not necessarily from other members of their species. At least one of the parents has the potential to be a founding member of a new species, provided it can find a mate with the *same* DNA "accent." Differences in (**G**+**C**)% have the potential to initiate the speciation process creating first "incipient species" with partial reproductive isolation, and then "species" that, by definition, are fully reproductively isolated. To see how this might work, we consider the chemistry of chromosome alignment at meiosis [9].

Crick's Unpairing Postulate

In 1922 Muller suggested that the pairing of genes as parts of chromosomes undergoing meiotic synapsis in the gonad might provide clues to gene structure and replication [10]:

> "It is evident that the very same forces which cause the genes to grow [duplicate] should also cause like genes to attract each other [pair] ... . If the two phenomena are thus dependent on a common principle in the make-up of the gene, progress made in the study of one of them should help in the solution of the other."

In 1954 he set his students an essay "How does the Watson-Crick model account for synapsis?" [11]. The model had the two DNA strands "inward-looking" (i.e. the bases on one strand were paired with the bases on the other strand). Crick took up the challenge in 1971 with his "unpairing postulate" by which the two strands of a DNA duplex would *unpair* to expose free bases in single-stranded regions [12]. This would allow a search for sequence similarity (homology) between two chromosomes (i.e. between two independent duplexes). Others later proposed that the single-stranded regions would be extruded as stem-loops. The "outward-looking" bases in the loops would be available to initiate the pairing process [13–15].

Thus, for meiotic alignment, maternal and paternal chromosomal homologues should mutually explore each other and *test* for "self" DNA complementarity, by the "kissing" mechanism noted in Chapter 6 [16–18]. Under this model (Fig. 8-1), the sequences do not *commit* themselves, by incurring strand-breakage, until a degree of complementary has been recognized. The mechanism is essentially the same as that by which tRNA anticodon loops recognize codons in mRNAs, except that the stem-loop structures first have to be extruded from DNA molecules that would normally be in classical duplex form. In all DNA molecules examined, base-order supports the formation of such secondary structures (see Chapter 5). If sufficient complementarity is found between the sequences of paternal and maternal chromosome homologues (i.e. the genomes are "reprotypically" compatible), then crossing over and recombination can occur (i.e. the "kissing" can be "consummated").

The main adaptive values of this would be the proper assortment of chromosomes among gametes, and the correction of errors in chromosome sequences (see below and Chapter 14). "Kissing" turns out to be a powerful metaphor, since it implies an exploratory interaction that may have reproductive consequences.

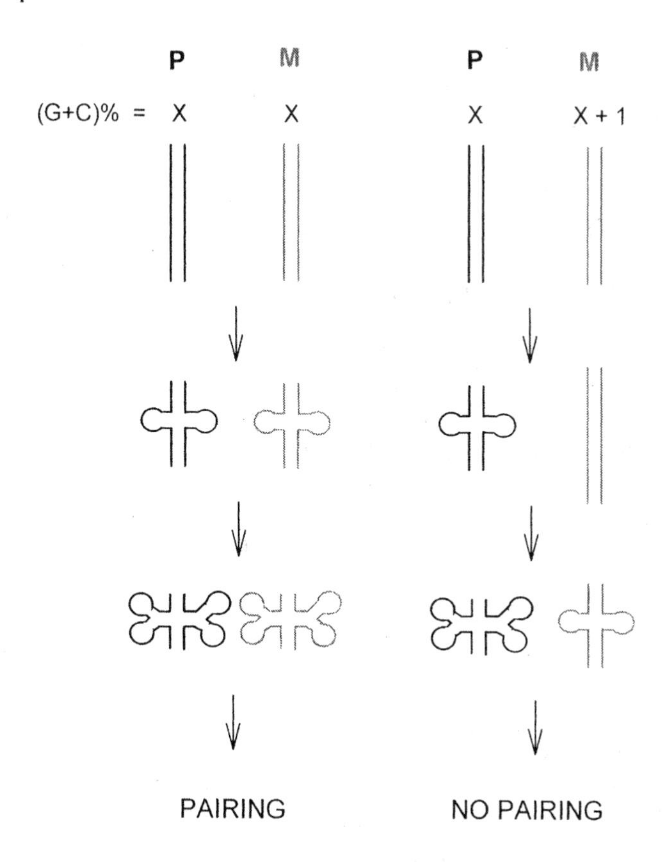

Fig. 8-1. The exquisite sensitivity of stem-loop extrusion from duplex DNA to differences in base composition can prevent the initiation of pairing between homologous DNA sequences. At the left, paternal (P) and maternal (M) duplexes have the same (**G**+**C**)% value (X). As negative supercoiling progressively increases, the strands of each duplex synchronously open to allow formation of equivalent stem-loop secondary structures so that "kissing" interactions between loops can progress to pairing. At the right, paternal and maternal duplexes differ slightly in (**G**+**C**)% (X, and X + 1). The maternal duplex of higher (**G**+**C**)% opens less readily as negative supercoiling increases, so strand opening is not synchronous, "kissing" interactions fail, and there is no progress to pairing. In this model, chromosome pairing occurs *before* the strand breakage that accompanies recombination (not shown). Even if strand breakage were to occur first (as required by some models), unless inhibited by single-stranded DNA-binding proteins the free single strands so exposed would rapidly adopt stem-loop conformations. So the homology search could still involve kissing interactions between the tips of loops

The model predicts that, for preventing recombination (i.e. creating reproductive isolation), a non-complementarity between the sequences of potentially pairing strands, in itself, might be less important than a non-complementarity associated with sequence differences that change the pattern of stem-loops. This implies differences in the *quantities* of members of the Watson-Crick base pairs in single strands (i.e. a parity difference). This is because parity between these bases would be needed for optimum stem formation. Parity differences should correlate with differences in stem formation, and hence, different stem-loop patterns, as will now be considered.

(G+C)% Controls Pairing

What role does the (**G**+**C**)% "accent" play in meiotic pairing? From calculated DNA secondary structures, it has been inferred that small fluctuations in (**G**+**C**)% have great potential to affect the extrusion of stem-loops from duplex DNA molecules and, hence, to affect the pattern of loops which would then appear (Fig. 5-2). A very small difference in (**G**+**C**)% (reprotypic difference) would mark as "not-self" a DNA molecule that was attempting to pair meiotically with another DNA ("self"). This would impair the kissing interaction with the DNA [19, 20], and so would disrupt meiosis and allow divergence between the two parental lines, thus initiating a potential speciation event.

The total stem-loop potential in a sequence window can be analysed quantitatively in terms of the relative contributions of base composition and base order, of which base composition plays a major role (see Chapter 5). Of the various factors likely to contribute to the base composition-dependent component of the folding energy of an extruded single stranded DNA sequence, the four simplest are the quantities of the four bases.

Two slightly more complex factors are the individual bases, from each potential Watson-Crick base pair, that are present in lowest amounts. For example, if the quantities of **A**, **G**, **C** and **T** in a 200 nucleotide sequence window are 60, 70, 30 and 40, respectively, then what may be referred to as "$\mathbf{AT}_{min}$" would be 40, and the corresponding "$\mathbf{GC}_{min}$" would be 30. These numbers would reflect the upper limit on the number of base pairs that could form stems, since the quantity of the Watson-Crick pairing partner that was *least* would place a limit on the possible number of base pairs. This value might be expected to correlate *positively* with folding stability.

Conversely, the excess of bases without a potential pairing partner (in the above example **A-T** = 20 and **G-C** = 40) might provide an indication of the maximum number of bases available to form loops. Since loops tend to destabilize stem-loop structures, these "Chargaff difference" values might be expected to correlate *negatively* with folding stability.

Although the bases are held in linear order, a vibrating single-stranded DNA molecule has the potential to adopt many structural conformations, with Watson-Crick interactions occurring between widely separated bases. Accordingly, pairing can also be viewed as if the result of random interactions between free bases in solution. This suggests that the two products of the quantities of pairing bases could be important (60 x 40, and 70 x 30, in the above example). The products would be maximal when pairing bases were in equal proportions in accordance with Chargaff's second parity rule. Thus, [50 x 50] > [60 x 40] > [70 x 30]). The product values (2500 > 2400 > 2100) might provide an index both of the absolute quantities of the members of a Watson-Crick base pair, and of their relative proportions.

In an attempt to derive formulae permitting prediction of folding energy values directly from the proportions of the four bases, Jih-H. Chen [21] examined the relative importance of eight of the above ten factors in determining the base composition-dependent component of the folding energy (FORS-M; see Chapter 5). These factors were **A**, **G**, **C**, **T**, $\mathbf{AT}_{min}$, $\mathbf{CG}_{min}$, **A** x **T**, and **G** x **C** (where **A**, **C**, **G**, and **T** refer to the quantities of each particular base in a sequence window). The products of the quantities of the Watson-Crick pairing bases (**A** x **T**, and **G** x **C**) were found to be of major importance, with the coefficients of **G** x **C** (the strongly interacting **S** bases), greatly exceeding those of **A** x **T** (the weakly interacting **W** bases). Less important were $\mathbf{AT}_{min}$ and $\mathbf{CG}_{min}$, and the quantities of the four bases.

All ten parameters were examined in an independent study, which confirmed the major role of the product of the quantities of the **S** bases in a segment (Table 8-1) [22].

Number of	Individual bases				Base products		Minimum bases		Chargaff differences		
predictors	A	T	C	G	A x T	G x C	AT_{min}	GC_{min}	A - T	G - C	r^2
1	-	-	-	-	-	+	-	-	-	-	0.92
2	-	-	+	-	-	+	-	-	-	-	0.96
3	-	-	+	+	-	+	-	-	-	-	0.98
4	-	-	+	+	-	+	-	-	+	-	0.98

Table 8-1. Relative abilities of ten base composition-derived parameters to predict the base composition-dependent component of the folding energy of single-stranded DNA. Best predictors are marked as plus. With 1 predictor, the product of **G** and **C** has a correlation of 0.92, as assessed by the r^2 value (see Appendix 1). Adding further predictors (rows 2, 3 and 4) does not greatly improve on this value

Of particular importance is that it is not just the absolute quantities of the **S** bases, but the *product* of the multiplication of these absolute quantities. This should amplify very small fluctuations in (**G**+**C**)%, and so should have a major impact on the folding energy of a segment and, hence, in the pattern of stem-loops extruded from the duplex DNA in a chromosome engaging in a "kissing" homology search for a homologous chromosome segment.

If stem-loops are of critical importance for the initiation of pairing between segments of nucleic acids at meiosis, then differences in (**G**+**C**)% could strongly influence the establishment of meiotic barriers, so leading to speciation. But barriers may be transient. Having served its purpose, an initial barrier may be superseded later in the course of evolution by a more substantial barrier (see Figure 7-4). In this circumstance evidence for the early transient barrier may be difficult to find. However, in the case of different, but related, virus species (allied species) that have the potential to coinfect a common host cell, there is circumstantial evidence that the original (**G**+**C**)% barrier has been retained.

Mutational Meltdown

Modern retroviruses, such as those causing AIDS (HIV-1) and human T cell leukemia (HTLV-1), probably evolved by divergence from a common ancestral retrovirus. Branching phylogenetic trees linking the sequences of modern retroviruses to such a primitive retroviral "Eve" are readily constructed, using either differences between entire sequences, or just (**G**+**C**)% differences [23]. The fewer the differences, the closer are two species on such trees.

Unlike most other virus groups, retroviruses are diploid. As indicated in Chapter 2, diploidy entails a considerable redundancy of information, a luxury that most viruses cannot afford. They need compact genomes that can be rapidly replicated, packaged and dispersed to new hosts. However, different virus groups have evolved different evolutionary strategies.

The strategy of retroviruses is literally to mutate themselves to the threshold of oblivion ("mutational meltdown"), so constituting a constantly moving target that the immune system of the host cannot readily adapt to. To generate mutants, retroviruses replicate their nucleic acids with self-encoded enzymes (polymerases) that do not have the error-correcting ("proof-reading") function that is found in the corresponding enzymes of their hosts. Indeed, this is the basis of AIDS therapy with AZT (azidothymidine), which is an analogue of one of the nucleotide building blocks that are joined together (polymerized) to form linear nucleic acid molecules ("polymers;" see Chapter 2). AZT is recognized as foreign by host polymerases, which eject it. But retroviral polymerases cannot discriminate, and levels of mutation (in this case termination of the nucleic acid sequence) attain values above the oblivion threshold ("hypermutation") from which it is impossible to recover ("error catastrophy"). Below the threshold, there is a most effective mechanism to counter mutational damage.

The retroviral counter-mutation strategy requires that two complete single-strand retroviral RNA genomes be packaged in each virus particle (i.e. diploidy). Each of these genomes will be severely mutated but, since mutations occur randomly, there is a chance that each genome will have mutations at different sites. Thus, in the next host cell there is the possibility of recombination (cutting and splicing) between the two genomes to generate a new genome with many less, or zero, mutations [24].

The copackaging of the two genomes requires a process analogous to meiotic pairing. On each genome a "dimer initiation" nucleotide sequence folds into a stem-loop structure. "Kissing" interactions between the loops precede the formation of a short length of duplex RNA (Fig. 6-8), so that the two genomes form a dimer. This allows packaging and, in the next host, recombination can occur.

Coinfecting Viruses

What if two diploid viruses both infected the *same* host cell, thus releasing four genomes into an environment conducive to recombination? In many cases this would be a most favorable circumstance, since there would now be four damaged genomes from which to regenerate, by repeated acts of recombination, an ideal genome. Thus, it would seem maladaptive for a virus with this particularly strategy to evolve mechanisms to prevent entry of another virus ("superinfection") into a cell that it was occupying, at least in the early stages of infection [25].

This presupposes that a co-infecting virus will be of the *same* species as the virus which first gained entry. However, HIV-1 and HTLV-1 are retroviruses of allied, but distinct, species. They have a common host (humans) and common host cell (known as the CD4 T-lymphocyte). When in the course of evolution these two virus species first began to diverge from a common ancestral retroviral species, a barrier to recombination had to develop as a condition of successful divergence. Yet, these two virus types needed to retain a common host cell in which they had to perform similar tasks. This meant that they had to retain similar genes. Many similar gene-encoded functions are indeed found. Similar genes implies similar sequences, and similar sequences implies the possibility of recombination between the two genomes. Thus, coexistence in the same host cell could result in the viruses destroying each other, as distinct species members, by mutually recombining (shuffling their genomes together).

Without a recombination barrier *each virus was part of the selective environment of the other*. This should have provided a pressure for genomic changes that, while not interfering with conventional phenotypic functions, would protect against recombination with the other type. If **(G+C)**% differences could create such a recombination barrier (while maintaining, through choice of appropriate codons, the abilities to encode similar amino acid sequences), then such differences would be selected for. When we examine the **(G+C)**% values of each of these species there is a remarkable difference. HIV-1 is one of the lowest **(G+C)**% species known (i.e. it is **AT**-rich). HTLV-1 is one of the highest **(G+C)**% species known (i.e. it is **GC**-rich). This might be regarded as just a remarkable coincidence save for the fact that, in some other situations where two viruses from different but allied species occupy a common host cell, there are also wide differences in **(G+C)**% [3, 19]. As set out above, these **(G+C)**% differences alone should suffice to prevent recombination.

Polyploidy

The plant which gives us tobacco, *Nicotiniana tabacum*, is a tetraploid which emerged some six million years ago when the two diploid genomes of *Nicotiniana sylvestris* and *Nicotiniana tomentosiformis* appeared to fuse. *Nicotiniana tabacum* is designated an *allo*tetraploid (rather than an *auto*-tetraploid) since the two genomes were from different source species (Greek: *allos* = other; *autos* = same). The two species are estimated to have diverged from a common ancestral species 75 million years ago. As allied species they should have retained some sequence similarities; so within a common nucleus in the tetraploid there should have been ample opportunity for recombination between the two genomes. Yet, the genomes have retained their separate identities. This can be shown by backcrossing to the parental types. Half the chromosomes of the tetraploid pair at meiosis with chromosomes of one parent type. Thus, recombination of the other chromosomes of the tetraploid with chromosomes of that parent type is in some way prohibited. In 1940 Goldschmidt noted [26]:

> "Clausen ... has come to the conclusion that *N. tabacum* is an allotetraploid hybrid, one of the genomes being derived from the species *sylvestris*, the other from *tomentosa*. By continuous backcrossing to *sylvestris* the chromosomes derived from *sylvestris* can be tested because they form tetrads with the *sylvestris*-chromosomes. They have been found to be completely different genetically [from *tomentosa*]. The idea of reaction system [pattern] thus becomes less generalized [less nebulous] and actually applies to the whole architecture of individual chromosomes."

The separateness of the genomes in allotetraploids is reflected in meiosis where, although there are four potentially pairing copies of each chromosome (four homologous chromosomes), each parental type only pairs with its *own* type. The two sets of chromosomes differ in a key sequence characteristic, their (**G+C**)% [27].

Gene Duplication

The survival of a duplicate copy of a gene depends on a variety of factors, including (i) natural selection favouring organisms where a function encoded by the gene is either increased or changed (i.e. there is either *concerted* or *divergent* gene evolution), (ii) a recombination-dependent process known as gene conversion, and (iii) a recombination-dependent process that can lead to copy-loss (see Fig. 8-2). These *intra*genomic recombinations can occur when there is a successful search for similarity between DNA strands. This is likely to be greatly influenced by the (**G+C**)% environment of the original gene and the (**G+C**)% environment where the duplicate copy locates.

Once a (**G+C**)%-dependent speciation process has begun, factors other than (**G+C**)% are likely to replace the original difference in (**G+C**)% as an *inter*genomic barrier to reproduction (i.e. a barrier to recombination between diverged paternal and maternal genomes within their hybrid, if such a "mule" can be generated; Fig. 7-4). In this circumstance, (**G+C**)% becomes free to adopt other roles, such as the prevention of recombination *within* a genome (*intra*genomic recombination). This can involve the differentiation of regions of relatively uniform (**G+C**)%, that Japanese physicists Akiyoshi Wada and Akira Suyama referred to as having a "homostabizing propensity" and Giorgio Bernardi and his coworkers named "isochores" (Greek: *iso* = same; *choros* = group) [28, 29]. These have the potential to recombinationally isolate different parts of a genome.

Thus, the attempted duplication of an ancestral globin gene to generate the α-globin and β-globin genes of modern primates might have failed since sequence similarity would favour recombination between the two genes and incipient differences (early sequence divergence) could have been eliminated ("gene conversion;" Fig. 8-3). However, the duplication appears to have involved relocation to a different isochore with a different (**G+C**)%, so the two genes became recombinationally isolated to the extent that initially the sequences flanking the genes differed in (**G+C**)%. Later the new gene would have increased its recombinational isolation by mutating to acquire the (**G+C**)% of its host isochore. As a consequence of the differences in (**G+C**)% the corresponding mRNAs today utilize different codons for corresponding amino acids, even though both mRNAs are translated in the *same* cell using the *same* ribosomes and *same* tRNA populations. So it is most unlikely that the primary pressure to differentiate codons arose at the translational level.

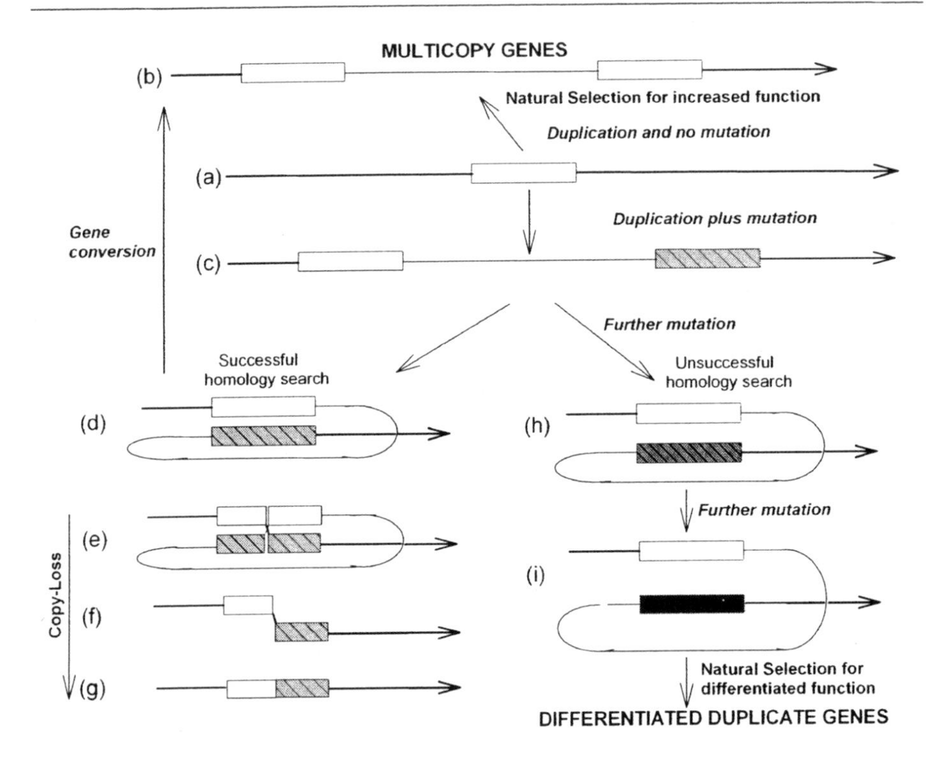

Fig. 8-2. Model for possible outcomes of a gene duplication. The duplication from *(a)* can result in identical multicopy genes *(b)* that confer an ability to produce more of the gene product. If this is advantageous, then the multicopy state will tend to be favored by natural selection. If unmutated (white box in *(b)*) or only slightly mutated (light grey striped box in *(c)*), there are not sufficient differences between the duplicates to prevent a successful homology search *(d)*. This allows the mutation *(c)* to be reversed to *(b)* by the process known as gene conversion (see Fig. 8-3). This maintains identical copies, so allowing concerted evolution of the multicopy genes to continue. However, the recombination necessary for gene conversion can also result in removal of a circular intermediate *(e, f)*, and restoration of the single copy state *(g)*. The risk of copy-loss due to recombination *(d-g)* can be decreased by further mutation (dark grey striped box in *(h)*). This will decrease the probability of a successful homology search. Being protected against recombination (i.e. preserved), the duplicate is then free to differentiate further by mutation (black box in *(i)*). If the product of the new gene confers an advantage, then the duplicate will be further preserved by natural selection (divergent gene evolution). In the general case, mutation facilitating recombinational isolation *(h) precedes* mutation facilitating functional differentiation *(i)* under positive Darwinian selection

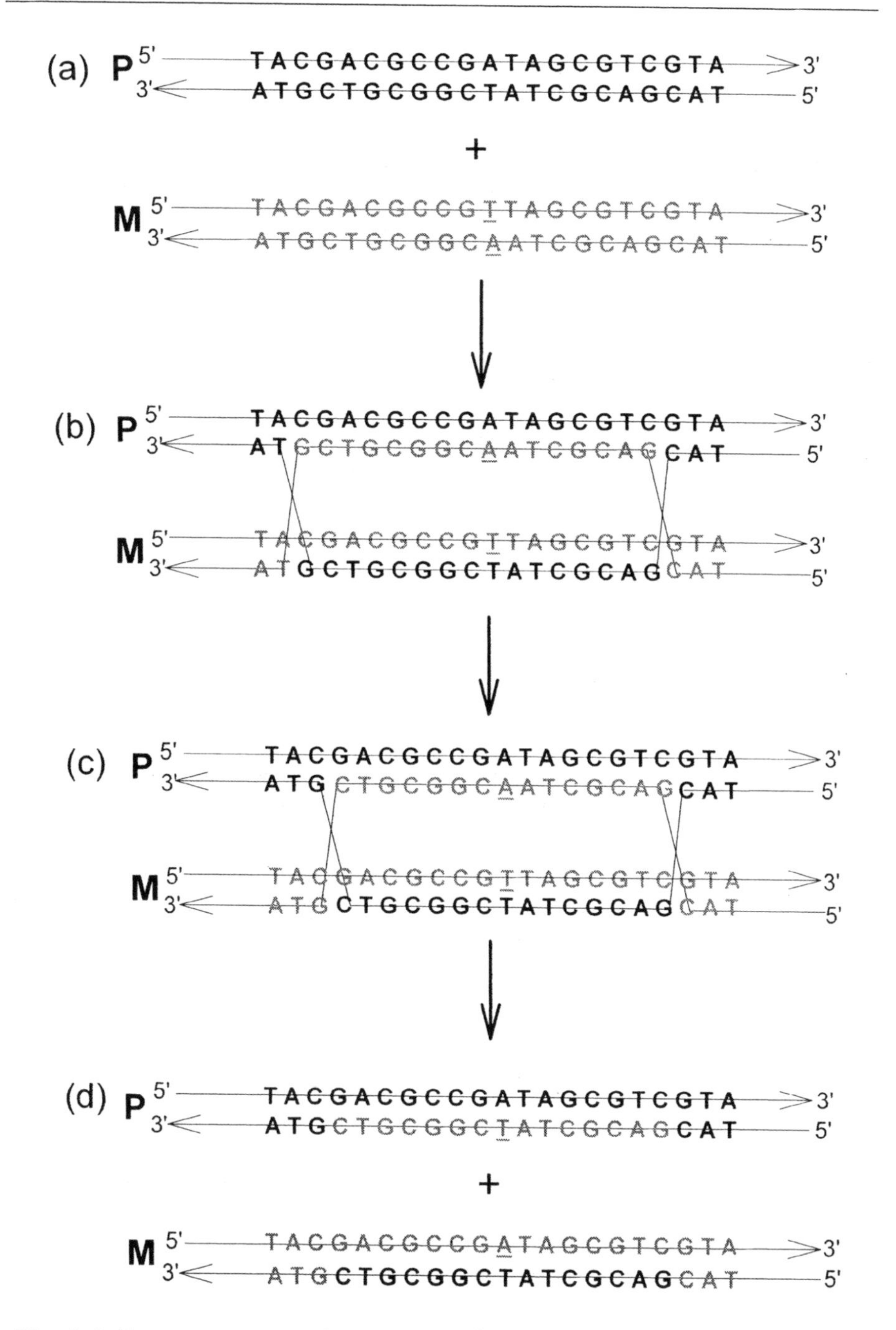

Fig. 8-3. Recombination repair and gene conversion. *(a)* Sequences 3.1 and 3.3 of Chapter 3 are taken as paternal (P) and maternal (M) versions (alleles) of a particular gene, with a single base-pair difference (underlined in the grey

maternal version). *(b)* Following a successful homology search, exchange of single-strands results in a P segment becoming part of the bottom duplex, and an M segment becoming part of the top duplex, without permanent strand breakage ("paranemic joint"). Because of the single base-pair difference, the recombination complex has an **A** mismatched with an **A** (top duplex), and a **T** mismatched with a **T** (bottom duplex). *(c)* Cross-over points can migrate without any permanent strand breakage (shown at left). *(d)* Strand breakage and resealing (ligation) occur, so that single-strand segments are exchanged between the duplexes (*recombination*). Without guidance, the "repair" of each mismatch has a 50% chance of restoring the *status quo* as in *(a)* (i.e. replacing the **A-A** non-Watson-Crick base pair with an **A**-**T** Watson-Crick base-pair in the top duplex, and replacing the **T**-**T** non-Watson-Crick base-pair with a **T-A** Watson-Crick base pair in the bottom duplex). In the alternative shown here, the *status quo* is restored to the top duplex (an **A** is mutated to **T**), but in the bottom duplex the **T**-**T** non-Watson-Crick base-pair is replaced with an **A**-**T** Watson-Crick base-pair (i.e. a **T** is mutated to an **A**). Thus, there has been *conversion* of the sequence of the original M allele to that of the P allele. There has been a loss of heterozygosity (as in *(a)*) and a gain of homozygosity (as in *(d)*). In this example, gene conversion involves copies of homologous genes (alleles) on different chromosomes. However, gene conversion can also involve homologous genes (non-allelic "paralogues") on the same chromosome (see Fig. 8-2). Note that, in Chapter 4, sequence 3.1 (P above) is shown to form a stem-loop with the central bases being located in the loop (sequence 4.4). Since the single base-pair difference between P and M versions is in this loop, then the M version has the potential to form a similar stem-loop. Because the loops differ slightly, during the initial homology search loop-loop "kissing" interactions might fail and prohibit subsequent steps. However, cross-over points can migrate (e.g. *(b)* to *(c)*), so that if crossing over is prohibited in one region there is some possibility of a migration from a neighboring region that would reveal mismatches. Thus, multiple incompatibilities (base differences) are most likely to inhibit the pairing of homologous chromosomes and the repairing of multiple mismatches

Each isochore would have arisen as a random fluctuation in the base composition of a genomic region such that a copy of a duplicated gene that had transposed to that region was able to survive without recombination with the original gene for a sufficient number of generations to allow differentiation between the copy and its original to occur. This would have provided not only greater recombinational isolation, but also an opportunity for functional differentiation. If the latter differentiation were advantageous, organisms with the copy would be favoured by natural selection. The regional base compositional fluctuation would then have "hitch-hiked" through the generations by virtue of its linkage to the successful duplicate (i.e. the copy would have been positively selected). By preserving the duplicate copy from recombination with the original copy, the isochore would, in turn, have itself been preserved by virtue of its linkage to the duplicate copy.

When functional differentiation of a duplicate is necessary for it to be selected (divergent evolution), there is the danger that, *before* natural selection can operate, recombination-mediated gene conversion will reverse any incipient differentiation, or intragenic recombination between the copies (paralogues) will result in copy-loss. In the case of duplicate eukaryotic genes that have diverged in sequence, Koichi Matsuo and his colleagues noted that divergence was greatest at third codon positions, usually involving a change in (G+C)% [30–33]. Thus, there was a codon bias in favour of the positions of *least* importance for the functional differentiation that would be necessary for the operation of natural selection. Where amino acids had not changed, different gene copies used *different* synonymous codons. It was proposed that the (G+C)% change was an important "line of defence" against homologous recombination between the duplicates. Thus, recombinational isolation of the duplicate (largely involving third codon position differences in (G+C)%) would *protect* (preserve) the duplicate so allowing time for functional differentiation (largely involving first and second codon position differences), and hence, for natural selection to operate. In the general case, isolation would *precede* functional differentiation, not the converse. (G+C)% differentiation, largely involving third codon positions, would precede functional differentiation, largely involving first and second codon positions under positive Darwinian selection.

From all this it would be predicted that, if a gene from one isochore were transposed to an isochore of different (G+C)%, and its ability to recombine with its *allele* were advantageous, then the gene would preferentially accept mutations converting its (G+C)% to that of the new host isochore (i.e. organisms with those mutations would be genetically fitter and thus likely to leave more fertile offspring than organisms without the mutations). Indeed, there is evidence supporting this. The sex chromosomes (X and Y) tend not to recombine at meiosis except in a small region (the "pseudoautosomal" region; see Chapter 14). Transfer of a gene from a non-recombining part of a sex chromosome to the pseudoautosomal region forces the gene rapidly to change its (G+C)% value [34].

For various reasons (e.g. large demand for the gene product), certain genes are present in multiple identical copies. But, in the absence of some restraint, copies that are initially identical will inevitably diverge in sequence [3]. So how can multicopy genes (e.g. rRNA genes) preserve their similarity to each other? To prevent divergence through the generations (i.e. to allow "concerted evolution"), they should mutually correct each other to eliminate deviant copies. This is likely to occur by a recombination-dependent process – "gene conversion" (Figs. 8-2, 8-3; see Chapter 10). Thus, multicopy genes should all be, either in the same isochore, or in isochores of very close (**G**+**C**)%, so that recombination can occur.

Isochores Early

Before DNA sequencing methods became available, "isochores" were described as DNA segments that could be identified on the basis of their distinct densities in samples of duplex DNA obtained from organisms whose cells had nuclei (eukaryotes). The method involved physically disrupting DNA by hydrodynamic sheering to break it down to lengths of about 300 kilobases. The fragments were then separated as bands of distinct densities by centrifugation in a salt density gradient. The densities could be related to the average (**G**+**C**)% values of the segments, since the greater these values, the greater the densities. This way of assessing the (**G**+**C**)% of a duplex DNA segment distinguished one large segment from another, and largeness became a defining property of isochores.

Isochores, as so defined, were not identified in bacteria, which do not have distinct nuclear membranes (prokaryotes; see Chapter 10). Since prokaryotes (e.g. bacteria) and eukaryotes (e.g. primates) are considered to have evolved from a common ancestor, does this mean that the ancestor had isochores that were subsequently lost by prokaryotes during or after their divergence from the eukaryote lineage (isochores-early)? Or did the ancestor not have isochores, which were therefore freshly acquired by the eukaryotic lineage after its divergence from the prokaryotic lineage (isochores-late)? If prokaryotes could be shown to have isochores, then this would favour the isochores-early hypothesis.

Indeed, prior to modern sequencing technologies, physical methods demonstrated small segments of distinct (**G**+**C**)% in the genomes of prokaryotes and their viruses. The 48 kb duplex genome of phage lambda (see Chapter 5) was extensively sheered to break it down to subgenome-sized fragments. These resolved into six distinct segments, each of relatively uniform (**G**+**C**)%, by the density method [35], and into thirty four "gene sized" segments by another, more sensitive, method (thermal denaturation spectrophotometry) [36].

With the advent of sequencing technologies, in 1984 Mervyn Bibb and his colleagues were able to plot the average (**G**+**C**)% values of every third base for small windows in the sequences of various bacteria (Fig. 8-4) [37]. Three plots were generated, the first beginning with the first base of the sequence (i.e. bases in frame 1, 4, 7, etc.), the second beginning with the second base of the sequence (i.e. bases in frame 2, 5, 8, etc.), and the third beginning with the third base of the sequence (i.e. bases in frame 3, 6, 9, etc.). In certain small regions (**G**+**C**)% values were relatively constant within each frame. *These regions of constant (**G**+**C**)% corresponded to genes.*

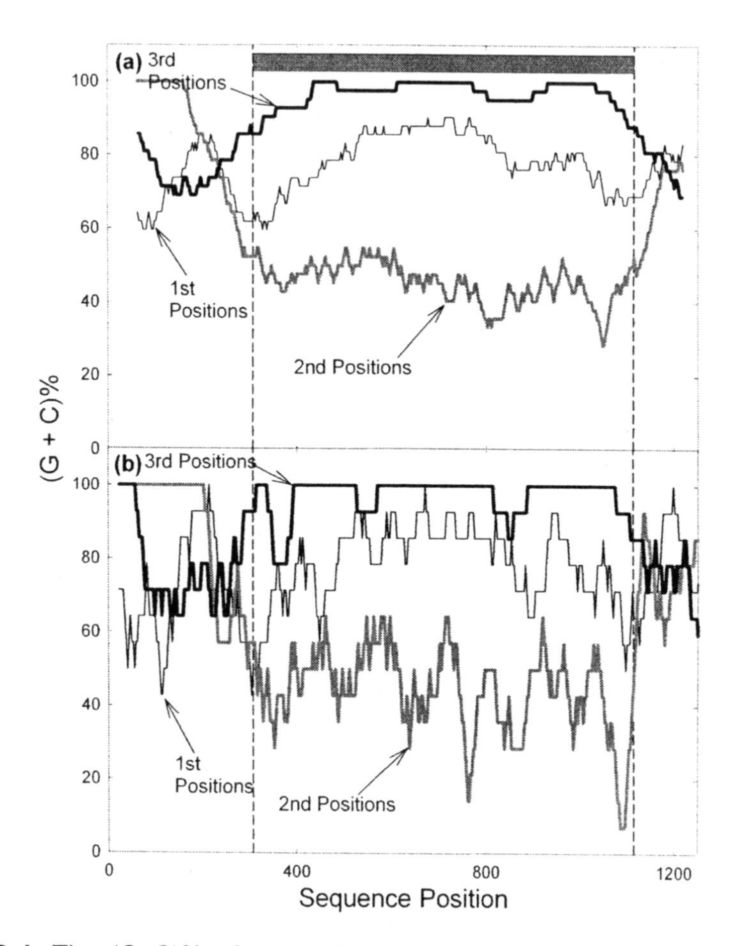

Fig. 8-4. The (**G**+**C**)% of every third base tends to be both distinct from neighboring bases, and constant, in the region of a gene. Bases were counted in windows of 126 bases *(a)* or 42 bases *(b)*, in one of three frames, either beginning with base 1 (frame 1, 4, 7, etc.), or with base 2 (frame 2, 5, 8, etc.), or with base 3 (frame 3, 6, 9, etc.). Thus, for calculating average (**G**+**C**)% values, each frame takes into account either 42 bases *(a)* or 14 bases *(b)*. The frames are here named according to the positions of bases in the codons of the gene (encoding the enzyme aminoglycoside phosphotransferase in the **GC**-rich bacterium *Streptomyces fradiae*). Vertical dashed lines and the grey rectangle indicate the limits of the rightward-transcribed gene. Note that the relative constancy of (**G**+**C**)% is most for the third codon position (mainly independent of the encoded amino acids), and least for the second codon position (most dependent on the encoded amino acids). The fluctuation in values at the second codon position is more apparent when a window size equivalent to 14 codons is used *(b)* than when a window size equivalent to 42 codons is used *(a)*. This figure was redrawn from ref. [37]

Thus, individual genes have a relatively uniform (G+C)% and each codon position makes a distinctive contribution to that uniformity. This is not confined to bacteria. Wada and Suyama noted that, whether prokaryotic or eukaryotic, "every base in a codon seems to work cooperatively towards realizing the gene's characteristic value of (G+C) content." This was a "homostabilizing propensity" allowing a gene to maintain a distinct (G+C)%, relatively uniform along its length, *which would differentiate it from other genes in the same genome* [38]. Thus, each gene constitutes a homostabilizing region in DNA.

Stated another way, if large size is excluded as a defining property, many bacteria have isochores. When isochores are defined as DNA segments of relatively uniform (G+C)% that are coinherited with specific sequences of bases, then bacteria have isochores. To contrast with the classical isochores of Bernardi, these are termed "microisochores," and their length is that of a gene, or small group of genes (see Chapter 9). Thus, classical eukaryotic isochores ("macroisochores") can be viewed as constellations of microisochores of a particular (G+C)%. The proposed antirecombination role of (G+C)% would required that, unless they represent multicopy genes, microisochores sharing a common macroisochore (i.e. they have a common (G+C)%) have other sequence differences that are sufficient to prevent recombination between themselves [39]. Within an organism, genes with similar (G+C)% values may sometimes locate to similar tissues, so that there is a tissue-specific codon usage tendency [31].

Since both prokaryotic (e.g. bacterial) and eukaryotic (e.g. primate) lineages have some form of isochore, this appears most consistent with the isochores-early hypothesis. While not endorsing a particular role for (G+C)%, this underlines the fundamental importance of (G+C)% differences in biology. Let metaphors multiply! A given segment of DNA is coinherited with a "coat" of a particular (G+C)% "color." A given segment of DNA "speaks" with a particular (G+C)% "accent," (and hence has a distinct potential vibrational frequency; see Fig. 5-2). A fundamental duality of information levels is again manifest.

The Gene as a Unit of Recombination

As will be further considered in Chapter 9, it is likely that differences in (G+C)% serve to isolate recombinationally both genes within a genome, and genomes within a group of species (a taxonomic group). The power to recombine is fundamental to all life forms because, for a variety of reasons, it is advantageous (see Chapter 14). However, the same power threatens to homogenize (blend) genes within a genome, and to homogenize (blend) the genomes of members of allied species within a taxonomic group (i.e. genus). This would countermand evolution both within a species and between species. *Thus, functional differentiation, be it between genes in a genome, or between genomes in a taxonomic group (speciation), must, in the general case, be preceded (or closely accompanied) by the establishment of recombinational barriers.* Species have long been defined in terms of recombinational barriers (see Chapter 7). In some contexts, genes are defined similarly. A species can be defined as a unit of recombination (or rather, of antirecombination with respect to other species). So can a gene.

Most definitions of the "gene" contain a loose or explicit reference to function. Thus, biologists talk of a gene encoding information for tallness in peas. Biochemists talk of the gene encoding information for growth hormone (a protein), and relate this to a segment of DNA (see legend to Fig. 10-1). However, before it can function, information must be *preserved.* Classical Darwinian theory proposes that function, through natural selection, is itself the preserving agent. Thus, function and preservation go hand-in-hand, but function is more fundamental than preservation. In 1966 biologist George Williams in the USA, an originator of the "selfish gene" concept, seemed to argue the converse when arriving at a new definition.

The function of any multipart entity, which needs more than one part for this function, is usually dependent on its parts not being separated. Preservation can be more fundamental than function. Williams proposed that a gene should be defined entirely by its property of remaining intact as it passes from generation to generation. He identified recombination as a major threat to that intactness. Thus, for Williams, "gene" meant any DNA segment that has the potential to persist for enough generations to serve as a unit for natural selection; this requires that it not be easily disruptable by recombination. The gene is a unit of recombination (or rather, of antirecombination with respect to other genes) [40].

> "Socrates' genes may be with us yet, but not his genotype, because meiosis and recombination destroy genotypes as surely as death. It is only the meiotically dissociated fragments of the genotype that are transmitted in sexual reproduction, and these fragments are further fragmented by meiosis in the next generation. If there is an ultimate indivisible fragment it is, by definition, 'the gene' that is treated in the abstract discussions of population genetics. Various kinds of suppression of recombination may cause a major chromosomal segment or even a whole chromosome to be transmitted entire for many generations in certain lines of descent. In such cases the segment, or chromosome, behaves in a way that approximates the population genetics of a single gene. ... I use the term *gene* to mean 'that which segregates and recombines with appreciable frequency'. ... A gene is

> one of a multitude of meiotically dissociable units that make up the genotypic message."

Despite this, Williams did not invoke any special chromosomal characteristic that might act to facilitate preservation. Pointing to "the now discredited theories of the nineteenth century," and lamenting an opposition that "arises ... not from what reason dictates, but from the limits of what the imagination can accept," his text *Adaptation and Natural Selection* made what seemed a compelling case for "natural selection as the primary or exclusive creative force." No other agency was required. This tendency, which can be infectious, to bolster the scientific with the *ad hominem* in otherwise rational discourse, will be considered in the Epilogue.In contrast, we have here considered intergenomic and intragenomic differences in (**G+C**)% as an agency, essentially independent of natural selection, which preserves the integrity of species and genes, respectively.

Within a species individual genes differ in their (**G+C**)%. Relative positions of genes on the (**G+C**)% scale are usually preserved through speciation events. If, in an ancestral species, gene *A* was of higher (**G+C**)% than gene *B*, this relationship has been sustained in the modern species that resulted from divergences within that ancestral species. Accordingly, when the (**G+C**)% values of the genes of one of the modern species are plotted against the corresponding (**G+C**)% values of similar (orthologous) genes in the other modern species, the points usually fit a close linear relationship (c.f. Fig. 2-5).

Species with intragenomic isochore differentiation can themselves further differentiate into new species. In this case, a further layer of intergenomic (**G+C**)% differentiation would be imposed upon the previous intragenomic differentiation. Again, when a sufficient degree of reproductive isolation had been achieved this initial barrier between species would usually be replaced by other barriers, thus leaving (**G+C**)% free to continue differentiating in response to intragenomic demands. However, (**G+C**)% is never entirely free. It can itself be constrained by demands on gene function (i.e. natural selection) that primarily affect first and second codon positions. Furthermore, as we shall see next, in extreme environments, natural selection can make direct demands on (**G+C**)%, which might then conflict with its role as a recombinational isolator.

Thermophiles

There are few environments on this planet where living organisms are not found. Hot springs, oceanic thermal vents, and radioactive discharges of nuclear reactors, all contain living organisms ("extremophiles"). Fortunately, since heat and radiation are convenient ways of achieving sterilization in hospitals, none of these organisms has been found (or genetically engineered to become) pathogenic (so far). Thermophiles are so-called because they thrive at high temperatures. Proteins purified from thermophiles may show high stability at normal temperatures, a feature that has attracted commercial interest (i.e. they have a long "shelf life"). Hence, the full genomic sequences of many prokaryotic thermophiles (bacteria and archaea) are now available.

Some thermophiles normally live at the temperature of boiling water. Nucleic acids in solution at this temperature soon degrade. So how do nucleic acids survive in thermophiles? The secondary structure of nucleic acids with a high (**G+C**)% is more stable than that of nucleic acids with a low (**G+C**)%. This is consistent with Watson-Crick **G-C** bonds being strong, and **A-T** or **A-U** bonds being weak (see Table 2-1). Do thermophiles have high (**G+C**)% DNA?

In the case of genes corresponding to RNAs whose *structure* is vital for RNA function, namely rRNAs and tRNAs, the answer is affirmative. Free of coding constraints (i.e. they are not mRNAs), yet required to form part of the precise structure of ribosomes where protein synthesis occurs, genes corresponding to rRNAs appear to have had the flexibility to accept mutations that increase **G+C** (i.e. organisms that did not accept such mutations perished by natural selection, presumably acting against organisms with less efficient protein synthesis at high temperatures). The **G+C** content of rRNAs is directly proportional to the normal growth temperature, so that rRNAs of thermophilic prokaryotes are highly enriched in **G** and **C** [41–43]. Yet, although optimum growth temperature correlates positively with the **G+C** content of rRNA (and hence of rRNA genes), optimum growth temperature does *not* correlate positively with the overall **G+C** content of genomic DNA, and hence with that of the numerous mRNA populations transcribed from the genes in that DNA (Fig. 8-5a). Instead, optimum growth temperature correlates positively with **A+G** content (Fig. 8-5b; see Chapter 12) [44].

The finding of no consistent trend towards a high genomic (**G+C**)% in thermophilic organisms has been interpreted as supporting the "neutralist" argument that variations in genomic (**G+C**)% are the consequences of mutational biases and are, in themselves, of no adaptive value, at least with respect to maintaining duplex stability [43, 45]. However, the finding is also consistent with the argument that genomic (**G+C**)% is too important merely to follow the dictates of temperature, since its primary role is related to other more fundamental adaptations.

The stability of duplex DNA at high temperatures can be achieved in ways other than by an increase in **G+C** content. These include association with small basic peptides (polyamines) and relaxation of torsional strain (supercoiling) [46, 47]. Thus, there is every reason to believe that, whatever their (**G+C**)% content, thermophiles are able, both to maintain their DNAs in classical duplex structures with Watson-Crick hydrogen-bonding between opposite strands, and to adopt any necessary extruded secondary structures involving intrastrand hydrogen-bonding (i.e. stem-loops). This will be further considered in Chapter 9.

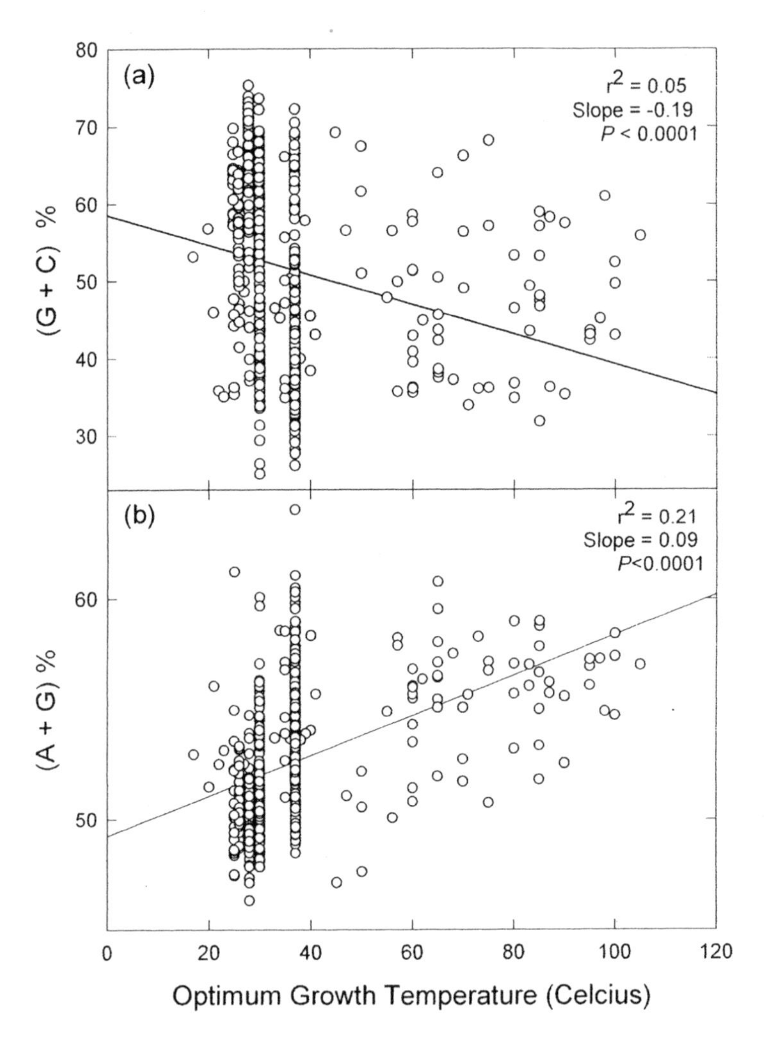

Fig. 8-5. Variation of average *(a)* (G+C)%, and *(b)* (A+G)% of the genes of prokaryotic species (bacteria and archaea), with the temperature required for optimal growth of each species. The large number of data points corresponding to ambient temperatures (20-30°C) and human blood temperature (37°C) reflect the fact that relatively few thermophiles have been sequenced at this time. Note that, whereas in *(a)* only 5% of the variation between points can be explained by growth temperature (r^2 = 0.05), in *(b)* 21% can be explained on this basis (r^2 = 0.21; see Appendix 1)

Saltum?

Darwin held that biological evolution reflected the accumulation of frequent very small variations, rather than few intermittent large variations. That Nature did not work by means of large jumps was encapsulated in the Latin phrase "*Natura non facit saltum.*" However, Huxley, while supporting most of Darwin's teachings, considered it more likely that evolution had proceeded in jumps ("*Natura facit saltum*"). According to the arguments of this chapter, both are correct. Within some members of a species small variations in the genome phenotype (i.e. in **(G+C)%**) accumulate, so that these members become progressively more reproductively isolated from most other members of the species, initially *without* major changes in the conventional phenotype. As it accrues, reproductive isolation increasingly favors rapid change in the conventional phenotype, often under the influence of natural selection. So, when their appearance is viewed on a geological time scale, new species can seem to "jump" into existence.

The rate increase reflects better preservation of frequent phenotypic micromutations rather than of infrequent phenotypic macromutations (i.e. of "hopeful monsters," to use Goldschmidt's unfortunately term). In other words, while there is *continuity* of variation at the genotype level, as far as speciation is concerned variants (mutant forms) seem to emerge *discontinuously* at the phenotype level. Being infrequent, and hence unlikely to find a member of the opposite sex with the same change, organisms with macromutations are not the stuff of evolution.

Summary

Single strands extruded from duplex DNA have the potential to form stem-loop structures that, through exploratory loop-loop "kissing" interactions, may be involved in the homology search preceding recombination. The total stem-loop potential in a sequence window can be analyzed quantitatively in terms of the relative contributions of base composition and base order, of which base composition, and particularly the *product* of the two **S** bases (**G** x **C**), plays a major role. Thus, very small differences in **(G+C)%** should impair meiotic pairing, resulting in hybrid sterility and the reproductive isolation that can initiate speciation (i.e. because their hybrid is sterile, the parents are, in an evolutionary sense, "reproductively isolated" from each other). In chemical terms, Chargaff's species-dependent component of base composition, **(G+C)%**, may be the "holy grail" responsible for reproductive isolation (non-genic) as postulated by Romanes, Bateson and Goldschmidt. Once a speciation process has initiated, other factors (often genic) may replace **(G+C)%** as a barrier to reproduction (preventing intergenomic recombination *between* species). This leaves **(G+C)%** free to assume other roles, such as preventing intragenomic (e.g. intergenic) recombination *within* a species. Thus, many organisms have "macroisochores," defined as long segments of relatively uniform (**G+C**)% that are coinherited with specific sequences of bases. These may facilitate gene duplication. Indeed, each gene has a "homostabilizing propensity" to maintain itself as a "microisochore" of relatively uniform (**G+C**)%. Protection against inadvertent recombination afforded by differences in (**G+C**)% facilitates the duplication both of genes, and of genomes (speciation). George Williams' definition of a gene as a unit of recombination rather than of function is now seen to have a chemical basis, and many of the old disputes among evolutionists (e.g. *saltum* versus *non-saltum*) appear reconcilable.

Part 4 Conflict within Genomes

Chapter 9

Conflict Resolution

> "There would appear to be no reason why the recombinator should not be a sequence which codes in the normal way for amino acids, and that it would therefore be within the genes rather than between them".
>
> Robin Holliday (1968) [1]

The tasks of evolutionary bioinformatics are to identify the forms of information that genomes convey, and show how potential conflicts between different forms are reconciled. Apparent redundancies (e.g. diploidy; Chapter 2), and beliefs in the existence of "neutral" mutations (Chapter 7), and of "junk" DNA (Chapter 12), tended to support the view that there is much vacant genome space, and hence "room for all" in the journey of the genes through the generations. Suggestions that there might be conflicts between different forms of information were not taken too seriously. However, when genomic information was thought of in the same way as the other forms of information with which we are familiar (see Chapters 2-4), it became evident that apparent redundancies might actually play important roles – error-detection and correction, and much more. The possibility of conflict could no longer be evaded. The essential argument of this book is that many puzzling features of genomes can best be understood in such terms, as will be emphasized in this and subsequent chapters.

The idea of competition at different levels can be traced back to Galton [2]. In 1876 he noted that "the limitation of space in the stirp [germ-line] must compel a limitation ... in the number of varieties of each species of germ" [now interpreted as gene]. Furthermore, there was the possibility of gene interactions:

> "Each germ [gene] has many neighbours: a sphere surrounded by other spheres of equal sizes, like a cannon ball in the middle of a heap of them, when they are piled in the most compact form, is in immediate contact with at least twelve others. We may there-

> fore feel assured, that the germs may be affected by numerous forces on all sides, varying with their change of place, and that they must fall into many positions of temporary and transient equilibrium, and undergo a long period of restless unsettlement, before they severally attain the positions for which they are finally best suited."

Two Levels of Information

DNA molecules engage in various protein-mediated transactions (transcription, replication, recombination, repair), which require that specific DNA sequences be recognized by specific proteins. Included among these recognition sequences in DNA might be sequences that themselves encode proteins. Thus some sequences would both encode, and be recognized by, proteins. This was noted by British geneticist Robin Holliday in 1968 in the context of a specific sequence recognized by proteins mediating recombination that he referred to as a "recombinator" sequence [1].

In this light, in 1971 Israeli geneticist Tamar Schaap distinguished the potential for two kinds of information in a DNA sequence, one "active" and one "passive" [3]:

> "If indeed recognition sites are located intragenically, the DNA must contain two kinds of information: the information transcribed into mRNA and translated into polypeptides, henceforth named 'active', and the information serving to distinguish particular regions of the DNA molecule, henceforth called 'passive'."

Recognizing that the coding needs of the two kinds might differ, Schaap continued:

> "It seems therefore logical to assume that the two kinds of information are carried by different codes using the same letters: the four nucleotides, which are read in triplets in the active information code, form "words" of a different length and/or different structure in the passive information code. A word in the passive code may, for example, consist of a nucleotide sequence in which only every second or third nucleotide is essential to the message."

Thus, an individual base mutation should have the potential to affect either one or both coding functions. Although Schaap used the term "passive," an underlying function could be affected by mutations in the passive information code. A mutation in Holliday's recombinator sequence, for example, might impair DNA recognition by recombination proteins, and hence impair recombination-dependent functions. So recombination could be impaired by mutations both in the DNA *encoding* the recombination proteins themselves, and in the DNA *targets* (substrates) of those recombination proteins. The former mutations would tend to be localized to the regions of the corresponding genes. The latter mutations would tend to be generally distributed.

Codon Bias

Schaap noted that the needs of passive information might dictate that certain codons be utilized preferentially (i.e. there would be codon bias):

> "Were homologous [synonymous] codons fully equivalent, the establishment of any particular one would have been most easily understood as the result of a random process … . However, non-equivalence of homologous codons may stem from their participation in sequences forming recognition sites. Consequently, the choice of a particular codon rather than its homologs may be a function of the selective value of recombination, or initiation of replication or transcription in its vicinity. Thus, the establishment of specific codons at particular sites may well be the result of natural selection rather than a random process."

A consequence of this was that a mutation, even though not changing an amino acid, was deemed unlikely to be neutral in its effect on the organism:

> "The dual information carried by DNA can also explain the so-called neutral mutations … . The difference between the selective values of an active and an inactive recognition site may be greater than the one between two polypeptides differing in one amino acid. In such cases, the passive rather than the active information will be the criterion for selection, causing the establishment of mutations with no apparent selective advantage."

Hence, selection would be on the genome phenotype, rather than on the conventional (classical) phenotype:

> "The establishment of homologous [synonymous] and neutral mutations, usually attributed to random processes, can be accordingly understood as the result of natural selection acting at the level of the DNA rather than that of the polypeptide."

In the 1970s nucleic acids from various species were first sequenced. Codon bias emerged as a phenomenon that was both general and species specific. There was a tendency for all the coding regions of genes within a species to use the same sub-set of the possible range of codons (Table 9-1).

Sextet = A Quartet plus a Duet

Arginine	Leucine	Serine
CGU	CUU	UCU
CGC	CUC	UCC
CGA	CUA	UCA
CGG	CUG	UCG
AGA	UUA	AGU
AGG	UUG	AGC

Quartets

Threonine	Proline	Alanine	Glycine	Valine
ACU	CCU	GCU	GGU	GUU
ACC	CCC	GCC	GGC	GUC
ACA	CCA	GCA	GGA	GUA
ACG	CCG	GCG	GGG	GUG

Duets

Lysine	Asparagine	Glutamine	Histidine	Glutamic
AAA	AAU	CAA	CAU	GAA
AAG	AAC	CAG	CAC	GAG

Aspartic	Tyrosine	Cysteine	Phenylalanine
GAU	UAU	UGU	UUU
GAC	UAC	UGC	UUC

Odd Numbers

Isoleucine	Methionine	Tryptophan	Terminators
AUU	AUG	UGG	UAA
AUC			UAG
AUA			UGA

Table 9-1. Codon sets (sextets, quartets, duets). Codons within a set are not uniformly employed. Species, and to a lesser extent genes within a species, differ in their choice of the most usual codon to encode a particular amino acid (codon bias). The 61 codons for amino acids are grouped into 20 sets of between one and six synonymous members. The code includes eight quartets and twelve duets, the isoleucine trio, the single codons of methionine and tryptophan, plus the three codons that signal termination of a protein sequence. Changes in first codon positions always change the nature of the encoded amino acids, with leeway only in the sextet sets. Changes in second codon positions always change the nature of the encoded amino acids, with leeway only in the case of the serine sextet. Changes in third codon positions fail to change the nature of the encoded amino acid, with leeway only in the duets (e.g. **CAA** to **CAU**), and odd numbered codon sets (e.g. **AUU** to **AUG**).

The issue of which codon was employed in a particular circumstance was considered in France by one of the founders of evolutionary bioinformatics, the USA-expatriate Richard Grantham. He observed in 1972 [4] that codon usage was not random in microorganisms, "suggesting a mechanism against [base] composition drift". Observing that "little latitude appears left for 'neutral' or synonymous mutations", he was led to his "genome hypothesis", which specified that undefined adaptive genomic pressure(s) caused changes in base composition and hence in codon usage [5]:

> "Each ... species has a 'system' or coding strategy for choosing among synonymous codons. This system or dialect is repeated in each gene of a genome and hence is a characteristic of the genome."

Grantham sensed that the coding strategy was of deep relevance to an organism's biology:

> "What is the fundamental explanation for interspecific variation in coding strategy? Are we faced with a situation of continuous variation within and between species, thus embracing a Darwinian perspective of gradual separation of populations to form new species ...? This is the heart of the problem of molecular evolution."

He further pointed to the need to determine "how much independence exists between the two levels of evolution" (that of the genome phenotype and of the classical phenotype) and considered "it is too easy just to say most mutations are neutral."

Grantham posed good questions, but left them unanswered. "Easy," non-adaptive, "neutralist," explanations won general support. Codon biases were held to reflect mainly biases in the generation or repair of mutations. It was, however, conceded that an adaptive factor influencing codon bias might be the need for highly expressed genes to be efficiently translated [6].

Adaptive factors such as the need to translate an abundant mRNA efficiently can, indeed, influence codon usage, but this cannot explain the generality of codon bias, since only in certain species is translation rate-limiting in the process of maximizing the number of descendents (i.e. maximizing biological fitness). Furthermore, while mutational biases (i.e. a tendency to mutate unidirectionally from one base to another) are possible [7] (see also Chapter 15), this explanation is a "default" (fall-back) explanation that thrives mainly because other explanations for codon-bias have not been forthcoming.

This book argues that genomic factors, such as **GC**-pressure, purine-loading pressure and fold pressure, play an important, and sometimes dominant, role in determining which codons, from a range of codons (from as many as six alternatives; Table 9-1), are preferred to encode a particular amino acid in a particular species. In general, codon-bias is not a primary phenomenon, but is *secondary* to underlying phenomena.

If, for reasons set out in Chapter 8, a particular species has a high genomic (**G**+**C**)% (which affects both coding regions and non-coding regions), there will be a bias in favour of **GC**-rich codons, rather than **AT**-rich codons. Thus instead of the **RNY** periodicity (see Chapter 7), coding sequences will tend to adopt a **GNC** periodicity (e.g. ...**GNC**,**GNC**,**GNC**,**GNC**...). Similarly, coding sequences in **AT**-rich, low (**G**+**C**)%, genomes will tend to adopt an **ANT** periodicity (e.g. ...**ANT**, **ANT**, **ANT**, **ANT**, ...). If accommodation to high **GC** is insufficient, then an encoded amino acid may have to change to one of the "GARP" amino acids. If accommodation to low **GC** is insufficient, then an encoded amino acid may have to change to one of the "FYMINK" amino acids (see Chapter 8).

GC-Pressure Versus Protein-Pressure

Since it is often independent of the encoded amino acid, the third codon position should be freer than the other codon positions to respond to pressures other than those related directly to protein synthesis. As such, it serves as a "control" against which to compared protein pressures on codons (i.e. the need to encode a particular amino acid at a particular position). Thus, it allows a contrast to be drawn between the conventional phenotype and the genome phenotype.

A way of showing this, elegantly exploited by Akira Muto and Syozo Osawa in Japan, is to plot separately the *average* base composition of *each* codon position of the *set* of all the sequenced genes of a species, against the average base composition of the genome of that species [8]. Figure 9-1 shows the three plots that result in the case of the (**G**+**C**)% of 1046 bacterial species [9, 10]. Each point here refers to a codon position in an individual bacterial species.

The first thing to note is the wide difference in (**G**+**C**)% among species, which ranges from around 20% to 80% in bacteria (see X-axis). We shall see in Chapter 11 that it is most appropriate to regard values around 50% as the norm, so that below 50% there is "downward **GC**-pressure", and above 50% there is "upward **GC**-pressure." However, it is sometimes convenient to think of bacterial species as responding to a pressure progressively to increase their overall (**G**+**C**)%. In this respect, all three codon positions participate positively in the species "response" to **GC**-pressure (i.e. the slopes in Figure 9-1 are all positive). However, third codon positions are freer to respond than first and second codon positions (i.e. the plot for third codon positions has the steepest slope).

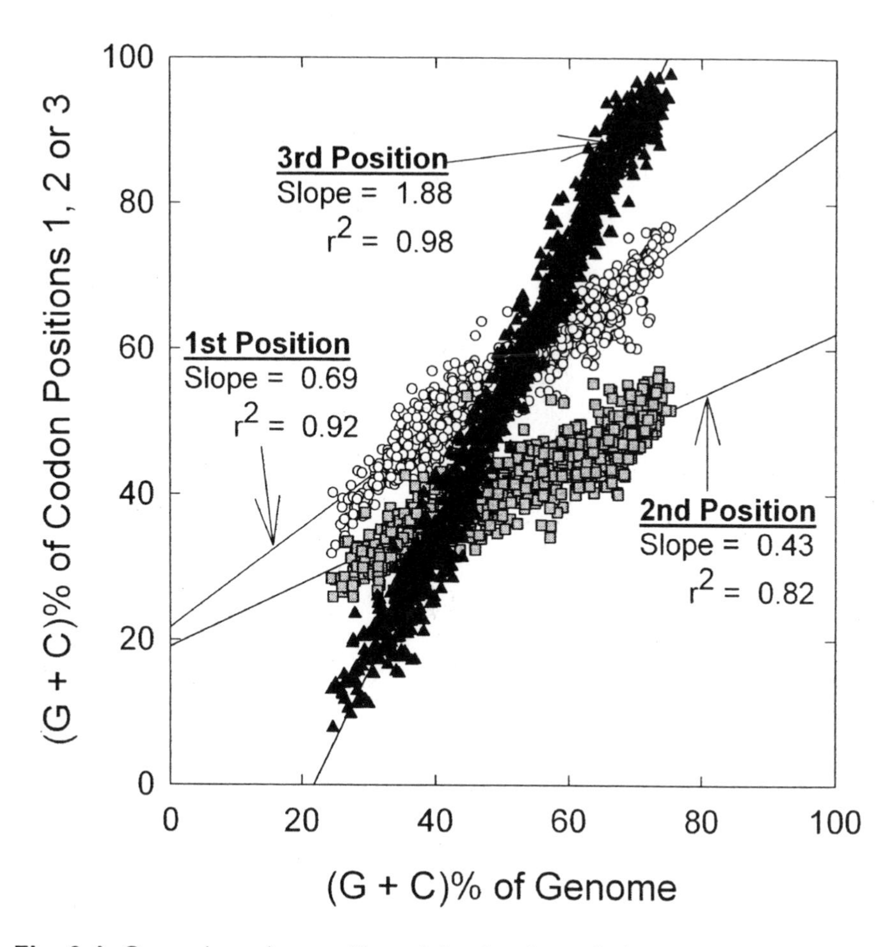

Fig. 9-1. Genomic codon position plots showing relative responses of different codon positions to **GC**-pressure in the 1046 species of bacteria represented in GenBank in the year 2000 by four or more genes. For each species there are 3 values corresponding to the average (**G**+**C**)% of first codon positions (open circles), the average (**G**+**C**)% of second codon positions (grey squares), and the average (**G**+**C**)% of third codon positions (black triangles). These three values are plotted against an estimate of average genomic (**G**+**C**)% derived from the coding regions of all sequenced genes of the species. Note that the slope values (of the regression lines) increase from 0.43 (second codon positions) to 1.88 (third codon positions). These slopes so obviously differ from zero that *P* values are not shown. It might appear that the points for third codon positions fit their line better than the points for first and second codon positions fit their respective lines. However, the standard errors of the mean (see Appendix 1) are 2.54 (first codon position plot), 2.54 (second codon position plot) and 3.74 (third codon position plot). Similar plots are obtained with other taxonomic groups of organisms

Values for (**G**+**C**)% for third codon positions cover a wide range, with some species having values below 10%, and others having values approaching 100% (see Y-axis). Thus, for genomes that are at the extremes of (**G**+**C**)% (low or high), the third codon position appears to amplify either a low average genomic (**G**+**C**)% by exceeding the first and second codon positions in its **AT**-richness, or a high average genomic (**G**+**C**)% by exceeding the first and second codon positions in its **GC**-richness. In some bacterial species, every third base in every codon is almost exclusively **A** or **T**. In others it is almost exclusively **G** or **C**. First and second codon positions are constrained by the pressure to encode proteins, some of which may be species-specific proteins with distinctive amino acid compositions; here the range of (**G**+**C**)% values (see Y-axis) is much narrower than in the case of third codon positions.

Species Versus Genes

Whereas Figure 9-1 shows the variation of base composition for different codon positions among entire *genomes* (i.e. among species), Figure 9-2 shows this for different codon positions among the sequenced *genes* of two bacterial species, one of very low, and the other of very high, genomic (**G**+**C**)%. Whereas each point in Figure 9-1 represents an average value for *all* sequenced genes of a species (i.e. each point represents a species), each point in Figure 9-2 represents an average value for an individual gene *within* a species (i.e. each point represents a gene).

Being at the extremes, the (**G**+**C**)% values for all codon positions tend to be low in the low (**G**+**C**)% species (so that all points tend to be in the bottom left corner of the figure), and high in the high (**G**+**C**)% species (so that all points tend to be in the top right corner of the figure). Nevertheless, there is still considerable differentiation among codon positions. Of particular note is that, whereas slopes for genic first and second codon positions are steep, slopes for genic third codon positions are almost *horizontal*. Thus, within each of these two genomes, genes differ from each other in (**G**+**C**)% mainly in their first and second codon positions. This would partly reflect the needs of the encoded proteins. In contrast, within a genome all genes have approximately the *same* (**G**+**C**)% in their third codon positions. The (**G**+**C**)% values of third codon positions would seem to be dedicated to the roles of keeping genomic (**G**+**C**)% low in the low (**G**+**C**)% species, and high in the high (**G**+**C**)% species.

This emphasises the independence of third codon positions, which appear to serve exclusively the needs of **GC**-pressure (i.e. the needs of the species), not the needs of the genes that contain them. Whatever the gene, its third codon position (**G**+**C**)% value is essentially the *same* as that of other genes of the same species. Thus, for species at the extremes of genomic (**G**+**C**)% (either low or high), the distinctiveness of the (G+C)% of a particular gene is largely determined by the (G+C)% of first and second codon positions. Yet, despite the tendency towards uniformity by virtue of constant third codon position (G+C)% values, there is some differentiation of genes (into "microisochores;" see Chapter 8). The range of this differentiation is limited. In the low (G+C)% species the range varies from 5-30%. In the high (G+C)% species the range varies from 60-80% (see X-axis of Fig. 9-2).

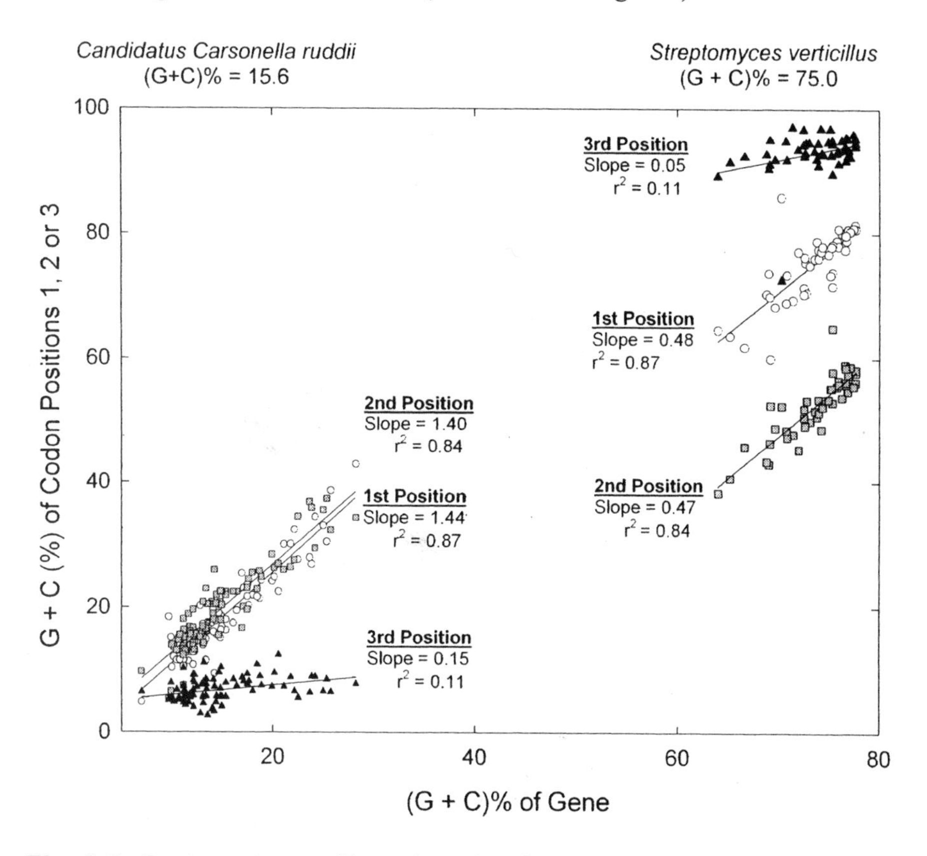

Fig. 9-2. Genic codon position plots showing responses to **GC**-pressure of different codon positions within individual genes for a low genomic (**G+C**)% species (15.6% average **G+C**), and for a high genomic (**G+C**)% species (75.0% average **G+C**). Each point represents the average (**G+C**)% value for a particular codon position in all the codons of a particular gene (i.e. there are three points for each gene, represented by a circle, a square, and a triangle

Most species have intermediate genomic (G+C)% values. One might predict that plots for genic third codon positions would again be almost horizontal, so that the needs of the species would again dominate. This is not so. In fact, *the very opposite is observed*. Slopes for genic third codon positions are *steeper* than slopes for genic first and second codon positions. The gut bacterium *E. coli*, provides a good example.

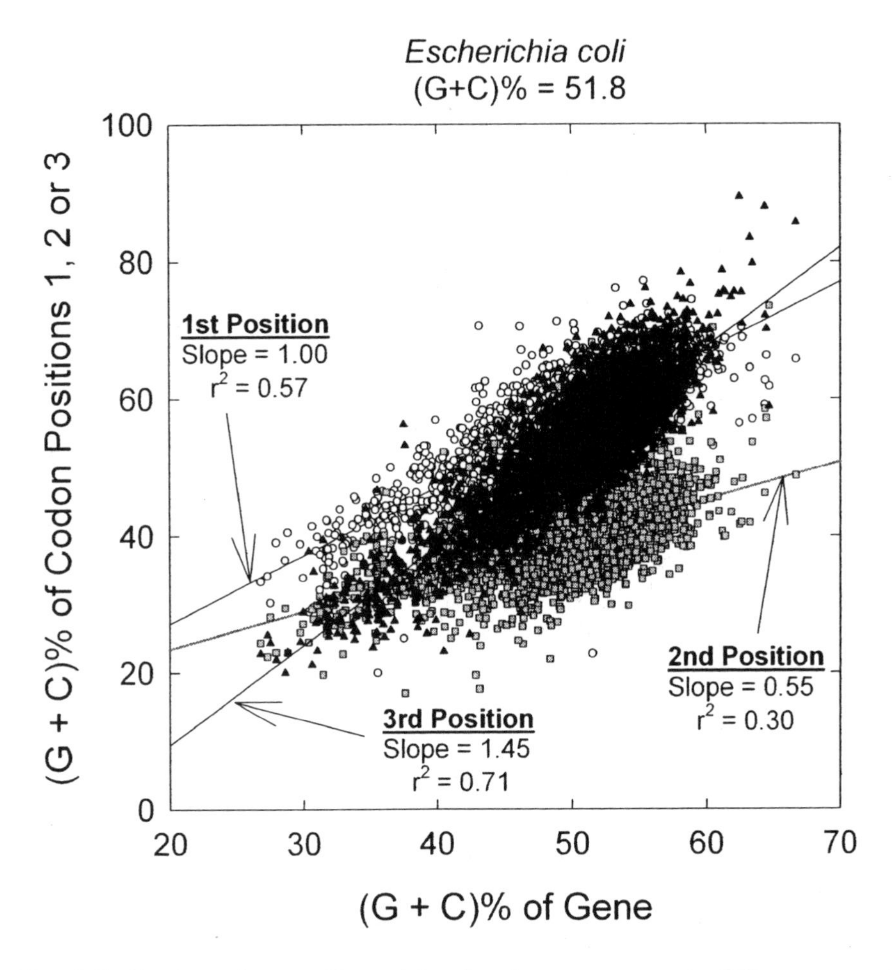

Fig. 9-3. Genic codon position plots showing responses to **GC**-pressure of different codon positions for each of the 4290 genes of an intermediate genomic (**G+C**)% species (*E. coli*). Each point represents the average (**G+C**)% value for a particular codon position in all the codons of a particular gene (i.e. there are three points for each gene). The standard errors of the mean (see Appendix 1) are 4.20 (first codon position plot), 4.07 (second codon position plot) and 4.53 (third codon position plot)

Figure 9-3, which shows average codon position (**G+C**)% values for *each* of the genes of *E. coli*, is like Figure 9-1, which shows average values for the *set* of all the sequenced genes of each of a wide range of bacterial species. Furthermore, the genes (microisochores) differ from each other over a *wide* range of (G+C)% values (25% to 70%). Thus, the dispersion of (G+C)% values among *genes within a genome* (Fig. 9-3) matches the dispersion of (G+C)% values among *genomes within a taxonomic group* bacteria; Fig. 9-1).

This similarity between the sets of codon position plots for genes and sets of codon position plots for genomes was noted in 1991 by Wada and his co-workers [11]. The plots are essentially context-independent (i.e. the same for genes and for genomes). Wada suggested that the plots "might be universal ones and the constraint parameters might have general biological meanings in relation to the DNA/RNA and protein functions." Thus, whatever was causing the dispersion among genes might also be causing the dispersion among genomes (e.g. the need to avoid intergenic or intergenomic recombination; see Chapter 8).

Species Win at (G+C)% extremes

In species with extreme genomic (G+C)% values, the slope values for genic first and second codon positions exceed those for third positions (Fig. 9-2), whereas the opposite holds for a species with intermediate genomic (G+C)% values (Fig. 9-3). Accordingly, at some genomic (G+C)% values between extreme and intermediate, the patterns must switch. This switch-over is shown when the three slope values taken from genic codon position plots for numerous individual bacterial species (of which Figs. 9-2 and 9-3 are examples) are plotted against the genomic (G+C)% values for those species. The switches occur at about 37% and 68% (Fig. 9-4).

Because they cover a wide range of genomic (G+C)% values, bacterial species are ideal for studies of this nature. However, similar results are found in "higher" organisms. Figure 9-5a shows that the low (G+C)% genome of the malaria parasite *P. falciparum* gives essentially the same result as the low (G+C)% genome of a bacterium (Fig. 9-2). Figure 9-5b shows that the intermediate (G+C)% genome of rice (*Oryza sativa*), gives essentially the same result as the intermediate (G+C)% genome of a bacterium (Fig. 9-3).

Hence, it is generally true that species with extreme genomic (G+C)% values (low or high) sustain those values largely by maintaining constant, extreme, third codon position (G+C)% values, which may be dubbed "gene-independent" in that they are only slightly influenced by the gene that contains them. Values for first and second codon positions, that specify amino acids, are partly gene-dependent, and do not so obviously contribute to the extreme (G+C)% values of the species. The situation changes dramatically in species with intermediate genomic (G+C)% values. All three codon positions assume gene-dependent (G+C)% values, and this is most evident in the case of third codon positions (Figs. 9-3, 9-5b).

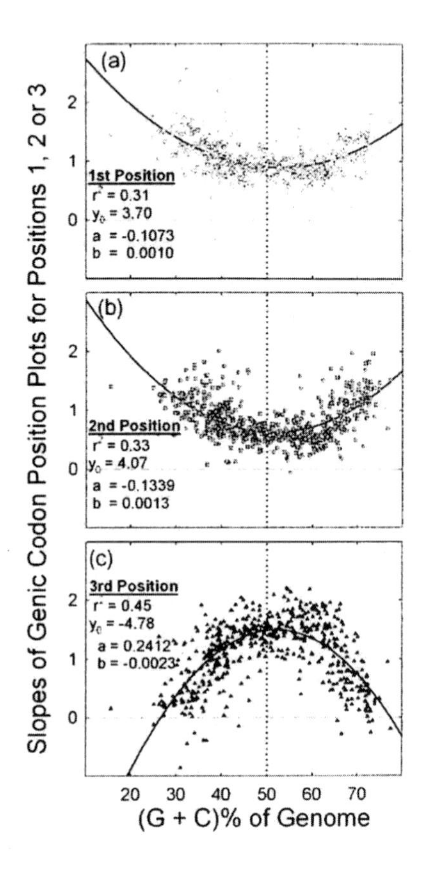

Fig. 9-4. Slopes of genic codon position plots for first and second codon positions (*a*, *b*) are low in species of intermediate genomic (**G**+**C**)%, and high in species with extreme genomic (**G**+**C**)%. Conversely, slopes for third codon positions (*c*) show the opposite trend. Slope values are from individual genic codon position plots for each of 546 bacterial species (each with at least 20 sequenced genes). Slope values were fitted to second order regression curves (see Appendix 1). In species of intermediate genomic (**G**+**C**)%, differentiation of genes for protein-encoding (mainly dependent on first and second codon positions) is reflected in slope values of 0.9 (or less) for these positions *(a, b)*. It is unlikely that, in species of extreme genomic (**G**+**C**)%, differentiation of genes for protein-encoding would be greater than in species of intermediate genomic (**G**+**C**)%. Yet, in species of extreme genomic (**G**+**C**)%, slope values are around 1.5 in plots for first and second codon positions *(a, b)*. The extra 0.6 slope units may reflect the need of genes in extreme genomic (**G**+**C**)% species to differentiate their (**G**+**C**)% values (to prevent intergenic recombination). This differentiation, as reflected in first and second codon position values, would be less in species of intermediate genomic (**G**+**C**)%, since here third codon positions *(c)* play the major role in achieving genic (**G**+**C**)% differentiation

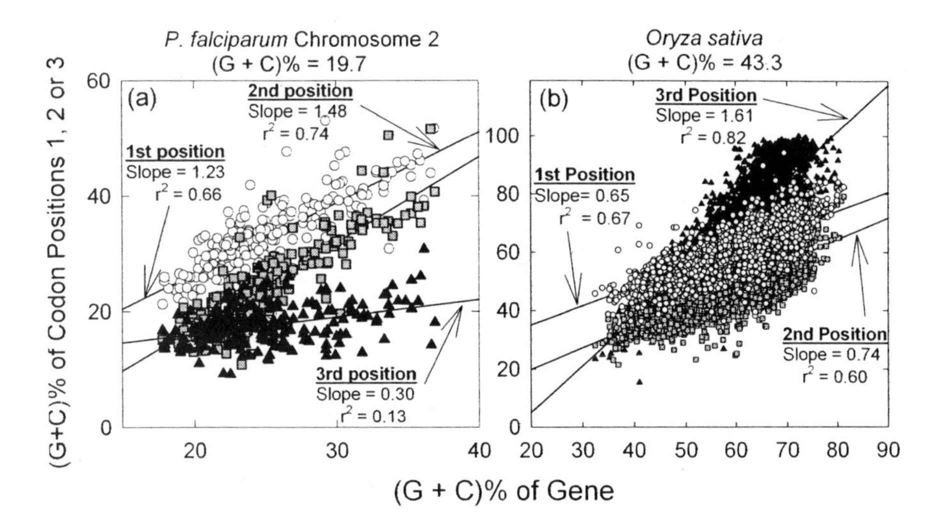

Fig. 9-5. Genic codon position plots of eukaryotes resemble those of prokaryotes of the same average genomic (**G+C**)%, as exemplified by *(a)* 205 genes from the second chromosome of the malaria parasite (*P. falciparum*; (19.7% average **G+C**) and *(b)* 3111 genes from the genome the rice plant (*Oryza sativa*; 43.3% average **G+C**)

So, if first and second codon positions within a gene have low (**G+C**)% values, then the corresponding third codon positions will have even lower (**G+C**)% values (i.e. they are richer in **A** and **T**). If first and second codon positions within a gene have high (**G+C**)% values, then the corresponding third codon positions will have even higher (**G+C**)% values (i.e. they are poorer in **A** and **T**). Accordingly, a low (**G+C**)% gene has many **AT**-rich codons such as **TTY** (generic codon for phenylalanine). Here **TTT** is preferred over **TTC**. On the other hand, a high (**G+C**)% gene has many **GC**-rich codons such as **GGN** (generic codon for glycine). Here **GGC** and **GGG** are preferred over **GGT** and **GGA**. Thus, third codon positions support and extend the (**G+C**)% values of first and second codon positions. Or, as stated by Wada and Suyama in 1985:

> "The average (**G+C**)-content of each gene is altered mainly by base substitution at the third site of the codon and ... from the positive correlation between ... [the three] sites, which is revealed when they are averaged over the entire length of a gene [see Fig. 8-4], every base in a codon seems to work cooperatively toward realizing the gene's characteristic value of (**G+C**)-content."

In this way, each gene in a species of intermediate genomic (**G**+**C**)% has come to occupy a distinct **GC**-niche (microisochore) amongst its fellow genes. In contrast to genes in species of extreme genomic (**G**+**C**)%, genes in species of intermediate genomic (**G**+**C**)% collectively span a wide (**G**+**C**)% range [12].

Intermediate (G+C)% Allows Compromise

Why are species with extreme genomic (**G**+**C**)% values so different? It is intuitively apparent that a sequence with only one base would have less ability to transmit information than a sequence with four bases. In fact, elementary information theory shows that information transmitting potential is maximized when the four bases are present in equal proportions (i.e. when Chargaff's second parity rule is obeyed; see Chapter 11). Thus, departures from 50% (downwards or upwards) are likely to progressively compromise the ability of genomes to transmit *further* information. The trade-off seems to be that, at extreme genomic (**G**+**C**)% values (below 37% and above 68%), third codon positions serve the information needs of the species, rather than of individual genes within that species. Why *further* information? It is likely that departures from 50% *themselves* have information content, which, at the extremes, happens to be of more relevance to the species than to individual genes (Chapter 8).

In species with intermediate genomic (**G**+**C**)% values, the contribution to genic needs is shared by all codon positions, with third positions making the greatest contributions (i.e. slope values are greatest for third positions; Figs. 9-3, 9-4c, 9-5b). In species with extreme genomic (**G**+**C**)% values, third positions do not contribute to genic needs, but, relative to species with intermediate genomic (**G**+**C**)% values, first and second codon positions *increase* their contributions (Fig. 9-4a, b). Thus, the slope values for the first and second codon positions in species with extreme genomic (**G**+**C**)% values are *greater* than in species with intermediate genomic (**G**+**C**)% values. This provides for some genic differentiation into microisochores in extreme (**G**+**C**)% genomes. However, the range of this genic differentiation (see X-axis values) is less than the range of genic differentiation in intermediate (**G**+**C**)% genomes. The position of a gene in that restricted range in extreme **GC**-species is partly determined by protein-pressure (the pressure to encode a particular protein) and partly determined by local *genic* pressure for (**G**+**C**)% differentiation as a microisochore distinct from other microisochores (genes).

While the nature and location of certain amino acids in a protein may be critical for protein function, the nature and location of certain amino acids encoded by genes in extreme **GC**-species is also strongly influenced by the general *genomic* pressure for that (**G**+**C**)%. Along similar lines, we shall see here and in Chapter 11 that sometimes base composition pressures cannot be accommodated without changing proteins, sometimes even to the extent of increasing their lengths, by inserting "placeholder" amino acids. The increases are sometimes evident as low complexity segments that primarily serve the needs of the encoding nucleic acid, not of the encoded protein.

Levels of Selection

The plots shown above illustrate the separate operations of *gene-level selection* and *genome-level selection* – subjects of much controversy and semantic confusion. The conflict between "selfish genes" and "selfish genomes" is reflected in the differential survival (reproductive success) of individuals (*individual level selection*) that, when the crunch comes (i.e. extreme genomic **(G+C)**% values), serves the species (i.e. "selfish genomes"). The quotation marks indicate that genes and genomes are not consciously selfish, but the dynamics of their evolution easily lends itself to this colorful implication.

Can natural selection act both on individuals and on species? There is sometimes talk of "species selection, in which whole species are the target of natural selection." It is even predicted that: "Species that split into new species faster than, or become extinct slower than, other species will become more common" [13]. But, what is usually the most important *biologically* is not the number of species, but the number of individuals within each of those species. A species can diverge into several new species, and the members of each can come to occupy distinct ecological niches that each supports, say, a million individuals. However, a group of species collectively comprising several million individuals cannot be considered more successful *biologically* than an individual species of, say, one hundred million individuals.

And how can a species *itself* become "more common" when the speciating (branching) event that creates it merely distinguishes one *individual* species from another *individual* species? It was impossible for the individual, Charles Darwin, to become more common. There was only one Charles Darwin. So, a species that is better able to diverge into new species than another species, may end up containing more individuals in its species collective (genus) than if it had not diverged. But here natural selection still has acted on individual members of each species, not on the species themselves. Individual *members* of a species, not individual species, become "more common" by virtue of natural selection

In passing it can be noted that species in which only a few of the total genes have been sequenced at the time of this writing (Figs. 9-2) appear just as informative in the present context as species in which many genes have been sequenced (Figs. 9-3, 9-5b). Evolutionary bioinformatics, the "new bioinformatics," often finds *many* incompletely sequenced genomes of greater value than a *few* completely sequenced genomes.

Dog Wags Tail

In the context of the "holy grail" of Romanes and Bateson (see Chapter 8), variations (mutations) of importance for achieving differentiation of species (reproductive isolation) are often independent of variations that affect genes. But **GC**-pressure acts both on intergenic DNA and genic DNA. To the extent that a gene does not "want" to change the nature of its products (RNA and/or protein) it must accommodate to the pressure for their continued faithful elaboration – for example, by accepting only a synonymous mutation that does not change the nature of an encoded amino acid. To this extent the synonymous mutation, although *structurally* "genic" (i.e. localized in a segment we call a gene), can be viewed as *functionally* "non-genic."

On the other hand, a mutation primarily affecting a gene could marginally affect the **(G+C)**% of a genome. If natural selection leads to an accepted genic change, meaning a changed gene product, then there may be some local increase or decrease in **(G+C)**%. Thus, while both functionally and structurally "genic," the mutation, albeit only marginally, would also be functionally "non-genic."

As part of a speciation event, two divergent lines may progressively differ in **(G+C)**% and the parts of genes under functional constraint (e.g. first and second codon positions) will tend to oppose this. Accordingly, the rates of mutation of first and second codon positions will usually be less than the rates of mutations of third codon positions, introns, and intergenic DNA (Fig. 2-5). Exceptions may be genes (i.e. nucleic acid) that encode proteins whose substrate (i.e. target) is *itself* nucleic acid. While proteins recognize nucleic acids generically by virtue of their common phosphate-ribose structure, many are not indifferent to the composition or order of bases. So, as nucleic acids change in **(G+C)**%, the proteins that recognize nucleic acids may need to adapt functionally to ensure optimal binding to their DNA substrate. They are on an adaptive treadmill. This implies that, when lines are diverging, the mutation rate in nucleic acid recognition proteins should be greater than in most other proteins, as is indeed found [14].

Thus, **GC**-pressure ("dog"), while causing mutation of both non-genic and genic DNA, particularly impresses the DNA of DNA-recognition enzymes ("tail"). Conversely, it might seem possible that mutations in genes encoding certain general DNA-recognition enzymes (e.g. DNA repair enzymes) might lead to directional changes in genomic **(G+C)**%. That the tail can wag the dog is indeed true in the case of certain "mutator" genes [7]. Since these genes can change **(G+C)**% generally, it would be tempting to consider them as candidate "speciation genes." However, such genes are likely to be exceptions rather than the rule. It is more likely that the small regions of genomic DNA encoding DNA-recognition proteins ("tail"), by virtue of their smallness, adapt to overall changes in the large lumbering genomic (**G**+**C**)% ("dog"), but usually do not themselves drive (**G**+**C**)% changes.

Certain dinucleotides differ less between species than other dinucleotides. In a high (**G**+**C**)% species (i.e. a low (**A**+**T**)% species) a high frequency of **SS** dinucleotides and a low frequency of **WW** dinucleotides is expected. However, dinucleotides of composition **SW** or **WS** are at intermediate frequencies, and vary less between species that differ in (**G**+**C**)%. Thus, enzymes which needed to recognize **SW** and **WS** dinucleotides might have to evolve less rapidly than enzymes that needed to recognize the **SS** and **WW** dinucleotides.

AG-Pressure Versus Protein-Pressure

The responsiveness of different codon positions in bacterial genomes to **AG**-pressure (purine-loading pressure) is shown in Figure 9-6. Following Figure 9-1, the average (**A**+**G**)% of individual codon positions of a species is plotted against the corresponding genomic (**A**+**G**)% of the species.

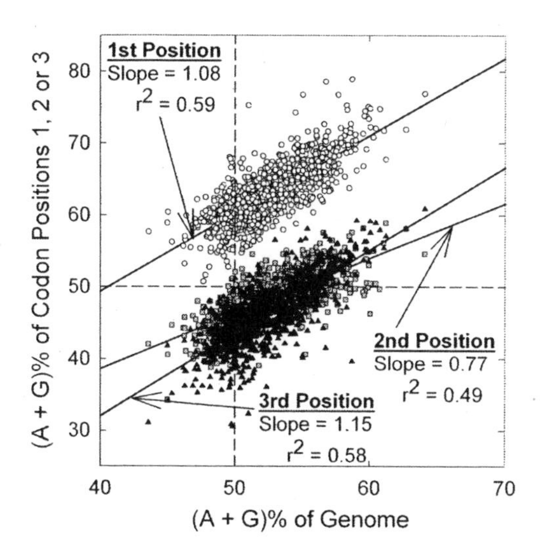

Fig. 9-6. Genomic codon position plots showing relative responses of different codon positions to **AG**-pressure in 1046 species of bacteria. For each species there are 3 values corresponding to the average genomic (**A**+**G**)% of first codon positions (open circles), the average genomic (**A**+**G**)% of second codon positions (grey squares), and the average genomic (**A**+**G**)% of third codon positions (black triangles). These three values are plotted against an estimate of the average (**A**+**G**)% of the genome derived from the coding regions of all the sequenced genes. Note that slope values increase from 0.77 (second codon positions) to 1.15 (third codon positions)

Here the range of genomic values is much less than in the case of **GC**-pressure (compare X-axis values with those of Fig. 9-1). In keeping with the **RNY**-rule (see Chapter 7), for all species the first codon position makes the greatest contribution to genomic purine-loading, and no species has a genomic (**A**+**G**)% value less than 50 for this codon position (i.e. no pyrimidine-loading). Few bacterial genomes achieve these high values in the second and third positions. Also in keeping with the **RNY**-rule, third codon positions are predominantly pyrimidine-loaded (values less than 50%). This purine-poor position would seem to be best poised to respond to **AG**-pressure. Indeed, from the slopes of the plots, third positions appear most responsive to **AG**-pressure so that, in species with high genomic (**A**+**G**)% values, third codon position values can exceed 50%. Thus, there is purine-loading, so that **RNY** becomes **RNR**, and the **RNY**-rule no longer holds. This may be construed as a conflict between purine-loading pressure ("**R**-pressure") and "**RNY**-pressure," which is settled in these species in favour of purine-loading pressure.

As in the case of the response to **GC**-pressure (Fig. 9-1), the response to **AG**-pressure is least in the second codon position (slope value of only 0.77). However, species with high genomic (**A**+**G**)% values have proteins enriched in amino acids encoded by purine-rich codons [15]. This may be construed as a conflict between purine-loading pressure and protein-encoding pressure, which is settled in these species in favour of purine-loading pressure.

GC-Pressure Versus AG-Pressure

Of course, the content of (**G**+**C**) in a sequence segment, is not independent of its content of (**A**+**G**), since both compositions include the base **G**. Indeed, there is a potential conflict between **AG**-pressure and **GC**-pressure. There are two ways to modulate genomic (**G**+**C**)% when the total number of bases is constant – either by changing the number of **G**'s, or by changing the number of **C**'s. If an increase in the number of **G**'s is matched by a decrease in the number of **C**'s, then there will be no overall change in (**G**+**C**)%. If one of these bases increases more than the other decreases, then there will be an overall increase. As genomic (**G**+**C**)% increases, replacing **A** with **G**, or **T** with **C**, would not affect the genomic **AG**-pressure (i.e. there would be no change in total purines). But if **T** were replaced with **G**, **AG**-pressure would increase as (**G**+**C**)% increases. Conversely, if **A** were replaced with **C**, **AG**-pressure would decrease as (**G**+**C**)% increases. The latter inverse relationship is observed in bacterial genomes, and in their viruses (bacteriophage), and in the several eukaryotes that have been examined. This indicates a preference for trading **A** for **C** as genomic **GC** content increases [16]. In a survey of 1046 bacterial species, 53% of the variation ($r^2 = 0.53$) could be explained in terms of the inverse relationship between (**A**+**G**)% and (**G**+**C**)% (Fig. 9-7).

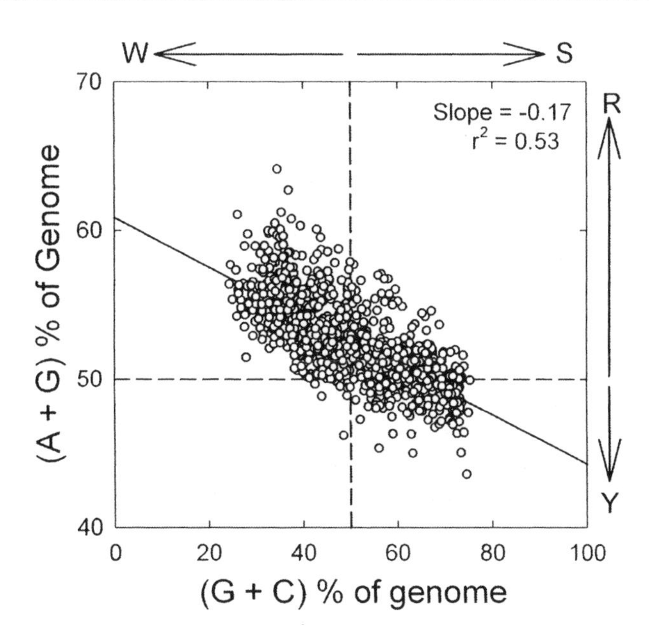

Fig. 9-7. Inverse relationship between average (**A**+**G**)% and average (**G**+**C**)% in the codons of 1046 species of bacteria. Each point represents an individual species. Horizontal arrows (top) indicate upward pressures towards high (**G**+**C**)% values (i.e. high **S** values) and downward pressures towards low (**G**+**C**)% values (i.e. high **W** values), when values depart from 50%. Vertical arrows (right) indicate pressures towards purine-loading (**R**), and pyrimidine-loading (**Y**), when values depart from 50%

To what extent do different codon positions contribute to this trading relationship? This can be examined by plotting the result, shown in Figure 9-7, for each codon position individually (Fig. 9-8). Peering beneath the fog of multiple data points it can be seen that, being already highly **Y**-loaded in accord with the **RNY**-rule, genomic third codon positions are less in a position to give up **A** and acquire **C** as (**G**+**C**)% increases. So third codon positions appear less responsive to this aspect of **GC**-pressure (slope of only -0.10) than the first and second codon positions. By the same token, being already highly **R**-loaded in accord with the **RNY**-rule, genomic first codon positions should best be able to trade **A** for **C**. But genomic *second* codon positions, the positions of most importance in determining the amino acid and least responsive to **GC**- and **AG**-pressures individually (Figs. 9-1 and 9-6), are more responsive than first positions in the trading relationship (slopes of –0.32 versus –0.23). In other words, at extreme (**G**+**C**)% values (low and high), **AG**-pressure is modulated by the introduction (when (**G**+**C**)% is low), or the removal (when (**G**+**C**)% is high), of amino acids corresponding to purine-rich codons. Do these amino acids necessarily serve protein function?

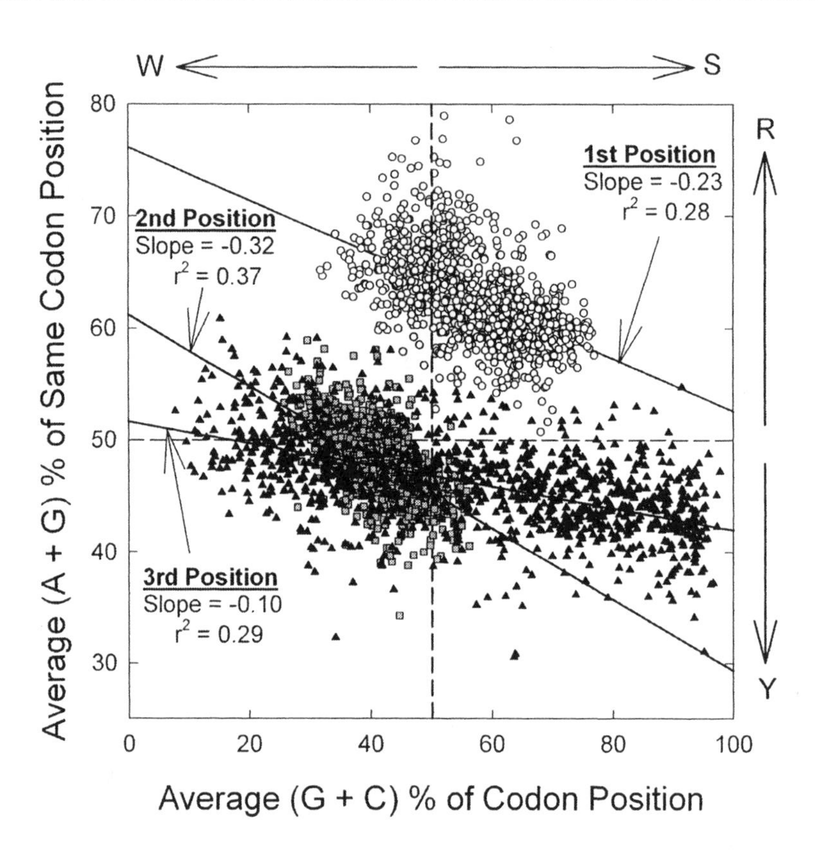

Fig. 9-8. Influence of **GC**-pressure on purine-loading at different codon positions. For each bacterial species, average (**A+G**)% values for each codon position are plotted against average (**G+C**)% values of the *same* codon position. Symbols are as in Fig. 9-1

AG-Pressure and "Placeholder" Amino Acids

The reciprocal relationship between **GC**-pressure and **AG**-pressure suggests that we should be careful when trying to attribute a given effect to just one of these pressures. For example, in animals a weak negative relationship between the (**G+C**)% of protein-encoding regions (exons) and the lengths of the corresponding proteins has been described [17]. This is seen quite clearly in the case of proteins encoded by the malarial parasite *P. falciparum*. In Fig. 9-9a (**G+C**)% values of exons are plotted against the corresponding lengths in kilobases. Note that 5% of the variation in (**G+C**)% can be explained on this basis ($r^2 = 0.05$; $P = 0.0009$). When the same lengths are plotted against (**A+G**)% the value increases to 7% ($r^2 = 0.07$; $P < 0.0001$; Fig. 9-9b). Thus, if you had to suggest which pressure might most likely be *primarily* related to protein length you would prudently pick **AG**-pressure.

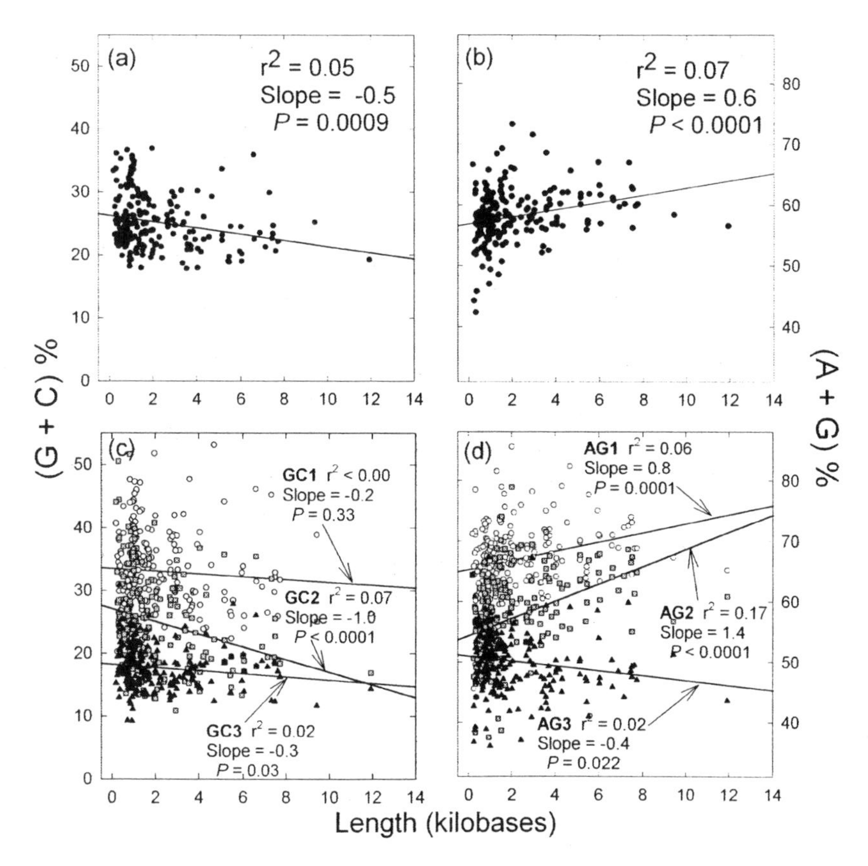

Fig. 9-9. Protein length is inversely proportional to (**G**+**C**)% and directly proportional to (**A**+**G**)% in 205 proteins encoded by chromosome 2 of *P. falciparum*. Each point refers to the protein-encoding region of a particular gene. *(a)* (**G** + **C**)% for all bases encoding each protein. *(b)* (**A**+**G**)% for all bases encoding each protein. Corresponding values for the three individual codon positions in each protein are shown for (**G**+**C**)% in *(c)*, and for (**A**+**G**)% in *(d)*. First codon positions (**GC1**, **AG1**; open circles), second codon positions (**GC2**, **AG2**; grey squares), and third codon positions (**GC3**, **AG3**; black triangles). Slopes are expressed in percentage units/kilobase of protein-encoding sequence.

The relationship with **AG**-pressure becomes more apparent when the base compositions of different codon positions are plotted against the lengths of the corresponding protein-encoding segments. In the case of **GC**-pressure (Figs. 9-9c), first and third codon position values change little as protein lengths increase. Second codon position values show the best correlation with protein lengths (slope = -1.0%/kilobase), and 7% of the variation can be explained on this basis ($r^2 = 0.07$).

In the case of **AG**-pressure (Fig. 9-9d) the **RNY**-rule prevails. In nucleic acid segments encoding both short and long proteins, first codon positions tends to be **R**-rich (i.e. high **(A+G)**%), and third codon positions tend to be **Y**-rich (i.e. low **(A+G)**%). The slopes, while modest, both support these trends (i.e. first codon position slopes are positive, indicating increasing **R**-loading, and third codon position slopes are negative, indicating increasing **Y**-loading).

Second codon positions appear much freer to vary their **(A+G)** % values as the lengths of protein-encoding regions increase (slope = 1.4%/kilobase), and 17% of the variation can be explained on this basis ($r^2 = 0.17$). Since second codon positions are most affected, it can again be inferred that base compositional pressures have caused amino acid-changing (non-synonymous) mutations to be accepted. Increased protein lengths appear to reflect the insertion of amino acids with **R**-rich codons, at the expense of **S**-rich (i.e. **GC**-rich) codons.

In Figure 9-9 slopes differ from zero because, on average, when their base compositions are presented as percentages, nucleic acid segments encoding large proteins differ from nucleic acid segments encoding small proteins. Why cannot small protein-encoding segments contain proportionately as many **R**-rich codons, as long protein-encoding segments, so that zero slopes result?

As will be further considered in Chapters 10 and 11, the probable explanation is that proteins tend to fold into a number of independent functional domains. Domains may involve from 60 to 300 contiguous amino acids. Small proteins tend to have just one domain. Large proteins tend to have many domains, and it is the space *between* domains that most readily accepts insertions of "placeholder" amino acids. These may not necessary play a role in the actual functioning of the protein. They are there *by default* because the corresponding genes have not "found" how to cater to the demands of nucleic acids in any way other than by adding interdomain codons.

By the same token, by virtue of their possession of more interdomain regions, genes encoding long proteins should more readily respond to pressures to mutate. For example, the preservation of a gene duplicate requires that it quickly mutate to avoid recombination with the source gene (see Fig. 8-2). This predicts that genes encoding long proteins should more readily form duplicates than genes encoding short proteins.

Thermophiles

We have already noted for prokaryotes that optimum growth temperature correlates both with **(A+G)**% and **(G+C)**%, but better with the former (see Chapter 8). Figure 9-10 shows this at the level of individual bases.

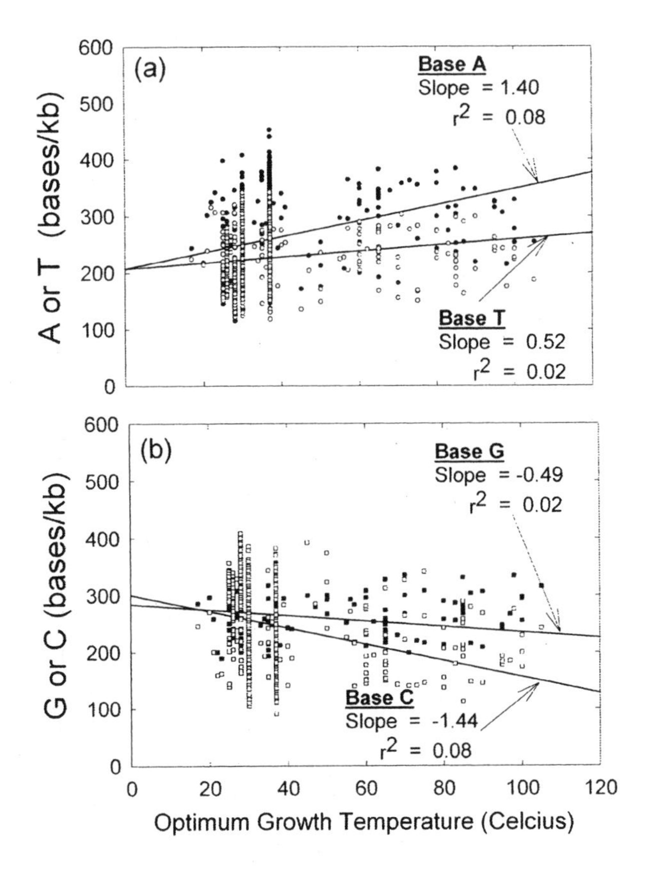

Fig. 9-10. Variation of the frequencies of bases in the DNA of the coding regions of genes of each of 550 prokaryotic species (eubacteria and archaebacteria) with the temperature require for optimal growth of that species. Each point represents a species. *(a)* The **W** bases: **A**, filled circle; **T**, open circle. *(b)* The **S** bases: **G**, filled square; **C**, open square. All slope values are significantly different from zero ($P < 0.001$). In addition, slope values for members of each pair are significantly different from each other; in *(a)* $P = 0.0005$, and in *(b)* $P = 0.0001$

Although an increase in **(G+C)**% can stabilize RNAs that must function by virtue of their structure in the cytoplasm, there is no consistent trend towards a high genomic **(G+C)**% in thermophiles. This may partly reflect a conflict with genomic **AG**-pressure, which increases with optimum growth temperature, and is accompanied by a small, but significant, decrease in genomic **(G+C)**% (see Fig. 8-5). Hence, a causal chain of evolutionary events might be: (i) need to adapt to high environmental temperature, (ii) purine-loading (to diminish RNA-RNA interactions; see Chapter 12), and (iii) decrease in **(G+C)**%. As might be expected from the decrease in genomic **(G+C)**% as optimum growth temperature increases, the increase in genomic (**A+G**)% mainly reflects an increase in base **A** at the expense of base **C**. Thus, the levels of the bases **T** and **G** remain relatively constant as the optimum growth temperature increases.

Since thermophiles strongly purine-load their RNAs, it might be suspected that the interdomain regions of their proteins would have many placeholder amino acids encoded by purine-rich codons. However, the demand of stability at high temperatures tends to exclude large interdomain regions from the proteins of thermophiles. Thus, the extra purines must usually be accommodated within intradomain regions [17, 18].

Summary

Genome phenotypes and classical phenotypes make potentially conflicting informational demands. The composition and order of bases in the nucleic acids of members of a particular biological species reflect the demands both of the genome phenotype (e.g. **GC**-pressure) and of the classical phenotype (e.g. protein-pressure – the pressure for a particular composition and ordering of amino acids in a protein). Base compositions of third codon positions, being less constrained by protein-pressure, can provide indices of non-classical pressures, so facilitating the identification and calibration of conflicts. Since changes in genetic fitness (number and reproductive health of descendents) can result from changes in the classical phenotype and/or the genome phenotype, a mutation appearing neutral with respect to the classical phenotype can nevertheless affect fitness by changing the genome phenotype. Within a group of species (taxonomic group), *genomic* (**G+C**)% values can cover a wide range that is most evident at third codon positions. However, within a genome, *genic* (**G+C**)% values can also cover a wide range that is, again, most evident at third codon positions. The dispersion of (**G+C**)% values among *genes within a genome*, matches the dispersion of (**G+C**)% values among *genomes within a taxonomic group*. This suggests that both dispersions are driven by a common factor – such as the need to avoid recombination in order to facilitate duplication, either of genes, or of genomes (e.g.. speciation). Third codon positions usually relate more to the genes that contain them than to the species. However, extreme (**G+C**)% values (low or high) constrain the potential of a genome to transmit further information. Third codon positions then tend to maintain a constant (**G+C**)%, thus relating more to the species than to the genes that contain them. Each gene in an intermediate (**G+C**)% genome has come to occupy a discrete (**G+C**)% niche (microisochore) amongst its fellow genes, which collectively span a wide (**G+C**)% range. Genes in extreme (**G+C**)% genomes rely mainly on first and second codon positions for differentiation as microisochores, which collectively span a narrow (**G+C**)% range.

Chapter 10

Exons and Introns

> "Ideas in the theory of evolution can be used in situations far removed from biology. Similarly, information theory has ideas that are widely applicable to situations remote from its original inspiration."
>
> Richard Hamming (1980) [1]

Organisms can be divided into those whose cells do not have a nucleus, the single celled "prokaryotes" (Greek: *pro* = before; *karyon* = nucleus), and those whose cells have a nucleus, the single- or multi-celled "eukaryotes" (Greek: *eu* = good or normal; *karyon* = nucleus). Prokaryotes include species of bacteria (eubacteria) and archaea (archaebacteria), the latter being a bacteria-like group sometimes found in extreme environments (e.g. hot springs). Eukaryotes (eukarya) include all species of animals and plants, both single-celled (protozoa, protophyta) and multi-celled (metazoa, metaphyta).

The etymology may suggest that organisms without a nucleus, somewhat like modern bacteria or archaea, evolved before organisms with a nucleus. However, as discussed in Chapter 8 when wondering whether isochores appeared early or late, prokaryotes and eukaryotes are believed to have evolved from a common ancestor, whose properties we can now only infer. It may appear parsimonious to imagine an ancestor without a nucleus. But parsimony may be in the mind of the beholder. Unlikely maybe, but perhaps the ancestor had a nucleus?

In the 1960s many eukaryotic RNAs were found to be first synthesized as "giant" RNAs, a large part of which was subsequently "wasted" as the RNAs were "processed" to generate mature cytoplasmic products. Considering a "gene" as corresponding to a unit of transcription, it appeared that a large part of a gene might contain redundant information. Eukaryotic ribosomal RNAs (rRNAs) were transcribed from DNA as long precursor RNAs that were subsequently processed by the removal of apparently functionless internal "spacer" sequences [2]. However, prokaryotic (bacterial) rRNA genes were more compactly organized. Did rRNA genes in the first organisms to evolve have the spacer sequences, which decreased later in prokaryotes ("spacers-early"), or were the long spacer sequences acquired later in eukaryotes ("spacers-late")?

A similar processing was found in the case of eukaryotic messenger RNAs (mRNAs). Again, bacterial mRNA genes were more compactly organized. By 1970 it had been shown that many freshly minted giant nuclear RNAs contained sequences corresponding to smaller mature cytoplasmic mRNAs. A simple model was that each giant RNA consisted of an mRNA sequence that was flanked by long redundant RNA segments [3]. It turned out, however, that instead of one DNA sequence segment compactly encoding one protein, as in most prokaryotic genes, eukaryotic genes often encode proteins in sequence segments that are interrupted by sequence segments that do not usually encode proteins. Thus, if a mature "sense" mRNA be represented as:

Maryhadalittlelambitsfleecewaswhiteassnow (10.1)

Then the corresponding giant precursor RNA would contain segments of, what would appear to be, "nonsense:"

Maryha*xqvhxmgeqz*dalittlela*qwxgtpscr*mbitsfleecewa*zgfxnyq*swhiteassnow (10.2)

Since an entire gene was transcribed, the internal RNA sequences (e.g. *xqvhxmgeqz*) derived from DNA "introns," had to be removed from the initial transcript. What remained in the processed RNA, the mRNA, was derived from DNA "exons" (Fig. 10-1) [4]. Thus, a protein-encoding gene consists of exons and introns. Since the "spacer" phenomenon had already been described for rRNA it should not have been a surprise that it also applied to mRNAs; but many, myself included, were surprised at its generality.

This chapter presents evidence that introns are a way of resolving intragenomic conflicts between different forms and levels of information. Cases of extreme intragenomic conflict, as when genes in "arms races" evolve rapidly under positive selection, demonstrate most clearly that "information theory has ideas that are widely applicable" [1].

Introns Interrupt Information

Although there were many attempts, it proved difficult consistently to associate exons with domains of protein structure or function [5, 6]. Introns interrupt genetic information *per se*, not just protein-encoding information. Thus, intron-encoded information is removed during transcript processing both from parts of mRNA precursors that encode protein, and from parts of mRNA precursors that do not encode protein (i.e. from 5' and 3' non-coding sequences). Certain special RNAs which, like rRNAs, are not translated to give a protein product (e.g. see Chapter 14), also have introns.

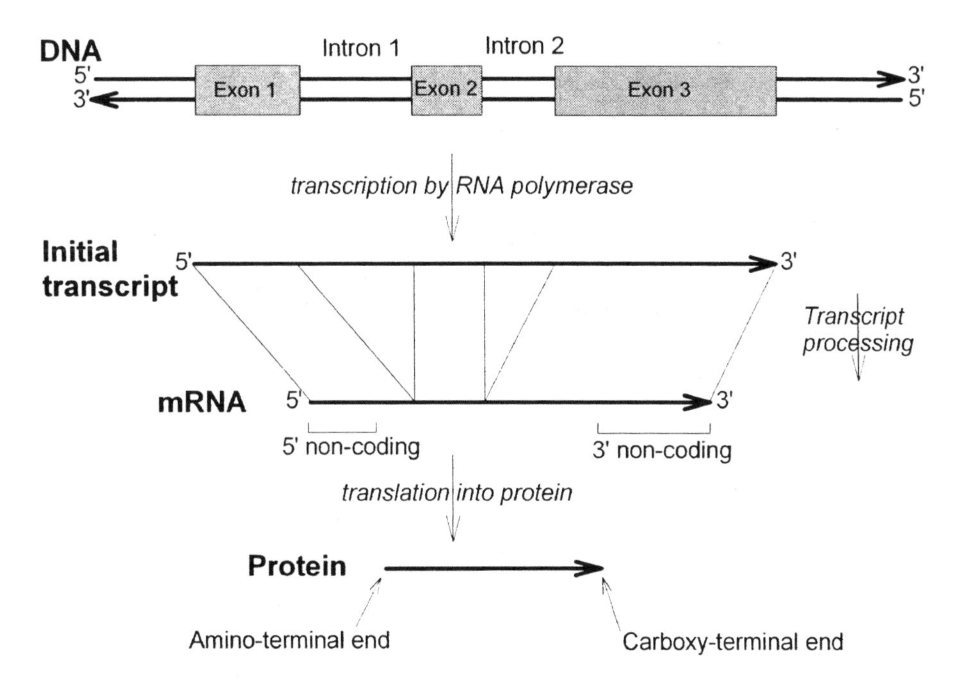

Fig. 10-1. Transcription and processing of mRNA from a eukaryotic gene. Sequences that correspond to the final mRNA are marked as boxes on the duplex DNA. The enzyme RNA polymerase initiates transcription at the left end of exon 1 using the bottom DNA strand as template, and terminates transcription at the right end of exon 3. Intron sequences are spliced out from the initial transcript, so that the mature mRNA consists of a 5' non-protein-coding region, a central protein-encoding region, and a 3' non-protein-coding region. The central region is translated into protein. A gene may be narrowly construed as corresponding to the segment of DNA that is transcribed. A broader definition would include other segments (e.g. the promoter region where RNA polymerase initially binds to DNA), which may lie outside the transcribed segment. Williams gives a different definition (see Chapter 8)

Exons have a narrow size range, with a peak at about 100 bases (Fig. 10-2) [7]. If coding, this would correspond to 33 amino acids. So proteins, often containing hundreds of amino acids, are usually encoded by long genes with many exons [8]. There appears to be a limit to the extent to which a genome will tolerate a region that both encodes protein, and is purine-loaded (see Chapter 6). When that limit is reached, protein-encoding and purine-loading functions are arrested, and can restart only after a "decent" interval – namely an intron interval. Introns are generally longer than the exons that surround them (see Fig. 2-5). Each exon being separated from its neighbours by introns, the sum of intron lengths tends to increase proportionately as the sum of exon lengths increases [9].

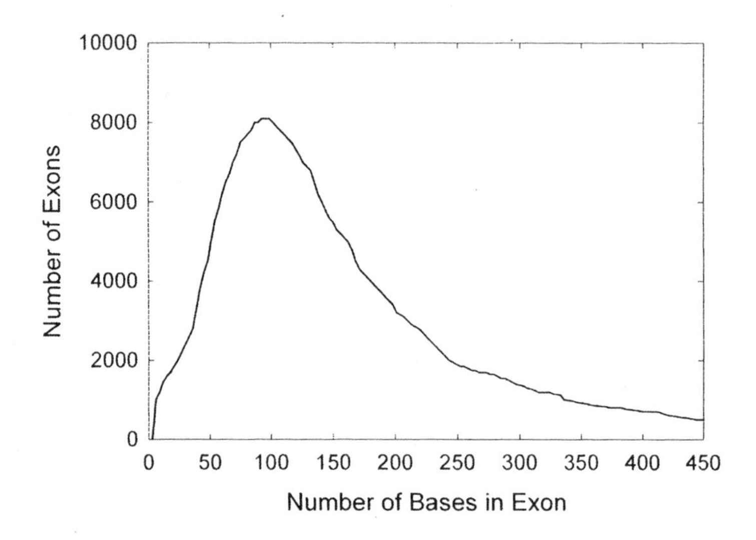

Fig. 10-2. Distribution of exon sizes in eukaryotic genes. The number of exons in different size categories increases to a maximum and then progressively declines. Most exons are about 100 bases in length. (Data are from the ExInt database)

What function(s), if any, do introns have? Did introns appear early or late in the evolution of biological forms? If introns can be dispensed with in bacteria, then perhaps they have no function. Alternatively, whatever function introns have, either is not necessary in bacteria, or can be achieved in other ways by bacteria. Since members of many bacterial species appear to be under intense pressure to streamline their genomes to facilitate rapid replication, if it were possible they should have dispensed with any preexisting introns and/or should have been reluctant to acquire them. On the other hand, if introns play a role, and/or do not present too great a selective burden, then organisms under less pressure for genome compaction might have retained preexisting introns, or might have acquired them.

An early origin of introns is suggested by the fact that the humans share the positions and sequences of many introns with a marine worm, indicating that their common ancestor had the same introns [10]. Introns did not first appear so that at some remote future date organisms with exons would be advantaged. Nature is not prescient. Although something playing a role at one point in time can come later to serve a quite unrelated role, in general evolution does not work this way. Sometimes a random event (genetic drift) provides an evolutionary toe-hold, but for something so widespread and drastic as introns there should be some immediate selective advantage.

Knowing the function of introns seemed critical for sorting this out. There were many ingenious suggestions. Some thought introns were just another example of the apparently useless "junk" DNA that appeared to litter the DNA of many organisms (see Chapter 12). Others thought that introns might have facilitated the swapping of protein domains to generate new proteins [4, 11], but that did not explain how introns initially arrived.

However, the notion of "message" sequences interrupted by "non-message" sequences is familiar to those working on noise affecting signal transmission in electrical systems. In these systems information scientist Richard Hamming pointed out that the non-message sequences can have an error-checking function that permits the receiver to detect and correct errors in the message sequence [1]. Could introns have a similar error-checking role [12]?

It appears that the order of bases in nucleic acids has been under evolutionary pressure to develop the potential to form stem-loop structures, which might facilitate "in-series" or "in-parallel" error-correction by recombination (see Chapter 2). This means that genomic sequences convey more than one level of information (see Chapter 9). Furthermore, as predicted in 1893 by the discoverer of DNA, Johann Miescher, a sexual process that brings molecules from separate sources together, could facilitate the mutual correction of errors (see Chapters 3 and 14). However, the need to participate in the process of error-detection and correction can result in redundancies and various constraints (see Chapter 4).

The error detection and correction process requires an alignment of two sequences, which itself depends on an initial "homology search." Thus, there must be sufficient similarity between two sequences for a successful homology search. One outcome of this is that segments of DNA link up ("recombine") with other segments. In the course of this swapping of segments, errors can be detected and corrected ("gene conversion;" Fig. 8-3). The process is referred to as recombination repair, which distinguishes it from a variety of other repair processes that will not be considered here [13]. The adaptive value of recombination repair is likely to be very great. So, if it *could* have arisen early in the evolution of primeval biological forms (perhaps in an "RNA world") prior to the evolution of protein-encoding capacity (i.e. if it were chemically feasible), then it *would* have arisen. Williams noted in 1966 [14]:

> "The existence of genetic recombination among the bacteria and viruses, and among all of the major groups of higher organisms, indicates that the molecular basis of sexuality is an ancient evo-

> lutionary development. Our understanding of the structure of the DNA molecule makes recombination at this level easy to visualize. In a sense sex is at least as ancient as DNA. ... I would agree, therefore, ... that sexual reproduction is as old as life, in that the most primitive living systems were capable of fusion and of combination and recombination of their autocatalytic particles. Modern organisms have evolved elaborate mechanisms for regulating this primitive power of recombination and for maximizing the benefits to be derived from it."

Given an early evolution of recombination, protein-encoding capacity could then have had to *intrude* into the genomes of biological forms already adapted for recombination. Would this intrusion have been readily accepted? Or would protein-encoding capacity have had to elbow its way forcibly into primeval genomes?

Protein Versus DNA

Although the degeneracy of the genetic code provides some flexibility as to which base occupies a particular position, there may still be a conflict between the needs of a sequence both to encode a protein and to respond to other pressures. Situations where protein-encoding and/or other pressures are extreme should be particularly informative in this respect.

Extreme protein-encoding pressure is apparent in the case of genes under very strong positive Darwinian selection. In Chapter 7 competition among speakers for the attention of an audience provided a simple metaphor for positive selection. In that case, speakers were positively selected if they could communicate rapidly to a fixed audience by overcoming idiosyncrasies of accent. Under biological conditions, however, "the audience" is not fixed. Positive selection often occurs under conditions where both "speakers" and "audience" are rapidly changing. This includes genes affected by "arms races" between predators and their prey.

For example, snake venom may decrease the rodent population (prey) until a venom-resistant rodent line develops and expands (i.e. a mutant line arises with this selective advantage). Now, while the rodent population expands, the snake population (predators) decreases because its members cannot obtain sufficient food (i.e. rodents). This decrease continues until a line of snakes with more active venom, which can overcome the resistance, develops and expands (i.e. a mutant line arises with this selective advantage). This population of snakes now expands, and the rodent population begins to fall again.

The cycle constitutes a biological arms race, and influences particular gene products. Parts of venom proteins which are important for toxicity are required to change so rapidly in response to this strong pressure from the environment (i.e. from rodents), that the corresponding genes can no longer afford the luxury of both encoding the best proteins (primary information) and attending to other pressures (secondary information). They must encode better proteins even at the expense of their abilities to respond to other pressures. Accordingly, under extreme positive selection pressure the rate of sequence change in protein-encoding regions is high (Fig. 10-3).

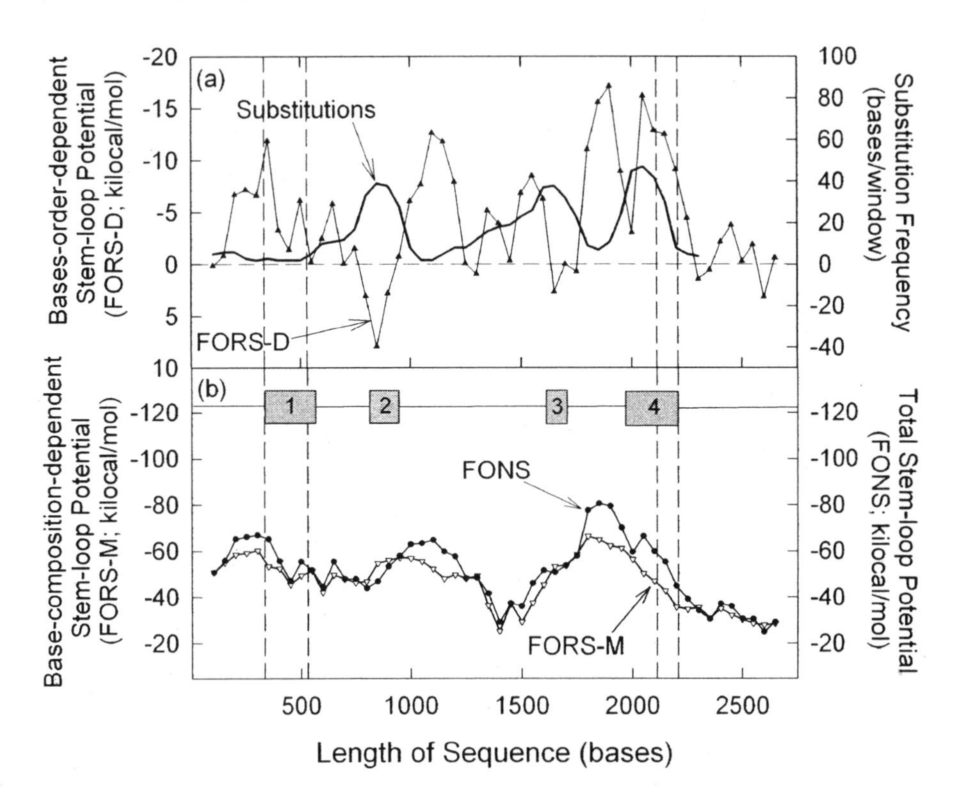

Fig. 10-3. High base substitution frequency and low base order-dependent stem-loop potential in exons of the rattlesnake gene encoding the basic subunit of venom phospholipase A_2 (PLA_2), which is under positive Darwinian natural selection. The distribution of base substitutions (continuous line in *(a)*) is compared with values for FORS-D (closed triangles in *(a)*), FORS-M (open triangles in *(b)*), and FONS (closed circles in *(b)*). Values were determined for overlapping 200 base windows, which were moved in steps of 50 bases. Substitutions are base differences relative to the rattlesnake PLA_2 acidic subunit gene. The two genes are likely to have arisen by duplication of a common ancestral gene. Boxes in *(b)* indicate the location of the four exons, with dashed vertical lines showing, consecutively, the beginning of exon 1, the beginning of the protein-coding part of exon 1, the end of the protein-coding part of exon 4, and the end of exon 4

When sequences of similar venom proteins (e.g. phospholipase A_2) from two snake species (or from duplicated genes within a species) are compared, *great* differences are found in the protein-encoding parts of exons (i.e. low sequence conservation). In contrast, *small* differences are found in introns, and in the 5' non-coding and 3' non-coding parts of exons (i.e. high sequence conservation; Fig. 10-3a). This is a dramatic reversal of the more usual situation where, in genes under classical negative ("purifying") Darwinian selection, exons are conserved much more than introns (i.e. introns display more variation; see Fig. 2-5).

What is being conserved in snake venom introns? Analysis of fold potential as it affects base order (FORS-D; see Chapter 5) reveals that base order-dependent stem-loop potential is low in exons (where sequence conservation is low) and high in introns (where sequence conservation is high; Fig. 10-3a). Base order-dependent stem-loop potential appears to have been conserved in introns; indeed, there is an inverse (reciprocal) relationship between base substitution frequency and base order-dependent stem-loop potential. When base order-dependent stem-loop potential is high, base substitution frequency is usually low (i.e. sequence conservation is high).

This suggests that the pressure to adapt the protein sequence (requiring non-synonymous codon changes) has been so powerful that base order has not been able to support base order-dependent stem-loop potential in the same exon sequence (Fig. 5-6). Instead, stem-loop potential is diverted to introns, which are appropriately conserved (fewer base substitutions than the surrounding exons). This is in keeping with the hypothesis that early in evolution protein-encoding potential was imposed on prototypic genomes that had already developed stem-loop potential. For this imposition to succeed without disturbing the general distribution of stem-loop potential, proteins had to be encoded in the fragments that we now call exons [15–17]. Thus, in the general case, introns were "early."

Another example of positive selection is the genome of the AIDS virus (HIV-1), which can be viewed as a predator, with us as its prey (see Chapter 8). Here, an inverse correlation between substitutions and base order-dependent stem-loop potential can be observed when the disposition of substitutions and fold potential along the genome are displayed (Fig. 10-4). At first glance the data appear as a confused jumble of lines. But when the paired values from along the sequence are plotted against each other, a significant inverse correlation emerges (Fig. 10-5c) [18, 19].

Thus, sequences varying *rapidly* in response to powerful environmental selective forces ("arms races") appear unable to order their bases to favor the elaboration of higher order folded structures (of a type that, in eukaryotes, might mediate meiotic chromosomal interactions and recombination repair; see Chapter 8). So, the encoding of nucleic acid stem-loop structure can be relegated either to non-protein-encoding regions, namely, introns, 5' and 3' non-coding regions, and non-genic DNA (the favored option in less compact genomes), or to less rapidly evolving protein-encoding sequences (or parts of such sequences) where there is some flexibility in codon or amino acid choice (the main option in compact genomes).

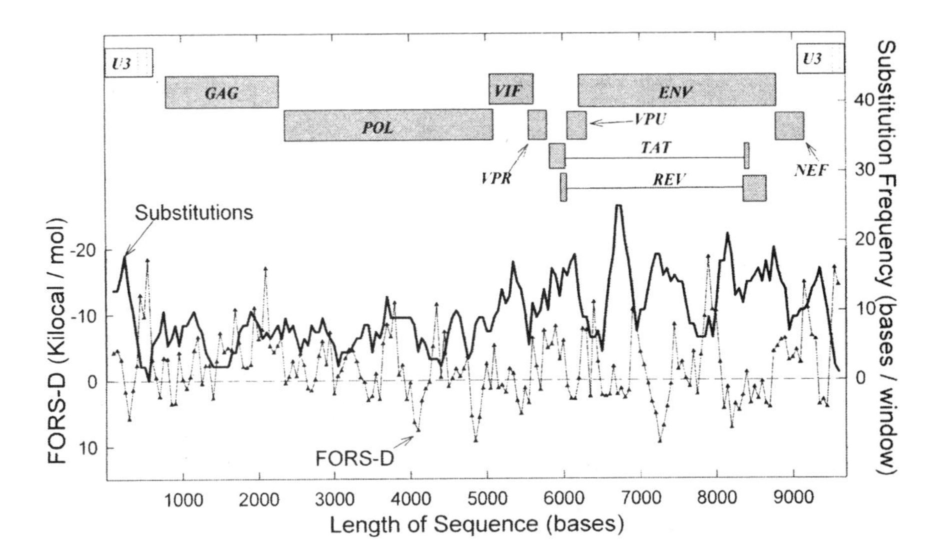

Fig. 10-4. High base substitution frequency and low base order-dependent stem-loop potential in regions of the AIDS virus genome that are under positive Darwinian natural selection. The various genes are shown as grey boxes, with their abbreviated names attached. Two genes, *TAT* and *REV*, each have two exons that are shown linked by continuous lines that represent introns. Thus, one gene's intron can be another gene's exon. The distribution of base substitutions for virus "subtype" HIVSF2 relative to virus "subtype" HIVHXB2 is shown as a continuous line. Values for base order-dependent stem-loop potential (FORS-D) are shown as filled triangles. All values are for 200 base overlapping windows, which were moved in steps of 50 bases

Functionally important regions are conserved in genes evolving slowly under classical negative selection (i.e. there is a low local base substitution frequency in protein regions under this selection pressure). If the rate of evolution has been slow, then there has been more time to arrive at an appropriate compromise with base order-dependent stem-loop potential. Thus, in the case of slowly evolving sequences, a relationship between base order-dependent stem-loop potential and sequence variability may be less evident.

On the other hand, the demands of faithful reproduction of a protein, with negative selection of individuals bearing mutations affecting its functionally most important parts, can leave the co-encoding of stem-loops not only to non-protein-encoding regions (e.g. introns), but also to regions encoding functionally less important, *and hence more variable*, parts of proteins (such as the protein surface, which cell water can readily access) [20, 21]. In this case, high base order-dependent stem-loop potential can correlate positively (not inversely) with high substitution rates (variability) when similar (homologous) sequences from different species are compared. Indeed, comparison of certain human and mouse oncogenes (*FOSB*) reveals a positive correlation between base order-dependent fold potential and substitution frequency [22].

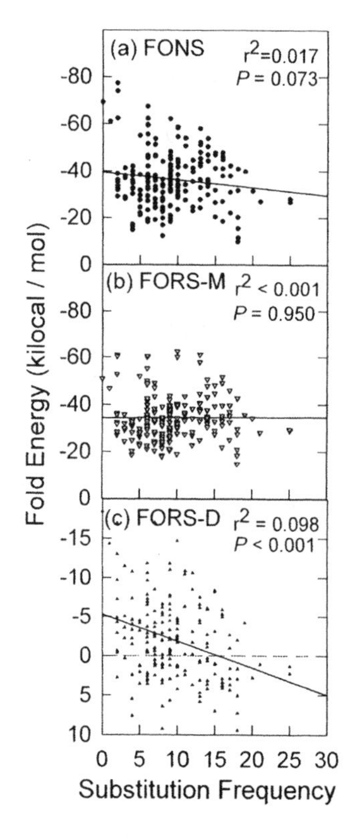

Fig. 10-5. In the AIDS virus genome there is *(c)* an inverse relationship between base order-dependent potential (FORS-D values) and substitution frequency, but *(b)* no detectable relationship between base composition-dependent stem-loop potential (FORS-M) and substitution frequency. Shown in *(a)* are FONS values, which are the sum of FORS-D and FORS-M values. Although only approximately 10% of the variation in substitutions can be explained by base order-dependent stem-loop potential (shown by the r^2 value in *(c)*), the downward slope is significantly different from zero (shown by the low *P* value in *(c)*)

Achilles Heels

The AIDS virus causes millions of deaths annually, and subtypes resistant to therapeutic agents have become more prevalent. Given the profiles shown in Figure 10-4, can you suggest potential targets for therapeutic attack – namely targets that are *least* likely to differ between different pathogens? Note that the region to the left of the *GAG* gene in the AIDS virus has the lowest base substitution rate (i.e. it is the most *conserved* part of the genome). This coincides with a major peak in base order-dependent stem-loop potential (indicated by high negative FORS-D values). This suggests that there is here a special need to conserve stem-loop potential [18]. Indeed, this is the location of the "dimer initiation" stem-loop sequence, which is necessary for the copackaging of two AIDS virus genomes as part of a process that resembles meiotic pairing (see Chapter 8). This may yet prove to be the Achilles heel of the AIDS virus, once the problem of "flushing out" latent forms from host genomic reservoirs is solved [23, 24].

We humans may also have an Achilles heel. Our genomes are rich in palindromes conferring stem-loop potential (Fig. 5-2), which is usually to our advantage (see Chapters 8 and 14). But sometimes the palindromes involve oligonucleotide repeats, which can be disadvantageous, as will be discussed in Chapter 11 (Fig. 11-8). This is particularly so when the repeats are AT-rich and consequently can readily adopt cruciform stem-loop configurations.

Palindromic regions containing AT-rich repeats are prone to recombine, at times when homologous chromosomes may not be precisely aligned, by kissing interactions with other palindromic regions containing AT-rich repeats. Thus, there may be "cut-and-paste" translocations (transpositions) between non-homologous chromosomes ("illegitimate recombination"), which can be detected in normal sperm samples [25]. Whether pathological results occur depends partly on the frequency of the translocations. This, in turn, depends on the lengths of the regions containing AT-rich repeats, which varies among individuals (polymorphism). It is likely that, by virtue of this Achilles heel, many individuals with long palindromic regions containing AT-rich repeats have been eliminated by natural selection.

Mirror Repeats

In Chapter 4 we encountered "inverted repeats" with palindromic properties *at the duplex level* (see also Fig. 2-4). For example a "top" single-strand in a duplex might read:

5' **AAAAACCCGGGTTTTT** 3' (10.3)

Here **AAAAACCC** in the 5' half of the top strand is repeated on the *complementary* strand, where it pairs with the **GGGTTTTT** sequence in the 3' half of the top strand. Such sequences appear to serve a DNA level function, since they facilitate the extrusion of stem-loop secondary structures (Fig. 5-2). Single-strands also contain "direct repeats" that might, for example, read:

5' **AAAAACCCAAAAACCC** 3' (10.4)

Here **AAAAACCC** is repeated. Single-strands can also contain "mirror repeats" that might read:

5' **AAAAACCCCCCAAAAA** 3' (10.5)

Here **AAAAACCC** can be considered the "mirror" of **CCCAAAAA**. In these two cases, (10.4) and (10.5), the repeats occur in the *same* strand. Direct repeats and, especially, mirror repeats, have the potential to oppose local stem-loop formation, and so to oppose any DNA level function that stem-loops might serve.

Remarkably, mirror repeats are found at particular locations in exons. Their locations correlate with the boundaries of various structural elements in proteins; indeed, mirror repeats can *predict* where such structural elements will occur [26]. In this case it appears that a conflict between protein and DNA has been won by protein. By preventing stem-loop formation mirror repeats should prevent local recombination and thus preserve the local integrity of the DNA encoding a structural element in a protein.

We know that a protein, by "insisting" (through natural selection) on having a particular amino acid at a particular position in its sequence, requires that the corresponding gene have a suitable codon at a particular position in its sequence. Now we see that a protein, by "insisting" (through natural selection) on having a particular structural element (e.g. alpha-helix, beta-strand), also requires that the corresponding gene have appropriately positioned mirror repeats (albeit often imperfect mirror repeats).

RNA Versus DNA

Usually a particular DNA sequence is transcribed into an identical RNA sequence, with the exception that RNA molecules have **U** (uracil) rather than **T** (thymine); but these are chemically similar bases. So it is not surprising that, in broad features, computer-derived secondary structures for an RNA molecule (using dinucleotide pairing energy tables for RNA bases), are similar to the structures derived for the corresponding DNA (using dinucleotide pairing energy tables for DNA bases; see Table 5-1). Yet there are genes with no protein product. The gene products are RNAs, which have specific functions dependent on the secondary (and higher order) structures they adopt (often selected for at the *cytoplasmic* level). If such stem-loop secondary structures also sufficed for function at the DNA level, then there might be no need for introns in genes for non-protein-encoding RNAs.

The fact that there are spacers or introns in such genes, implies that sequences generating the stem-loop secondary structures that suffice for function at the RNA level, do *not* suffice for, and may even *conflict* with, the sequences needed for stem-loop secondary structures that function at the DNA level. Since patterns of RNA stem-loops are influenced by the purine-loading of loops (the selective force for which probably operates at the cytoplasmic level; see Chapter 6), then purine-loading pressure (which would constrain stem-loop patterns in exons) should support stem-loop pressure in provoking the splitting of what might otherwise have been large exons.

Overlapping Genes

Strict adherence to the **RNY**-rule (see Chapter 7) would dictate that, of the three possible triplet reading frames in the "top" mRNA synonymous "coding" strand of DNA, the one that best fits the **RNY** pattern would be the actual reading-frame. However, sometimes it is expedient for genes to overlap, either entirely or in part, and in this case one of the genes, if transcribed in the same direction, can use another, non-**RNY**, reading-frame. This applies to some of the genes of the AIDS virus, which are all transcribed to the right (Fig. 10-4).

In some circumstances, genes transcribed in different directions may overlap. Thus, the "top" strand may be the coding strand of one gene, and the "bottom" strand may be the coding strand of another gene. Again, one reading frame is **RNY** and two are non-**RNY** (Fig. 7-2), and any of the three may be employed.

Whatever the transcription direction, in overlapping genes the region of the overlap can come under extreme protein-encoding pressure, and this might conflict with other pressures. Indeed, consistent with the argument made here, base order-dependent folding potential is constrained where genes overlap [27].

Simple Sequences

As in the above examples of genes under extreme protein-encoding pressure, genomes under another extreme pressure, **GC**-pressure, should also be highly informative. The genome of the most lethal malaria species, *Plasmodium falciparum,* satisfies this requirement, being under strong "downward

GC-pressure." It is one of the most **AT**-rich species known (i.e. very low (**G+C**)%; see Fig. 9-5a). Another unusual feature of *P. falciparum* is that many proteins are longer than their equivalent proteins (homologues) in species that have less extreme genomic (**G+C**)% deviations. This reflects the acquisition by the *P. falciparum* proteins of low complexity "simple sequence" segments that have no known function. Simple sequence at the protein level (i.e. runs of amino acids from a limited range of the twenty possible amino acids) is encoded by simple sequence at the nucleic acid level (i.e. runs of bases from a limited range of the four possible bases; see Chapter 11).

There are other unusual features of the *P. falciparum* genome. Unlike many eukaryotic genomes, there is poor correlation between the length of a gene and the combined lengths of the introns of that gene (Fig. 10-6a). Yet there is a close correlation between the length of a gene and the combined lengths of low complexity segments in that gene (Figs. 10-6b). In this respect the low complexity elements appear like introns; however, unlike introns, they are not removed during processing of the RNA transcript.

Furthermore, introns and low complexity segments are not interchangeable in that, as absolute intron length increases, there is little decline in length of low complexity segments in a gene (Figs. 10-6c). It is only when the lengths are expressed as a proportion of gene length that a reciprocal relationship emerges (Figs. 10-6d). Whereas in many eukaryotic genomes intron locations show no relationship to protein functional domains (since if splicing is accurate intron location is irrelevant to the protein), low complexity segments in *P. falciparum* must, of necessity, predominate between functional domains.

If low complexity segments have no, or minimal, function at the protein level, do they reflect a function at the nucleic acid level? Low complexity segments in *P. falciparum* are usually of high (**A+G**)%, and so they contribute to purine-loading. Introns, in contrast, tend to be of low (**A + G**)% (i.e. tending towards pyrimidine-loading). It will be shown in Chapter 11 that, when contributing to **AG**-pressure (purine-loading pressure), low complexity segments can countermand fold potential. But when **AG**-pressure is not extreme, low complexity segments do contribute to fold potential. In this respect, they do resemble introns.

Spacers and introns are likely to have arisen early in evolution because they are preferential sites for the encoding of the stem-loop structures in DNA that are necessary for initiating recombination and, hence, error-detection and correction. While, in extreme cases, by virtue of this function, introns are conserved more than exons (Fig. 10-3), in the general case, to facilitate the anti-recombination necessary for gene or genome duplication (speciation), introns evolve more rapidly (are conserved less) than exons (see Fig. 2-5 and Chapter 8).

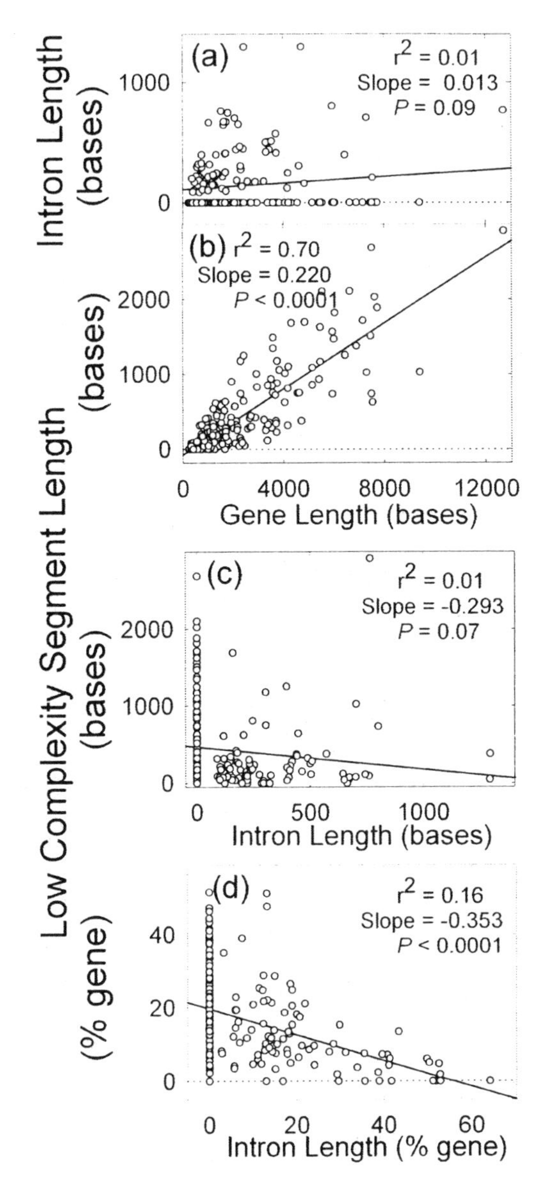

Fig. 10-6. In the genome of the malaria parasite, *P. falciparum*, low complexity segment length, not intron length, correlates best with gene length. Shown for chromosome 2 are relationships of lengths of introns *(a)*, and low complexity segments *(b)*, to the lengths of the corresponding genes, and relationships of either absolute *(c)*, or percentage *(d)*, lengths of low complexity segments to the lengths of introns in the same genes. *P* values indicate the probability that the slope values are not significantly different from zero (i.e. a low *P* value indicates high significance)

Multiple Pressures

The multiple, potentially conflicting, pressures affecting both the genome phenotype and the conventional (classical Darwinian) phenotype are summarized in Figure 10-7 and Table 10-1.

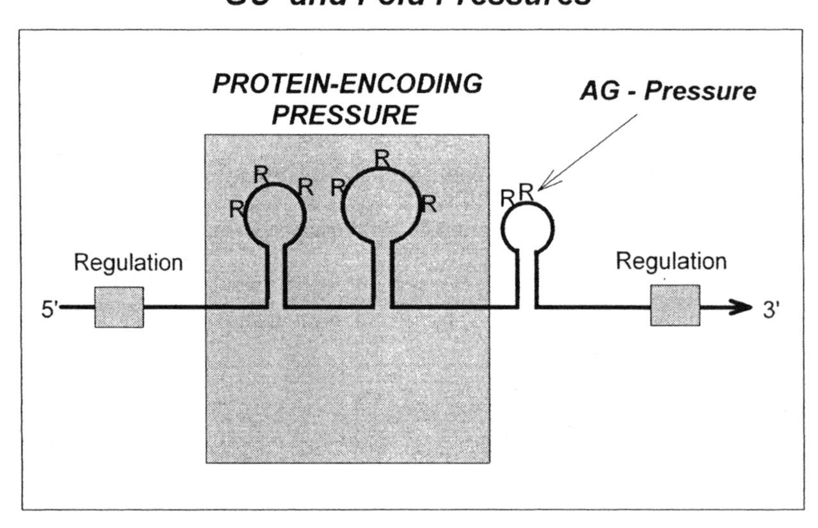

Fig. 10-7. Summary of potentially conflicting pressures operating at the mRNA level. The genome-wide pressures, **GC**-pressure and fold (stem-loop) pressure, influence the entire mRNA (shown as a thick horizontal arrow with loops enclosed in the large outer box). Purine-loading pressure (**AG**-pressure) is a local pressure that also influences the entire mRNA. Small grey boxes indicate potential sites for the binding of regulatory factors that sometimes preferentially locate to the 5' and 3' non-coding regions. The large grey box indicates the central, protein-encoding region, where "protein-pressure" is deemed to operate

The proteins of *P. falciparum* provide an extreme example of nucleic acid level pressures that affect protein sequence. However, proteins of all organisms are, to a degree, manifestations of nucleic acid level pressures. The correlation between gene length and content of low complexity segment (Fig. 10-6b) is probably general, with low complexity segments seeming to serve pressures for purine-loading, rather than for protein function. In a species where purine-loading was a dominant nucleic acid level pressure, proteins should be longer (and hence generally larger) than their homologous proteins in other species where purine-loading was less dominant. Since there is a reciprocal relationship between GC% and purine-loading (Fig. 9-7), then proteins should tend to be smaller in GC-rich species (Fig. 9-9), unless such GCrichness itself required addition of GC-rich low complexity segments. Since GC-rich species tend to have small proteins, it is likely that purine-loading is generally the dominant nucleic acid pressure in this respect (see Chapter 11).

Environmental selective factors	Selection for mutations which:	Primary effect on DNA function	Biological result	Observed features of modern DNA
Classical phenotypic selective factors	Change encoded proteins	Protein-encoding can constrain other functions	Change in classical phenotype	Changed base-pair in accordance with Chargaff's first parity rule
Competitors with more efficient translation, and intracellular pathogens	Purine-load RNAs	Purine-loading can constrain other functions	Efficient translation, and no "self" double strand RNA formation	Chargaff's cluster rule and Szybalski's transcription direction rule
Mutagens	Promote DNA stem-loop potential	Recombination promoted	Error detection and repair	Chargaff's second parity rule and genome-wide stem-loop potential
Classical phenotypic selective factors	Impair homology search between recently duplicated genes	Recombination between similar genes impaired	Gene duplication	Chargaff's (**G+C**)% rule
Differences in "reprotype" (Recombinationally "not-self" sexual partners)	Impair homology search between DNAs of species members whose sequences are diverging	Meiotic recombination impaired	Species duplication (speciation)	

Table 10-1. Summary of multiple, potentially conflicting, levels of information in genomes, and their relationships to Chargaff's rules

Nucleic acid pressures can affect the success of protein alignment analyses based on the scoring schemes described in standard bioinformatics texts. For example, since the amino acid tryptophan is highly conserved in proteins but glutamic acid is not, a tryptophan match between two protein sequences scores more than a glutamic acid match between two protein sequences [28]. Thus, two protein sequences may have a low similarity score because tryptophans do not match. Where there is tryptophan in one sequence, an alignment program may score a mismatch, or place a gap, in the other sequence. However, tryptophan may be present at a certain position in a particular protein sequence because **TGG** (the codon for tryptophan) plays some role at the nucleic acid level that is not needed in the homologous gene from another species.

Thus, gaps in the alignment of homologous protein sequences may occur if nucleic level pressures on the corresponding DNA sequences differ. To avoid this problem, protein alignment algorithms may depend on data from closely related organisms (e.g. "PAM matrices"), or use short "blocks" of amino acids (e.g. "BLOSUM matrices") corresponding to regions that are highly conserved at the protein level (e.g. active sites of enzymes). Trade-offs between competing pressures will be further explored in Chapter 11.

Summary

If genome space is finite with little, if any, DNA that is not functional under some circumstance, then potential conflicts between different forms of genomic information must be resolved by appropriate trade-offs. These trade-offs include the insertion into genes of spacers, introns, and simple sequence elements. The nature and extent of the trade-offs varies with the biological species. Study of trade-offs is facilitated in genes or species where demands are extreme (e.g. genes under positive selection pressure to adapt proteins, genes that overlap, and species under extreme downward or upward **GC**-pressures). Spacers and introns are likely to have arisen early in evolution because they are preferential sites for the stem-loop structures in DNA that are necessary for initiating recombination and, hence, error-detection and correction. Purine-loading pressure would have supported fold pressure in provoking the splitting into introns of what might otherwise have been large exons. From this perspective we can identify the Achilles heel of the AIDS virus as the dimer-initiation sequence that is essential for the copackaging of disparate genomes, so allowing recombination repair in a future host.

Chapter 11

Complexity

> "All perception and all response, all behaviour and all classes of behaviour, all learning and all genetics, all neurophysiology and all endocrinology, all organization and all evolution – one entire subject matter – must be regarded as communicational in nature, and therefore subject to the great generalizations or 'laws' which apply to communicational phenomena. We therefore are warned to expect to find in our data those principles of order which fundamental communication theory would propose."
>
> Gregory Bateson (1964) [1]

Some things are more complex than others. We feel intuitively that highly complex things are associated with more information than less complex things. But this will not suffice for *Homo bioinformaticus*. Information and complexity must be measured. When this is done, amazing "principles of order" emerge from large amounts of seemingly disparate data.

Scoring Information Potential

Imagine a DNA sequence that is a mix of just two bases, **A** and **T**. Given this sequence, and an appropriate code, you can "read" the information it contains. However, if the sequence consists of just one base, either **A** or **T**, this is not possible. A string of **A** bases, in the sense we have been discussing in this book, conveys little information. Sprinkle in a few **T**s and the potential of the sequence to convey information increases. Sprinkle in a few more and information potential increases further, reaching a maximum when **A**s and **T**s are at equal frequencies. Since the bases are alternatives to each other, we can express their frequencies in terms of just one of the bases. So, when **A%** is zero (i.e. **T%** is 100), information potential is zero. When **A%** is 50 (i.e. **T%** is 50), information potential is maximum. When **A%** is 100 (i.e. **T%** is zero), information potential is zero again.

Biological DNA sequences tend to display intrastrand parity between complementary bases, such as **A** and **T**, so that information potential is maximized (see Chapter 4). Thus, violations of Chargaff's second parity rule (i.e. violations of base equifrequency) tend to decrease the information potential of DNA sequences. "Violation" is a strong word, implying that something unwanted has happened. But whatever affects base parity in a DNA strand, might *itself* impart information of possible adaptive value to that strand. This might include information relating to secondary structure (Chapter 5), or to purine-loading (Chapter 6), or to DNA "accent" (Chapters 7 and 8). Thus, it is the *remaining* information potential that is decreased by a lack of parity.

This lack of parity can be quantitated simply in percentage terms (e.g. departures from 50% **A**), or can be related to conventional measures of information content – the binary digits or "bits" of the information scientist (see Chapter 2) [2]. Each unit scores as 1 bit in the case of a two unit sequence (e.g. a sequence containing an even mix of only two bases), 2 bits in the case of a four unit sequence (e.g. a sequence containing an even mix of four bases), and 4.322 bits in the case of a 20 unit sequence (e.g. a sequence containing an even mix of twenty amino acids; see Appendix 2). Thus, in a two base sequence, a segment containing only one of the bases would score as zero for each base, whereas a segment containing an even mix of both bases would score two for each base. In a two unit sequence, segments can be scored as to their information content on a scale from zero bits/base to two bits/base. Here we will mainly discuss measurement of information potential in percentage terms.

In various contexts, the quality measured can be referred to either as "uncertainty," "entropy," or "complexity." This also relates to "compressibility," since a simple sequence is easier to characterize in terms that allow a decrease in total symbol number (e.g. "**AAAAAAAAAA**" can be written in compressed form as "10**A**"). On this basis, one linear DNA molecule can be described as less complex (more compressible) than another. Furthermore, a given linear DNA molecule can be considered to consist of segments of varying complexity. Low complexity segments often have large deviations from the even parity of Chargaff's second parity rule (e.g. the frequencies of the Watson-Crick pairing bases approach 0% or 100%). High complexity segments often have small deviations from even parity (e.g. the frequencies of the Watson-Crick pairing bases approach 50%). However, a low complexity segment can have even parity. For example, **ATATATATAT** is of low complexity, but has even parity with respect to **A** and **T**, and **CGATCGATCGATCGAT** is of low complexity, but has even parity with respect to all four bases.

The New Bioinformatics

The "old" bioinformatics is largely concerned with aligning different sequences to determine the degree of similarity. If a newly discovered gene in biological species *A* is found to display a high degree of sequence similarity with a well-characterized gene in species *B*, then this may provide a clue to the function of the product of the new gene in species *A*. If many genes in species *A* show similar degrees of similarity to genes in species *B*, then it is possible that the genes are homologous (orthologous) and the species have diverged relatively recently (in evolutionary time) from a common ancestral species. Thus, the "old" bioinformatics is (i) a gene-product-orientated bioinformatics serving those who seek to identify and manipulate gene-products, and (ii) a phylogenetic bioinformatics serving those who seek to classify species on the basis of evolutionary relationships.

Confusing these tasks are low complexity segments, sometimes referred to as "simple sequences." Two genes that are not related evolutionarily may have *similar* simple sequences. These similarities are often sufficient to be picked up in alignment analyses. So simple sequences are distracters – an embarrassment that can generate spurious high-scoring alignments between genes that are not actually related. The parts of genes responsible for specific gene function usually correspond to specific sequences of high complexity. Accordingly, programs with names such as "Repeat masker" and "Seg" are widely used to identify low complexity segments and mask them prior to conducting alignment analyses (with programs with names such as "Blast," "Clustal," and "Genalign").

On the other hand there is evolutionary bioinformatics, which has as its primary concern the understanding of biological evolution. This "new" bioinformatics views simple sequences as having the potential to provide profound insights. The goals of the "old" and the "new" bioinformatics differ, but programs such as "Seg" can greatly aid the evolutionary bioinformaticist, as will be shown here.

Protein or Nucleic Acid Level Function?

Low complexity at the nucleic acid level correlates with low complexity at the protein level. For example, since the codon **TTT** in DNA encodes the amino acid phenylalanine (Table. 7-1), then a sequence containing only **T** bases [poly(d**T**)] would encode a protein containing only the amino acid phenylalanine (i.e. polyphenylalanine). On coming across a polyphenylalanine segment within a protein, we are faced with another chicken-and-egg problem. Would the polyphenylalanine have conferred some advantage to the function of the protein, with the automatic consequence that a **T**-rich nucleic acid sequence was selected in the course of evolution? Alternatively, did poly(dT) confer some advantage at the nucleic acid level with the automatic consequence that the protein sequence became phenylalanine-rich? Thus, a primary selective influence operating at one level could be the cause of low complexity at both levels.

Honghui Wan and John Wootton [3] noted for a wide range of organisms that **GC**-rich genomes and **GC**-poor (**AT**-rich) genomes have more low complexity segments in their proteins than genomes in which the quantity of **G+C** is approximately equal to the quantity of **A+T** (Figure 11-1).

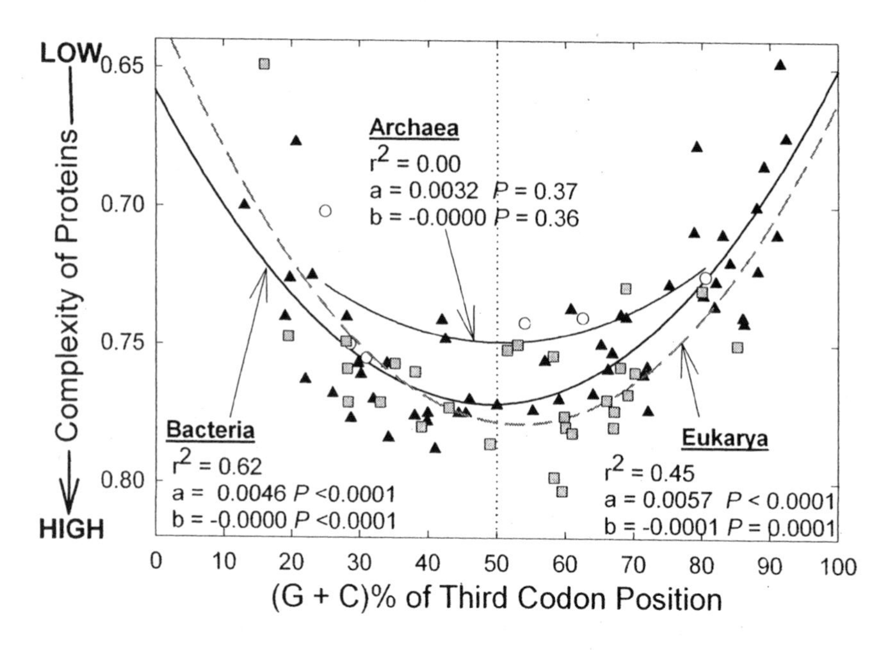

Fig. 11-1. Variation of protein complexity with species (**G+C**)% for three major groups of organisms: bacteria (triangles); archaea (circles); eukarya (squares). Each point corresponds to a particular species within a group. Values for protein complexity decrease from high to low on the Y-axis (note the *descending* scale). More high complexity proteins (fewer low complexity proteins) are found in species with 50% (**G+C**) and fewer high complexity proteins (more low complexity proteins) are found in species with extreme (**G + C**) percentages (low or high). This is exaggerated here by showing the base composition of the codon position that is least constrained by protein-pressure (third position). The points were fitted to curves (quadratic parabolas of general form $Y = Y_0 + aX + bX^2$) and parameters of the curves are shown in each figure with corresponding probabilities (*P*). The fit for archaea (archaebacteria) is not significant, but the curve seems to match those for bacteria (eubacteria) and eukarya (eukaryotes). The data are from a paper by Wan and Wootton [3], who measured the "median global complexities" of protein sequences from a variety of extensively sequenced genomes

While the cause of **GC**-pressure has been controversial, there is agreement that it operates genome-wide both on genes and on intergenic DNA. On the other hand, the pressure to encode a protein operates only on exons. Again, we are faced with a chicken-and-egg problem. If protein-pressure were driving the (**G**+**C**)%, then only exons should be affected. We might postulate that (**G**+**C**)% is somehow infectious and, once exons have arrived at a particular value, this spreads into introns and non-genic DNA. However, it is more likely that **GC**-pressure is itself primary (see Chapter 8). At the extremes – low or high – (**G**+**C**)% biases the amino acid composition of proteins towards certain amino acids, so generating low complexity segments in proteins (i.e. there may be departures from an even mix of twenty amino acids).

Nucleic acid pressures can be stronger than protein pressures, even to the extent that some amino acids may be mere "placeholders" (see Chapter 9). If a low complexity segment in a protein-encoding region of DNA primarily serves a nucleic acid level function, then the corresponding amino acid sequence in the encoded protein must follow this. On the other hand, if the low complexity segment primarily serves a protein-level function, then the nucleic acid sequence need only follow this with respect to the bases in codons that determine the nature of the encoded amino acid. These are the bases in first and second codon positions (see Chapter 7). Bases in third codon positions usually do not determine the nature of the encoded amino acid and so are often free to respond to pressures other than the pressure for having a particular amino acid at a particular position in a protein.

For example, although phenylalanine is encoded by **TTT**, it is also encoded by **TTC**. Accordingly, if selection operated at the protein level, a **TTC**-based polymer [poly(**TTC**)], might satisfy protein-encoding demands (protein-pressure) for polyphenylalanine just as well as a polymer containing a mixture of **TTC**s and **TTT**s, or just as well as a pure **TTT**-based polymer [poly(**T**)]. However, by virtue of the contained **C**, poly(**TTC**) might satisfy other pressures, such as **GC**-pressure, better than poly(**T**).

Base Pair Pressures

GC-pressure is one of a variety of pressures appearing to operate directly on nucleic acid sequences. By "base pair pressure" is meant a pressure due to a nucleic acid segment's content of a particular pair of bases (i.e. the segment's base *composition*, irrespective of base order). So we are not here concerned with 2-tuple (dinucleotide) frequencies (see Chapter 4). From four bases there are six combinations of two different bases, consisting of three mutually exclusive pairs (**GC**, **AT**; **AG**, **CT**; **AC**, **GT**). Implicit in values for (**G**+**C**)% in DNA segments are values for the reciprocating base pair, (**A**+**T**)%. Implicit in values for (**A**+**G**)% (purine-loading pressure) are values for the reciprocating base pair, (**C**+**T**)%. **AC**-pressure (reciprocating with **GT**-pressure) is currently an abstract entity of uncertain significance.

"Pressure" is a useful intuitive concept implying some force or tendency with a directional (vectorial) component. The terms "**GC**-pressure," "**AG**-pressure," and "**AC**-pressure" are used here to indicate evolutionary forces for *departures* from base equifrequencies. However, currently there is no strictly defined usage in the literature. Thus, in some contexts (see Chapter 9), the effect of "**GC**-pressure" (i.e. pressure for a change in (**G**+**C**)%) is considered to increase over the range 0% (**G**+**C**) up to 100% (**G**+**C**), while the effect of "**AT**-pressure" is considered to decrease correspondingly from 100% (**A**+**T**) down to 0% (**A**+**T**). Here we avoid the latter term (**AT**-pressure), which is implicit in the former (**GC**-pressure). In the present context the effect of "**GC**-pressure" is considered to increase both from around 50% (**G**+**C**) up to 100% (**G**+**C**) ("upward **GC**-pressure") and from around 50% (**G**+**C**) down to 0% (**G**+**C**) ("downward **GC**-pressure"). Similar considerations apply to **AG**-pressure and **AC**-pressure.

If you were "the hand of Nature" and wanted to increase two bases in the coding region of an existing mRNA from scratch, how would you go about it? Take purine-loading for example. The easiest option would be to take advantage of third codon position flexibility so that, despite the **RNY**-rule, all codons that can have a purine in this position acquire one (if they do not already possess one). Thus, in the case of leucine with six alternative codons of general form **YYY** and **YYR** (see Table 7-1), you would pick **YYR**. In the case of serine (**YYY**, **YYR** or **RRY**) there is room for some modulation according to the extent of the need for purine-loading. You could pick **YYR** or **RRY**. If selective use of codons with **R** in third positions did not suffice, or **RNY**-pressure was insistent, then you could see if the protein would remain functional when you changed from amino acids with **Y**-containing codons to amino acids with similar properties that had more **R** in their codons. For example, the small hydrophobic amino acid alanine (**RYN**) might be replaced by glycine (**RRN**), which is also small and hydrophobic.

Alternatively, you could try deleting an amino acid with a **Y**-rich codon to see if the protein could manage without it. In so doing you would achieve a relative enrichment for **R**. Or, you could insert an amino acid with an **R**-rich codon into some, perhaps innocuous, position in the sequence. Hopefully the protein would not mind. If the latter strategy succeeded, but the mRNA was still insufficiently purine-loaded, then you could try inserting more amino acids with **R**-rich codons next to the first one. Because many proteins consist of an assembly of independent functional domains, it is indeed possible to insert long segments of amino acids between these domains without interfering with overall protein function. Each inserted amino acid is a mere placeholder.

You could arrange that the amino acid composition of these inserted segments would be biased towards amino acids with **R**-rich codons. But, some amino acids with **R**-rich codons might be better than other amino acids with **R**-rich codons (i.e. would interfere least with the normal protein function). Which amino acids would be least offensive? Whatever amino acids you choose, the selective use of particular amino acids means that the sequence would be relatively simple. Would the order of the amino acids in the inserted, low complexity, simple sequence be important, or irrelevant? Now let's see how "Nature" has actually gone about the task. The genomes of Epstein-Barr Virus (EBV) and of *P. falciparum*, both under extreme **GC**-pressure (upwards and downwards, respectively), are most informative.

Epstein-Barr Virus Simple Sequence Element

EBV (human herpesvirus 4), the agent of infectious mononucleosis ("mono" or "glandular fever"), has various forms of latency in its human hosts. The initial infection may be asymptomatic. Most people are infected by the age of twenty. Almost certainly the virus is within you at this moment, but you will remain symptom-free. The latent state is associated with the transcription of only a few RNAs ("latency-associated transcripts") from the circular virus genome.

There are various types of latency, but all forms of latency have one RNA transcript in common – the transcript encoding the EBNA1 (Epstein-Barr Nuclear Antigen-1) protein. This contains a long simple-sequence element that can be removed experimentally without interfering with known protein functions. Since the EBV genome appears to be under pressure for compactness (i.e. there are few introns and little intergenic DNA), the simple-sequence element should only be present if, under some circumstance, it confers an adaptive advantage either at the protein level, or at the level of the encoding nucleic acid (either DNA or RNA). Indeed, an element that increases the length of a gene places the integrity of the gene in jeopardy, since it increases vulnerability to disruption by recombination (see Chapter 9). One function of the element appears to be the purine-loading of the corresponding mRNA to an extent that cannot be achieved, because of protein-encoding constraints, by the more complex (non-simple sequence) parts of the mRNA [4, 5].

The EBV genome has a relatively high **(G+C)**% and, consistent with the reciprocal relationship between **(G+C)**% and **(A+G)**% (Fig. 9-7), generally has a low **(A+G)**% in its genes. So the mRNA-synonymous "coding" DNA strands of most EBV genes are *pyrimidine-loaded*. The EBNA1-encoding gene is a notable exception. It is highly *purine-loaded*, mainly by virtue of a long low complexity segment (Fig. 11-2a). The reason for this special feature of the EBNA1-encoding gene will be considered in Chapter 12.

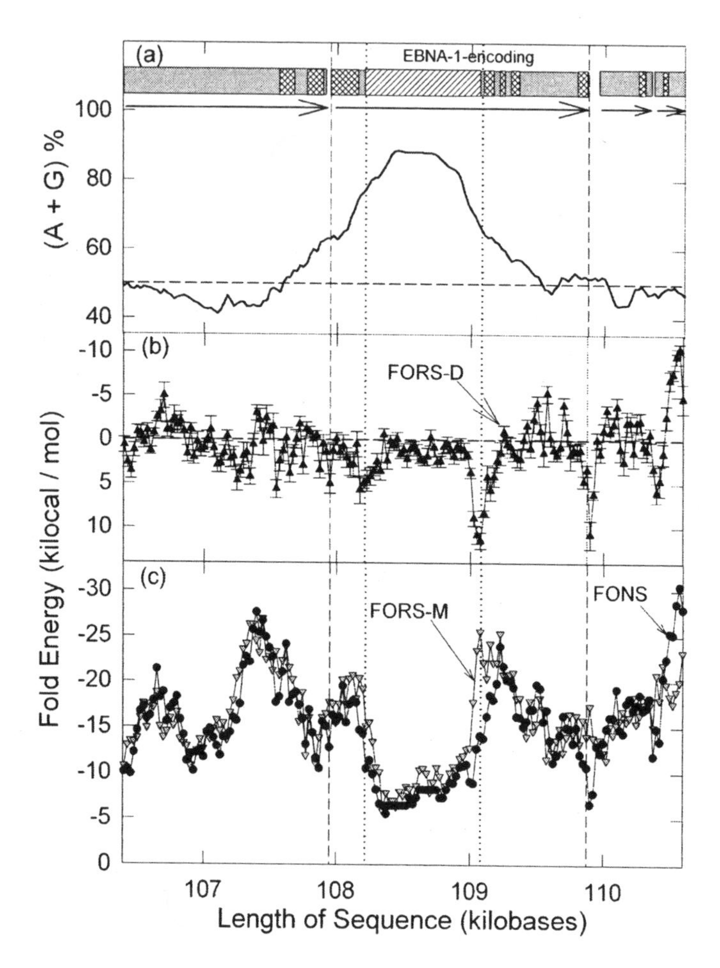

Fig. 11-2. The long low complexity element in the gene encoding the EBNA1 antigen of Epstein-Barr virus (EBV) is purine-loaded *(a)*, and has low folding energy *(b, c)*. In *(a)* four genes are shown as grey boxes with four horizontal arrows indicating transcription directions. Vertical dashed lines indicate the borders of the EBNA1-encoding gene. The region with the long low complexity, simple-sequence, repetitive element in the EBNA1 gene is shown as a white box with diagonal stripes, and is demarcated by vertical dotted lines. Other low complexity elements in the EBNA1 and neighboring genes are shown as crosshatched white boxes. In *(a)* purine-loading is shown as (**A**+**G**)% (continuous line). These values are greater than 50% in the EBNA1 gene (purine-loaded), and are less than 50% in neighboring genes (pyrimidine-loaded). In *(b)*, values for the base order-dependent component of the folding energy (FORS-D) are shown with standard errors. In *(c)*, values are shown for the base composition-dependent component of the folding energy (FORS-M), and for the total folding energy (FONS). Low complexity segments were detected using the Seg program

Different EBV genes, and hence the corresponding proteins, vary in their content of low complexity sequence. By eye you can tell from Fig. 11-2a that the EBNA1-encoding gene has about 70% low complexity sequence (see white boxes with diagonal or hatched lines). When values for percentage low complexity in individual EBV proteins are plotted against values for the various base pressures, points are widely scattered. However, they can be fitted to curves (quadratic parabolas) as shown in Figure 11-3.

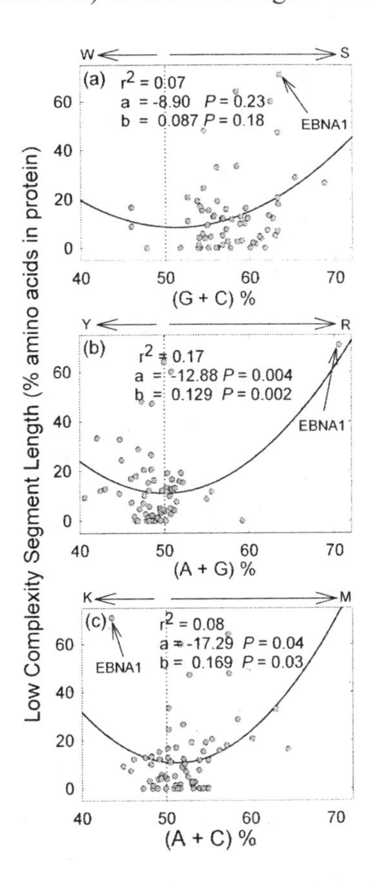

Fig. 11-3. Variation of low complexity segment length in EBV proteins with percentage content of *(a)* **G+C**, *(b)* **A+G**, and *(c)* **A+C**, in the corresponding genes. The point corresponding to the gene encoding EBNA1 protein is indicated with an arrow. The points were fitted to curves (quadratic parabolas) and parameters of the curves are shown in each figure. Horizontal arrows pointing to **W** and **S** indicate progressive enrichment for either the **W** bases (**A+T**) at low (**G+C**)%, or the **S** bases (**G + C**) at high (**G+C**)%. **Y** and **R** indicate the enrichment for either the **Y** bases (**C+T**) at low (**A+G**)%, or the **R** bases (**A+G**) at high (**A+G**)%. **K** and **M** indicate the enrichment for either the **K** bases (**G + T**) at low (**A+C**)%, or the **M** bases (**A+C**) at high (**A+C**)%

Most points are located near minima, at base compositions around 50%. Thus most genes have very little low complexity sequence (i.e. they are mainly of high complexity, with equifrequencies of bases that are irregularly ordered). The exceptional low complexity value, corresponding to the EBNA1-encoding gene, is at one extreme of the curves [6]. The scatter of points, either to the left or right of the vertical dotted line (marking 50%), shows the general tendency of EBV genes towards **S**-loading (i.e. high **G**+**C**; Fig. 11-3a), and **Y**-loading (i.e. low **A**+**G**; Fig. 11-3b). The maverick EBNA1 gene is both highly **S**-loaded and highly **R**-loaded. The plots become more informative when the base pressures exerted by different codon positions are plotted independently (Fig. 11-4).

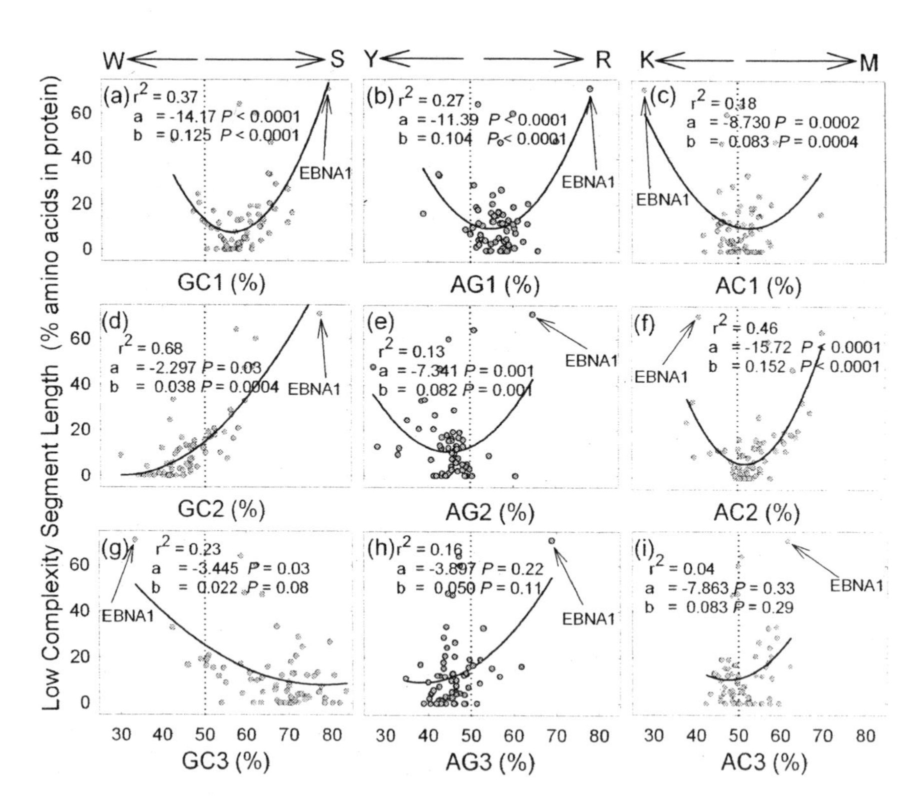

Fig. 11-4. Variation of low complexity segment length in EBV proteins with the content of either **G** + **C** *(a, d, g)*, **A** + **G** *(b, e, h)*, or **A** + **C** *(c, f, i)*, at first *(a, b, c)*, second *(d, e, f)* and third *(g, h, i)* codon positions (e.g. on the X-axis **GC**1 refers to the (**G**+**C**)% of first codon positions). Details as in Figure 11-3

By comparing the different plots, sense emerges from an apparently bewildering scatter of points. The amino acid-determining first and second codon positions tend to be the best predictors of the amount of low complexity segment (i.e. points for first and second codon positions best fit the curves, so that you would have more faith in predicting the degree of complexity from values for first and second codon positions, than from values for third codon positions). This can be determined by examining the various values shown in each figure.

Values for the square of the correlation coefficient (r^2) can range from zero to one and indicate how much of the observed variability (i.e. 0% to 100%) can be attributed to low complexity segment content (e.g. 68% in the case of the **G+C** content of the middle bases of codons as in Figure 11-4d, and 27% in the case of the **A+G** content of the first bases of codons as in Figure 11-4b). Similarly, the equations of quadratic parabolas have two key determinants (a and b); the probability values (*P*) associated with these (ranging from zero to one) indicate how likely it is that the match is *not* a good one (i.e. low *P* values indicate a good match of points to the curve). Third codon positions.– those least likely to relate to the encoding of amino acids – tend to have the highest *P* values (i.e. there is a relatively *poor* match of points to the curve; Figs. 11-4g,h,i).

With the caveat given in Chapter 7, that in some species third codon positions can affect the speed or accuracy of translation, third positions can be regarded as controls (i.e. they provide a frame of reference). Like non-genic DNA segments, *whether they are part of low or high complexity segments, they should be equally free to respond to pressures other than the encoding of amino acids*. Thus, the base compositions of third codon positions (X-axes) should be relatively independent of the quantity of low complexity segment (Y-axes). So, as expected, values for r^2 are low and values for *P* are high.

On the other hand, first and second codon positions, when in *high* complexity segments primarily play the amino acid-encoding role and *are restrained* from responding to other pressures. But when in *low* complexity segments first and second codon positions, if encoding mere placeholders, *would not be so restrained*. Consistent with this, the low complexity segment in the EBNA-1-encoding gene appears free to serve other pressures by virtue of the base content not only of third codon positions, but also of the first and second codon positions.

The flexibility of third codon positions would have been insufficient to meet the demands of pressures other than the encoding of proteins, and so first and second codon positions would have been *recruited* to assist. In simple-sequence elements, this assistance would not be to the detriment of the amino acid-encoding role of first and second codon positions, because the nature of the encoded amino acids would be largely irrelevant to the function of the protein.

We now go back to an earlier figure to consider secondary structure. In the DNA encoding the EBNA-1 repetitive element there are up to 800 purines per kilobase so that there are relatively few pyrimidines to pair with purines for formation of the stems of DNA stem-loop secondary structures. Thus, the potential for forming secondary structure is greatly decreased as measured by the FONS value (Fig. 11-2c). As discussed in Chapter 5, the latter value decomposes into a base composition-dependent component (FORS-M; Fig. 11-2c) and a base order-dependent component (FORS-D; Fig. 11-2b). Compared with neighboring regions, base order contributes poorly to the folding potential (assessed as negative kilocalories/mol) in the region of the repetitive element. Indeed, base order supplements base composition in *constraining* the folding potential (so most FORS-D values, being slightly below the horizontal dashed line, are positive; Fig. 11-2b).

Since secondary structure in DNA is likely to enhance recombination (see Chapter 8), this predicts decreased recombination between regions of the gene corresponding to the amino-terminal and carboxy-terminal ends of the EBNA1 protein. The gene would otherwise be more vulnerable to fragmentation by recombination, due both to the increase in gene length resulting from the presence of the low complexity element, and to the increased probability that a viral low complexity element will find a homologous recombination partner (i.e. another low complexity element with a similar sequence) elsewhere in its own, or the host cell's, genome.

Apart from the low complexity *repetitive* element in the EBNA1 protein, there are other *non-repetitive* low complexity elements (Fig. 11-2a). Similar elements are encoded in neighboring genes. Thus, the immediate upstream gene encodes two small elements. The two immediate downstream genes each have only one small element. However, in many respects the EBNA1-encoding gene is different from all other EBV genes, most of which are not expressed during latency. Indeed, there are more similarities with genes of the parasite responsible for the most malignant form of malaria, *Plasmodium falciparum*.

Malaria Parasites

Low complexity segments with biased amino acid composition are abundant in the proteins of *P. falciparum*. In Italy, Elizabetta Pizzi and Clara Frontali wondered whether these low complexity protein segments have special properties that provide a functional explanation for their existence [7]. They compared the properties of amino acids in low complexity sequences with the properties of amino acids in high complexity sequences. The amino acid biases in low complexity segments did not correlate with properties of individual amino acids (hydrophobicity, volume, flexibility). However, a correlation was found with the A content of the codons. There was a correlation, not at the protein level, but at the nucleic acid level.

In the extremely **AT**-rich malaria parasite genome the third codon position is dominated by intense downward **GC**-pressure, making it one of the most **AT**-rich eukaryotic species (see Fig. 9-5a). Being less influenced by the nature of the encoded amino acid, third codon positions have more flexibility to respond to **GC**-pressure and **AG**-pressures. However, this flexibility can be insufficient to meet the demands both of downward **GC**-pressure and of upward **AG**-pressure (purine-loading pressure). So first and second codon positions are recruited. This threatens to change the nature of the encoded amino acid, which could then disturb protein function. So, again, a solution is to create extra protein segments that lie between functional protein domains. These extra segments should neither play a role in, nor impede, normal protein function. The amino acids in these segments are mere place-holders. Thus, the low complexity segments represent evolutionary adaptations allowing genes, through the compositions of first and second codon positions (as well as third codon positions), to respond to extreme base compositional demands that could not be met by high complexity segments alone (despite their flexibility at third codon positions).

Consistent with this, as in the case of EBV, the quantity of low-complexity sequence correlates well with the base composition of first and second codon positions, but less well with that of third codon positions. Figure 11-5 shows that the percentage content of low complexity segment in malaria parasite genes relates to the (**G+C**)% of each gene in a way that is dependent on codon position. By eye you can tell, from the clustering of points relative to the X-axis, what average (**G+C**)% values for different codon positions are likely to be. Average (**G+C**)% values become progressively less as codon position changes from first position (Figs. 11-5a) to third position (Figs. 11-5c). The extreme value of the third position (around 18 % **G+C**) indicates that it has the most flexibility to respond to downward **GC**-pressure. This is largely *independent* of the percentage of low complexity segment (indicated by the low r^2 value, and high P values in Figure 11-5c). Thus, whatever their content of low complexity sequence, on the basis of their third codon position (**G+C**)% values genes contribute strongly to downward **GC**-pressure.

On the other hand, data points for first and second codon positions fit to curves with minima corresponding to genes with low percentages of low complexity segment (Figs. 11-5a, b). Most data points relate to the leftward limbs of the curves. For first and second codon positions the curves show that acquisition by certain genes of low (**G+C**)% values correlates with an increased content of low complexity segment. The more **AT**-rich a gene, the more low complexity segment it contains. In the case of second codon positions 18% of the variation can be explained on this basis ($r^2 = 0.18$).

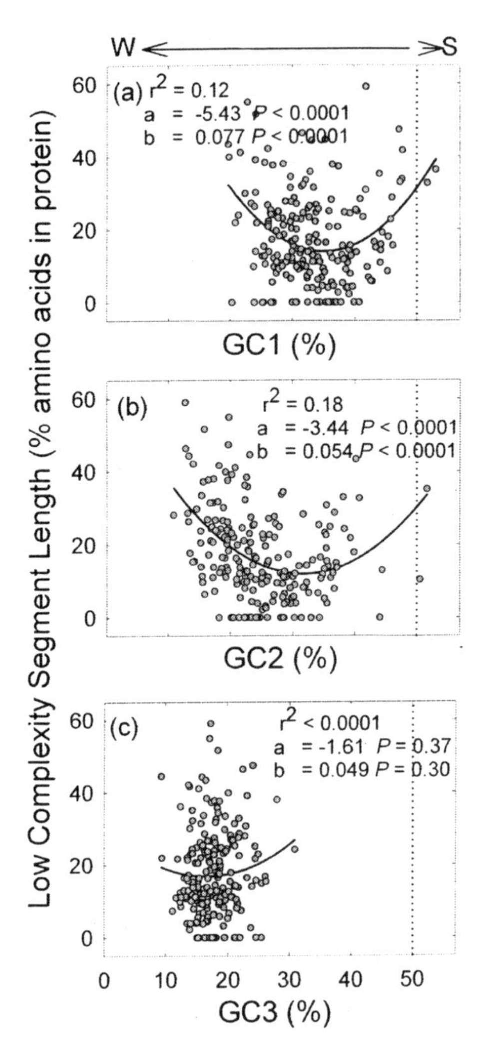

Fig. 11-5. Variation of low complexity segment length in the proteins encoded by chromosome 2 of *P. falciparum* with the (**G** + **C**) content of first *(a)*, second *(b)* and third *(c)* codon positions (expressed as **GC**1, **GC**2 and **GC**3 on the X-axis). Parameters of second order linear regression plots are shown in each figure. Horizontal arrows pointing to **W** and **S** indicate the enrichment for either the **W** bases (**A** + **T**) at low values of (**G**+**C**)%, or the **S** bases (**G** + **C**) at high values of (**G**+**C**)%

Similar results are seen in the case of **AG**-pressure (Fig. 11-6). Departure from equifrequency of bases (50% purine: 50% pyrimidine) is greatest in the case of first codon positions. Average (**A**+**G**)% values become progressively less as codon position changes from first position (Figs. 11-6a) to third position (Figs. 11-6c). This is consistent with the **RNY**-rule (purine for first codon base, pyrimidine for third codon base; see Chapter 7). Values for first and second codon positions can best be fitted to the curves; most data points relate to rightward limbs (Figs. 11-6a, b), indicating a positive correlation between a gene's purine content (i.e. upward **AG**-pressure) and its content of low complexity segment. In the case of second codon positions, 28% of the variation can be accounted for on this basis ($r^2 = 0.28$).

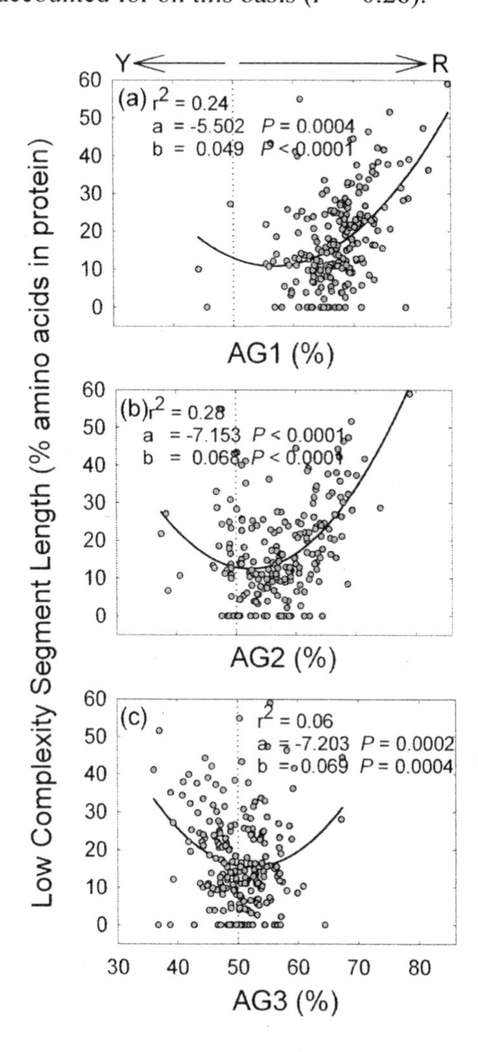

Fig. 11-6 Variation of low complexity segment length in proteins encoded by chromosome 2 of *P. falciparum* with the (**A** + **G**) content of first *(a)*, second *(b)* and third *(c)* codon positions. Horizontal arrows pointing to **Y** and **R** indicate the enrichment for either the **Y** bases (**C** + **T**) at low values of (**A**+**G**)%, or the **R** bases (**A** + **G**) at high values of (**A**+**G**)%

Values for third codon positions show poor correlations (i.e. $r^2 = 0.062$; Fig. 11-6c). So, as expected, the distinction between genes with low and high percentages of low complexity segment length is least for this position. Note that the correlations (r^2 values) in the case of purine-loading (Figs. 11-6a, b) are better than in the case of **GC**-loading (Figs. 11-5a,b). This is consistent with the notion that purine-loading is usually the dominant pressure with respect to increasing the length of a protein by insertion of low complexity elements (see Chapters 9 and 10).

Conflict with Fold Potential

The long low complexity sequence in the EBNA1 gene appears to countermand the potential of the encoding DNA duplex to extrude single-stranded DNA that adopts a stable stem-loop secondary structure (Fig. 11-2). Low complexity sequences in *P. falciparum* genes might be expected to act similarly – but not necessarily so. A prominent example is the gene encoding the circumsporozoite protein (CSP; Fig. 11-7). This is expressed on the form of the malaria parasite (sporozoite) that is initially injected by infected mosquitoes and enters liver cells. A host preimmunized against parts of the CSP might be able to nip the infection in the bud.

So we should not be surprised to learn that the two end segments of the protein (N-terminal and C-terminal segments) correspond to CSP antigens that vary between different individual parasites (i.e. are polymorphic) and are evolving rapidly under positive Darwinian natural selection. There is one low-complexity, non-repetitive element, and a central low-complexity simple sequence element encoding repeats of eight amino acids (the octapeptide NVDPNANP) or four amino acids (the tetrapeptide NANP; Fig. 11-7a). Downstream of the single exon CSP-encoding gene is a three exon gene encoding an unidentified (therefore referred to as a "hypothetical") protein. Here the upstream half of the small middle exon has a low complexity, non-repetitive, element.

The codons of the low complexity segments in the CSP-encoding gene tend to be **A**-loaded, and contribute to the generally high level of purine-loading (i.e. the curve in Fig. 11-7a is mainly above 50% (**A+G**) in the region of the gene). In regions corresponding to the N- and C-terminal segments this is achieved mainly by **A** bases, with a small contribution from **G** bases (the two bases are not shown separately in the figure). For the middle segment this is achieved entirely by **A** bases, and there is an *excess* of **C** compared with **G** (not shown in the figure). The inclusion of the low-complexity regions can be considered as a device to enhance purine-loading, as in the case of the EBNA1 antigen (Fig. 11-2). However, the fact that **G** works *against* this suggests other factors are in play.

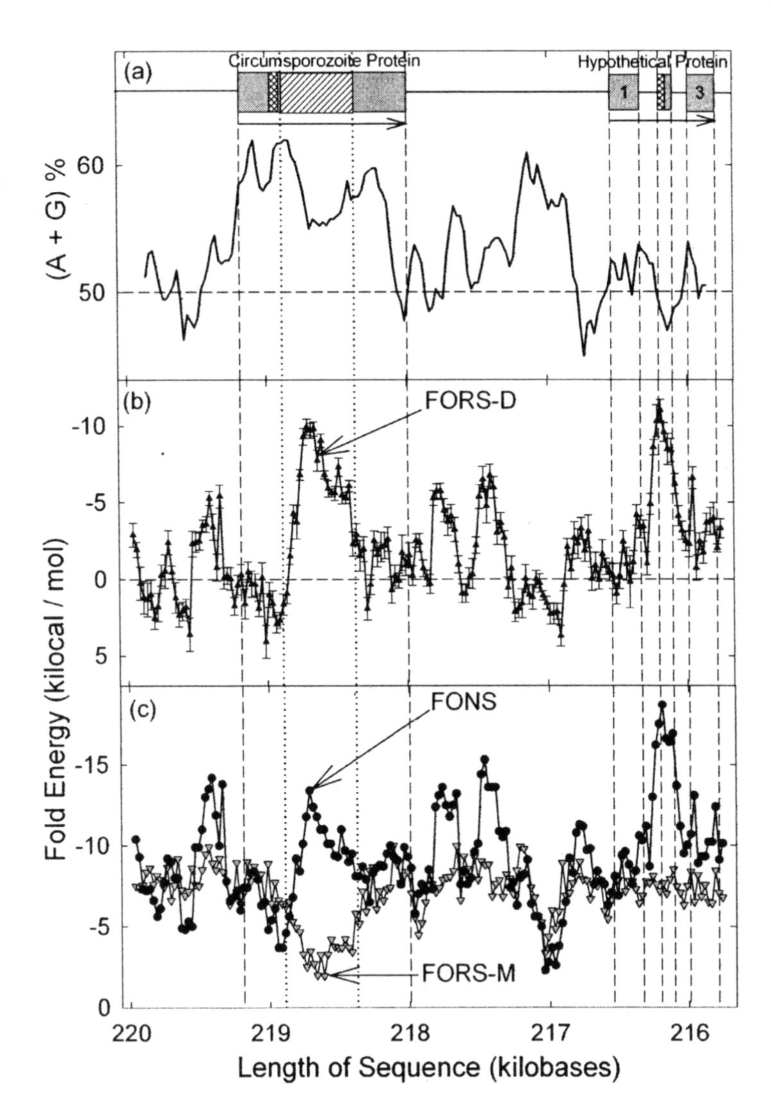

Fig. 11-7. The long low complexity element of the *P. falciparum* circumsporozoite protein (CSP) -encoding gene supports two nucleic acid level functions – purine-loading, and base order-dependent stem-loop potential. Details are as in Figure 11-2. Vertical dotted lines indicate the boundaries of the long low complexity, simple-sequence, repetitive element. Vertical dashed lines indicate the exon borders of the single exon CSP gene and of a three exon gene corresponding to an unidentified "hypothetical" protein. Note, from the genome sequence numbering on the X-axis, that the sequence was reversed so that the "bottom" strand is the analyzed strand (mRNA-synonymous coding strand) and genes formally designated in GenBank as transcribing to the left are shown as transcribing to the right

Figures 11-7b and 11-7c show the potential of windows in the segment to form secondary structure. The base order-dependent component (FORS-D) is impaired (i.e. tends to be positive) in regions of the CSP gene corresponding to the N- and C-terminal segments of the protein; this is characteristic of sequences under positive selection [8]. Consistent with the high purine-loading of the repetitive middle segment, the base *composition*-dependent component of the folding potential (FORS-M) is greatly *decreased* in the middle segment (Fig. 11-7c); but this is more than compensated for by an *increase* in the base *order*-dependent component (FORS-D; Fig. 11-7b). Hence, the natural sequence shows a large increase in overall stem-loop potential (FONS) in this region (Fig. 11-7c). This suggests that recombination between the two ends of the gene might be enhanced (see later).

In general, stem-loop potential in genes tends to be relegated to introns, especially in the case of genes under positive selection where introns may be more conserved than exons (see Chapter 10). In the case of the CSP gene, what might have been an intron appears to have been retained and expressed in the protein sequence, perhaps because of an adaptive value at the mRNA level (i.e. purine-loading) and/or at the protein level.

Similar considerations may apply to the second exon of the downstream gene. Whereas there is purine-loading (i.e. **AG**% greater than 50) both in the single exon CSP gene and in the first and third exons of the downstream gene (i.e. Szybalski's transcription direction rule is obeyed), the second exon is pyrimidine-loaded (i.e. **AG**% is less than 50). Much of the second exon consists of a low complexity element, which strongly supports base order-dependent stem-loop potential.

Roles of Low Complexity Segments

Since low complexity is often associated with departure from base-equifrequencies, it is not surprising that proteins with high percentages of low complexity segments correspond to the extremes of base pair percentages (Fig. 11-1). The issues are whether low complexity results from selective pressures, and if so, whether these pressures operate primarily at the nucleic acid level, or at the protein level. A sequence primarily generated in response to a nucleic acid level pressure might secondarily have acquired a protein level function, thus making it difficult to determine which was primary. A clear assignment of nucleic acid level roles can assist this task.

Unlike the first and second codon positions, the base composition of third codon positions tends not to correlate with the percentage of low complexity segment in proteins, especially in the case of **(G+C)**%. This suggests that, for this position, low and high complexity segments are equally free to respond to base compositional pressures. First and second positions are constrained by other pressures. In high complexity protein segments, mainly corresponding to conserved globular domains, these pressures include pressure to retain specific protein function (i.e. protein-encoding pressure). On the other hand, the encoding of amino acids by low complexity segments is largely determined by the base compositional pressures. Such segments mainly correspond to non-globular hydrophilic surface domains that do not play a critical role in a protein's structure and specific function.

Yet, some protein-level function might still be of value. In the case of the CSP gene, association with enhanced stem-loop potential (Fig. 11-7) would be expected to promote recombination and hence enhance CSP polymorphism, a feature conferring protection against host immune defenses. The long low complexity segment in the gene encoding the EBNA1 protein does not enhance stem-loop potential (Fig. 11-2), so probably functions solely as a means of purine-loading the mRNA to avoid formation of double-stranded RNA, which would alert the host's immune system (to be considered in Chapter 12).

Simple Sequence Repeats

The low complexity elements contain either non-repetitive, or repetitive, sequences of amino acids, the latter being prevalent in long low complexity elements. For example, the long low complexity element in the EBNA1 protein consists mainly of a repeat of two glycines and an alanine. If Nature's purpose is merely to purine-load, then there should be no particular virtue in having placeholder amino acids *ordered* in this way:

GlyGlyAlaGlyGlyAlaGlyGAlaGlyGlyAla... (11.1)

Surely, a disordered sequence might serve just as well – for example:

GlyAlaAlaGlyGlyGlyAlaGlyGlyAlaGlyGly... (11.2)

The ordering (periodicity) might somehow help the insert remain innocuous with respect to the function of the protein. Hence it would be selected for (i.e. natural selection would prefer viruses with ordered amino acids in the low complexity element, over viruses with disordered amino acids in the low complexity element). However, in some cases the ordering can be shown to depend on the "mechanical" (i.e. chemical) process by which the repeat was first generated. In fact, the process can get out-of-hand and the insert can grow so long that it begins to interfere with protein function, often by making the protein insoluble. Clear examples of this are found in the "trinucleotide expansion" diseases, which often affect central nervous system function, and include the human brain disorder known as Huntington's disease (Fig. 11-8).

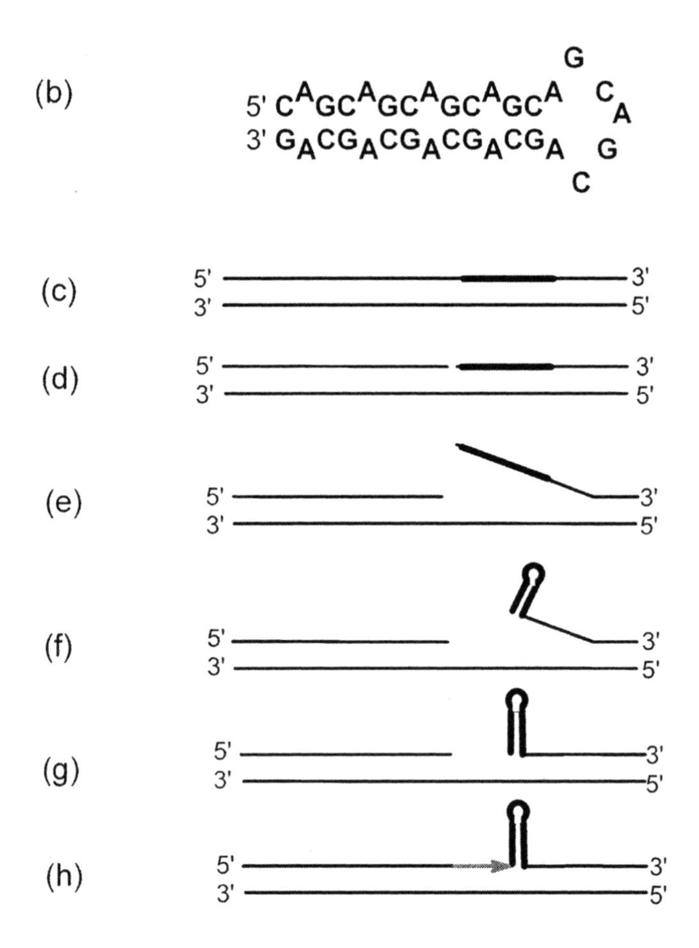

Fig. 11-8. A mechanism for trinucleotide repeat expansion in Huntington's disease. *(a)* Tandem repeats of **CAG** that encode polyglutamine. *(b)* Ability of the repeats to form a stem-loop structure by virtue of complementarity between the strongly pairing **C** and **G** bases. *(c)* A DNA duplex containing a tract of **CAG** repeats (thickened line) in the "top" mRNA-synonymous strand. This tract complements a tract of **CTG** repeats in the "bottom" mRNA-template strand. Initially both tracts would be short. *(d)* Single-strand breakage. *(e)* Temporary displacement of the top strand. *(f)* The **CAG** repeat adopts a stem-loop configuration. *(g)* Restoration of the duplex. *(h)* Restoration of top-strand continuity (grey arrow) by consecutive actions of a DNA polymerase (which introduces more top-strand **CAG** repeats using bottom-strand **CTG** repeats as templates), and DNA ligase. Thus the **CAG** pattern is reiterated. When this "repaired" duplex is later replicated, one of the child duplexes has an increased length of the **CAG** tract

In Huntington's disease there is a long repeat of the trinucleotide **CAG** in the "top" mRNA-synonymous strand of a protein-encoding exon. The corresponding mRNA thus contains a long **CAG** tract, which is in the codon reading frame (i.e. it is read as **CAG**, not as **AGC** or **GCA**; Fig. 11-8a). Since **CAG** is a codon for the amino acid glutamine, the protein contains a long polyglutamine tract. The **CAG** tract in the DNA is likely to have expanded because it can form an internal stem-loop structure (Fig. 11-8b). The stem can form because of the appropriate ordering of the **C** and **G** bases. This periodicity is conducive to stem-formation. The **A** bases are displaced and do not contribute to the stability of the structure. A possible mechanism for the expansion, believed to operate at the level of post-mitotic and post-meiotic haploid gametes, is shown in Figures 11-8c-h [9].

In normal humans the gene that, when changed, causes Huntington's disease, has a short **CAG** tract (less than 84 bases). This can be presumed to satisfy the normal need for purine-loading. However, as shown in Figure 11-8 the tract can increase in length. In fact, the longer it gets, the more likely it is to adopt the stem-loop structure, so facilitating a further expansion. Once it reaches a critical length (greater than 84 bases) disease features begin to appear. Initially, these features appear late in life, so that there may have been time to produce children. Yet, within a generation, the tract can expand, so in these children the tract may be longer than in the affected parent. This causes the disease to be more severe and to begin earlier in life.

It is easy to assume that Huntington's disease is caused by the insoluble protein aggregates that are observed in brain cells. Such aggregates consist of the polyglutamine tract-containing protein in association with a variety of other coaggregated normal proteins. The *same* set of normal proteins consistently coaggregate with the polyglutamine tract-containing protein [10]. Many members of this set show *specificity* for the polyglutamine-tract containing protein (i.e. they do not just coaggregate with any protein that happens to be aggregating). Why do, say, proteins A, Q and Y coaggregate, but not proteins G, L and V? This is a fundamental point that we will consider in Chapter 13.

Note, however, that the **CAG** tracts can adopt stem-loop secondary structures with long stems not only at the DNA level, but also at the RNA level. The lengths of double-stranded RNA so generated should be sufficient to trigger intracellular alarms, thus activating the immune system [11, 12]. Perhaps, double-stranded RNA is responsible for (or is a necessary cofactor in) the development of Huntington's disease? Consistent with this, in some other trinucleotide expansion diseases that affect the nervous system, (e.g. spinocerebellar ataxia) the expansions are not in protein-encoding regions. The pathogenic agents appear to be RNA transcripts [13]. We will consider this further in Chapter 12.

Many diseases associated with intracellular, and sometimes extracellular, protein aggregates are brain diseases that slowly progress over many years. Notable examples are the "prion" diseases, such as "mad cow disease," and Alzheimer's disease. An obvious explanation would be that the proteins are brain-specific. Yet, often the proteins are not brain-specific. The proteins are expressed in non-neural tissues, but aggregates are not seen and tissue function is not impaired. Perhaps of relevance is the fact that the brain is "immunologically privileged," being partially protected from the immune system by membranes ("blood-brain barrier"). This raises the possibility that the immune system can recognize cells with aggregates in non-neural tissues and destroy them. The cells can then be replaced by the division of neighboring cells or of "stem cells" (whose primary role is replacement of "worn out" cells). But somehow this is not possible in neural tissues (see Chapter 13).

Long-Range Periodicities

In Chapter 7 we encountered short-range **RNY** periodicities at the codon level. Indeed, in protein-encoding regions the level of every third base tends to remain constant (Fig. 8-4). Here we have encountered the local periodicities displayed by low complexity elements, whether they are in coding or non-coding regions. Are there long-range periodicities?

By virtue of their structure we sometimes encounter long-range periodicities in written texts. In this book every 10-20 pages the word "Chapter" appears in enlarged type with a number following it. This also applies to DNA texts. For example, as mentioned in Chapter 6, the DNA double-helix is normally slightly unwound (negative supercoiling); where other pressures do not countermand, every 10-11 Watson-Crick base pairs there are dinucleotide sequences ("bendability signals") that support this. The signals are weak but the periodicity becomes quite clear when long sequences are examined [14, 15]. The number of base pairs corresponds to about one turn of the double-helix (Fig. 2-1). Periodicities occurring at intervals of 200 bases, 400 bases, and even as long as 500 kilobases, have been related to aspects of chromosome structure [16, 17].

Summary

Protein segments that contain few of the possible twenty amino acids, sometimes in tandem repeat arrays, are referred to as containing "simple" or "low complexity" sequence. Many proteins of the malaria parasite, *P. falciparum*, are longer than their homologues in other species by virtue of their content of such low complexity segments that have no known function; these are interspersed among segments of higher complexity to which function can often be ascribed. If there is low complexity at the protein level then there is low complexity at the corresponding nucleic acid level (often seen as a departure from equifrequency of the four bases). Thus, low complexity may have been selected primarily at the nucleic acid level and low complexity at the protein level may be secondary. The amino acids in low complexity segments may be mere placeholders. In this case the amino acid composition of low complexity segments should be more reflective than that of high complexity segments on forces operating at the nucleic acid level, which include **GC**-pressure, **AG**-pressure and fold pressure. Consistent with this, for amino acid-determining first and second codon positions, low complexity segments show significant contributions to downward **GC**-pressure (revealed as decreased percentage of **G+C**) and to upward **AG**-pressure (revealed as increased percentage of **A+G**). When not countermanded by high contributions to **AG**-pressure, which locally decrease fold potential, low complexity segments can also contribute to fold potential. Thus they can influence recombination within a gene. These observations have implications for our understanding of malaria, infectious mononucleosis, and brain diseases in which protein aggregates accumulate.

Part 5 Conflict between Genomes

Chapter 12

Self/Not-Self?

> "Thrice is he armed that hath his quarrel just,
> But four times he who got his blow in fust."
>
> Josh Billings, His Sayings (1866) [1]

Given the "struggle for existence" an organism might hope to outshine its competitors by virtue of the excellence of its positive characters (e.g. greater speed, longer neck). But an organism that could also (i) identify, defend against, and attack its foes, and (ii) identify and support its friends, would seem to have an even greater advantage. To identify is to allow discrimination – friend or foe? – the former classification being non-dangerous (potentially unharmful), the latter classification being dangerous (potentially harmful). For this, friend and foe must have distinguishable characters and the organism must both detect these characters, and know which characters are likely to associate with non-danger (friend) and which characters are likely to associate with danger (foe). After this discrimination, there must then be a response (either non-alarm, or alarm) that, in turn, would lead to a variety of other responses of an adaptive nature (e.g. escape from, or destroy, the foe).

All this is a rather long-winded way of saying that successful organisms are street-smart and look out for number one! However, there are semantic pitfalls that can easily confuse. For example, it is easy to believe that the initial act of character recognition is the same as danger recognition. But an organism does not initially recognize danger, it recognizes and classifies characteristics (attributes, signals) in a source organism, and on the basis of this discrimination draws conclusions regarding the potential harmfulness of that source.

When grazing sheep hear a sound in the wood they know it is not from one of themselves. So the sound has arisen from not-self. The first discrimination is a self/not-self decision. In one step the sheep detect the sound and register it as not-self. The not-self signal may alone suffice to make them move away from the wood as a precautionary measure. If the source of the sound is subsequently seen with the attribute "shepherd," the sheep relax. If the attribute is "wolf," then alarms sound (i.e. bleating and running). The initial *recognition* (detect and register as not-self or self) and *classification* (friend or foe) events are usually quiet. The subsequent alarm is noisy – an amplification of the initial quiet signal. The alarm itself will be independently recognized and responded to; but this is secondary, and does not require the specificity of the quiet primary recognition event. It is true that an individual deaf sheep may respond only when it sees its companions running. So it is responding primarily to a general alarm signal. But the primary event in the sheep collective is the recognition of the not-self signal. The intuition to move away on this basis alone makes sense. Even if the source turns out to be the shepherd, one day they are going to be mutton, and the shepherd is going to have a hand in this!

An added complication, with which we are here most concerned, is that the foe may be within an organism's own body. Here, the characters recognized, and the sensors that do the recognizing, are likely to be chemical (e.g. receptors that should bind "not-self" molecular groupings, but should not bind "self" molecular groupings). Irrespective of its location, be it external or internal, the foe is part of the environment to which the organism has had to adapt through the generations. Likewise, the foe has had to counteradapt.

Thus, the full understanding of the genome of any biological species requires an understanding of the genomes of the species with which it has coevolved. A virus, for example, is not going to just sit around and let itself be tagged as "foe." It is going to exploit its power of rapid mutation to fool its host into believing it is "self." Step by step, the virus will mutate to look more and more like its host, and if each step confers an advantage it will continue on this path. If, however, as it continues it encounters progressively stiffer host defences, then it will discontinue. Thus, hosts whose immunological forces are poised to attack "near-self" versions of not-self (a subset of not-self), rather than not-self *per se* (the entire set of not-self – formidable in range), should be at a selective advantage (see Chapter 13).

Homology Search

Human sexual reproduction is an example of a multistep discriminatory process, the sensitivity of which increases step by step, culminating in a homology search that is the final arbiter of "friend" or "foe." First your mother, possibly discriminating among a number of suitors, chose your father. Then, one of his spermatozoa was recognized as "self" (a "friend" to the extent that it was from a member of the same species) by receptors on the surface of your mother's egg (ovum). Accordingly, your father's spermatozoon was able to enter your mother's ovum to form a zygote. Next, through their gene products, your parental genomes attempted to work with each other to enable the single zygotic cell to multiply and develop into you, an adult. For this there had to be an absence of not-self discrimination (i.e. any one maternally-encoded product might have failed to cooperate with one or more paternally-encoded products, leading to the developmental failure known as "hybrid inviability"). All this must have proceeded satisfactorily or you would not now be reading this.

But you should not be complacent. The production of gametes (spermatozoa, ova) tends to be an on-going process. Depending on your sex and stage of life, the chances are that at this very moment in your gonad your parental genomes are repeatedly completing the initial act of conjugation your parents themselves initiated decades earlier [2, 3]. The genomes appear to conjugate during the process known as meiosis, which results in the production of new gametes. In the meiotic minuet your 22 maternally-derived autosomal chromosomes pair with your corresponding 22 homologous paternally-derived autosomal chromosomes, and 1 maternally-derived sex chromosome (X-chromosome) pairs with 1 paternally-derived sex chromosome (X if you are female, Y if you are male). Between each pair of homologues there is a search for similarity ("homology search"). A high or moderate degree of similarity of base sequence registers as "self" (i.e. "friend") and the dance continues (the usual situation). However, below a certain threshold value for similarity, the base sequence registers as "not-self" (i.e. "foe"). The music stops and meiosis, and hence gamete production, fails. So, although unlikely, you could be sterile (a "mule").

The latter sterility ("hybrid sterility") should not be regarded as unnatural since, as related in Chapter 8, through variations in DNA sequences Nature confers sterility barriers on some species members; this allows branching into new species whose members, though natural selection, can explore phenotypes that were not possible within the confines of the parent species.

Antibody Response

The human embryo gains its nourishment from its mother by way of the placenta. But the embryo has a paternal component that is foreign to the mother. For this reason the evolution of placentation has required that there be barriers preventing an immunological response by the mother against the embryo. In theory, this would require that paternally-encoded immunogenic macromolecules not access the maternal immune system, and/or that maternal antibodies, which can bind those macromolecules, not be able to penetrate the placental barrier.

Initial formation of the zygote requires that the integrity of one of the mother's cells, the ovum, be breached by a paternal cell (spermatozoon). The ovum is usually permissive for this invasion by a specialized foreign (not-self) cell, as long as it is derived from a member of the same species. Should there be an initial intracellular self/not-self recognition event, the usual outcome would be that the paternal genome would be classified as friend rather than as foe (i.e. at this level of discrimination the paternal genome would be registered as likely to be unharmful).

However, the invading of female ova by foreign male spermatozoa of the same species, is a very special case. Organisms and their cells may become hosts to foreign organisms of a different species (viruses and bacteria). Molecules, or parts of molecules, which can mark an organism as self or not-self are referred to as antigens, or antigenic determinants. The immune recognition events elicited by these can occur extracellularly and, probably, also intracellularly.

An organism is said to have undergone a specific antibody response when, within its extracellular body fluids, there is an increase in the concentration of protein antibodies able to combine specifically with eliciting antigens. Antibodies function extracellularly and have an intracellular existence only within the cells of the lymphoid system where they are produced; they are rapidly secreted from these source cells and encounter antigens while remaining within the confines of the body of the source organism. This means that usually antibody molecules from one organism do not escape the body perimeter to locally assist its fellow organisms; an exception is milk (colostrum), which may transfer maternal antibodies to the newborn.

The antibody response to foreign (not-self) antigens has six consecutive steps. These six steps would appear fundamental to an internal biological defense system where there must be self/not-self discrimination and an adaptive response. 1. There is a randomization process whereby an organism acquires the potential to generate a wide repertoire of antibodies of varying specificities (i.e. with abilities to bind a range of antigens, be they self or not-self). 2. The potential to bind self-antigens is eliminated by the negative selection of self-reacting elements (i.e. the antibody repertoire is purged of self-reactivity, thus creating "holes" in the repertoire). 3. The repertoire is moulded to "dissuade" pathogens from mutating to take advantage of the "holes" (positive selection for "near-self"; see Chapter 13). 4. When the presence of a not-self antigen is detected by virtue of its reaction with specific antibody, then the concentration of that antibody increases. 5. The interaction between not-self antigen and specific antibody triggers alarms so that the host organism responds adaptively to eliminate the source of the not-self antigen. 6. The host organism learns from the encounter (i.e. acquires memory) so that, if challenged again by the same antigen, there is a better response (e.g. faster and quantitatively greater). These are six steps fundamental to processes of adaptive self/not-self discrimination, which might include processes for *intracellular* self/not-self discrimination.

We should note that, just as the above-mentioned sheep in the meadow responded immediately to the self/not-self discrimination signal without the potential harmfulness of the source being determined, so a not-self antigen at the surface of a bacterium that has been killed by some physical or chemical treatment usually suffices to trigger an adaptive immune response (steps 4-6) when dead bacteria are injected into a host organism. Detection and registration as not-self occur in one step because the repertoire has been pre-screened to avoid detection of self (steps 1-3). The response is not conditional on attribute recognition (i.e. classification as potentially harmful or non-harmful).

This trigger-happy tendency is fortunate, since it allows prophylactic immunization with attenuated or dead organisms, so conferring advanced protection against subsequent challenge with the live organism. Thus, our bodies do not allow bacteria to wander around like tourists. Even if likely to be non-harmful, bacteria, dead or alive, are recognized as not-self and responded to. Touring bacteria are not required to "break windows" before the host organism will respond by making specific antibodies.

Antibody Variable Genes

In principle, the ability to generate, purge, and mould, elements of an immunological repertoire (steps 1-3) could occur, together or independently, either during the life of an organism (i.e. during somatic time), or over multiple generations (i.e. during evolutionary time). In the latter case the organism would be born with an appropriately purged and moulded repertoire (i.e. the repertoire would be innate, having been inherited through the germ line). In the case of the antibody response there is an innate repertoire of variable region genes, but the generation of diverse antibody molecules from individual variable region genes involves further variations, so that essentially steps 1-3 all occur somatically.

The necessity for this is apparent from the fact that two diploid parents are likely to have independent germ-line histories, so that within their diploid child a paternally-donated repertoire would not have been purged of antibodies with the potential to react with maternal antigens, and a maternally-donated repertoire would not have been purged of antibodies with the potential to react with paternal antigens. This problem is eliminated if the purging process occurs during somatic time (i.e. "self" is defined afresh when a new organism is produced from two parents of different genetic constitutions). An additional advantage is that each individual is likely to have a unique repertoire, so that a pathogen that somehow evaded the repertoire of one host could not count on an equally easy ride in its next host.

Generation of an antibody repertoire during somatic time provides an opportunity to develop antibodies of high specificity for provoking antigens. The chemical bonding between antigen and specific antibody is strong. Thus the antibody repertoire can encompass both range (variety of different antibodies) and high specificity (strong bonding). On the other hand, an innate repertoire would be limited by genome size. To detect not-self there might be range, but this might be at the expense of specificity. Bonding between antigen and antibody might be weak. Perhaps this is one reason why immunologists did not readily transfer their thinking about antibodies, essentially an extracellular phenomenon of multicellular organisms, to the idea of some intracellular antibody equivalent (see below).

A parallel can be drawn in terms of friends and acquaintances. Most of us have a few "friends" with whom we are strongly bonded, and many "acquaintances" with whom we weakly bond. Remarkably, quite often in the grand scheme of things it is the acquaintances that make the difference. Mark Granovetter's study of social networks, *The Strength of Weak Ties*, argues this well [4]. We shall consider here the possibility that, in an intracellular environment, weak ties can provide a sufficient degree of specificity for self/not-self discrimination. The "crowded cytosol" (Chapter 13) greatly supports this process.

Prototypic Immune Systems

If some evolutionary adaptation is highly advantageous, then the "hand of Nature" is likely to have explored all possible ways to achieve it. For this, organisms must be in possession of any necessary *facilitatory* adaptations, and there must be no *countermanding* adaptations. Thus, if not countermanded by weight, wings can be highly advantageous. Necessary facilitatory adaptations are appendages. To evolve wings an appendage must "find" how to trap air effectively when moving in one direction, but not when returning to its original position by moving in the opposite direction. Birds, bats and flying insects have solved this problem in different ways. The underlying adaptive principle is the same.

In this light, when appreciating the advantages of adaptive immune systems as extensively studied in the higher vertebrates, we should be aware that organisms perceived as lower on the evolutionary scale might have similar adaptations that might differ from the adaptive immune systems we are familiar with, no less than the membranous wings of insects differ from feathers. For example, jawless vertebrates such as the lamprey have no immunoglobulin antibodies, but they have developed an analogous immune system, as have snails [5, 6]. How far down the evolutionary scale can we go before the presence a countermanding adaptation, or absence of some facilitatory adaptation, becomes prohibitory?

Unicellular organisms are likely to have evolved some 800 million years before multicellular organisms. In 1861 the German physiologist Ernst Brücke dubbed single cells "elementary organisms" implying that many multicellular level functions might have prototypic equivalents at the unicellular level [7]. For example, the functional separation of germs cells and somatic cells, resulting in what Weismann called "the continuity of the germ-plasm" (see Chapter 1), was originally construed as a phenomenon of multicellular organisms. Certainly, there is no record of it in the unicellular amoeba. To reproduce itself, the amoeba enters mitosis and evenly divides the contents of its solitary nucleus between the two daughter cells. However, in another group of protozoa, the ciliates, there may be two nuclei one of which can be construed as a "working nucleus" and the other as a non-working "archival nucleus." When so inclined, ciliates conjugate sexually and in this process working nuclei ("soma") are destroyed while the archival nuclei reciprocally exchange haploid products ("germ line"), and the regenerated archival nuclei then form new working nuclei.

Give such sophistication, it can hardly be doubted that, over 800 million years, some nucleic acid sequences would have "learned" how to break away from one cell and enter another. So unicellular organisms are likely to have been challenged by many viruses and to have had an abundance of time to adapt to this challenge. Indeed, it is possible that immune systems of multicellular organisms arose as extensions of immune systems pre-existing at the unicellular level. The likely characteristics of such prototypic immune systems can be deduced, and this might help us understand modern immune systems.

In a clonal unicellular population where asexual reproduction predominated, early self-destruction (i.e. apoptosis; see Chapter 13) would be a simple mechanism to prevent spread of an intracellular pathogen, such as a virus, from host to host. This would diminish the opportunity for virus multiplication within the host cell, so that fewer copies of the virus would be available to infect other cells. Since members of the host population would be genetically identical, this altruistic act by one individual might allow more copies of its genes to survive than if it were not altruistic (i.e. if it struggled to cope with the virus itself, and did not self-destruct). Thus natural selection would favor organisms with genes for altruism ("selfish genes;" see later).

However, even such a primitive defense would require specific antibody-like sensors and discrimination between self and not-self. If released from the cell these sensors would probably be lost by diffusion and hence would not be available to help either the cell of their origin or others in the vicinity. Thus, it is unlikely that a system for the secretion of antibody-like molecules would have evolved prior to the appearance of multicellular organisms with an extracellular space within a perimeter (skin) preventing loss by diffusion. It is more likely that, if it were at all feasible for it to evolve (i.e. if chemically possible), a system for intracellular self/not-self discrimination would have evolved first.

Although at the time of this writing still speculative, such a sensory system could consist of the multiplicity of structurally distinct macromolecules that are normally present within cells, of which we emphasize here two populations – "immune receptor" proteins and "immune receptor" RNAs (Fig. 12-1). Many of these would have distinct properties necessary for cell function (catalytic, structural, transporting, templating, etc.). Because of these properties both molecular populations would be variable in type and number (i.e. they would be qualitatively and quantitatively differentiated; for example, multiple heterogenous mRNA species each encode a different protein product).

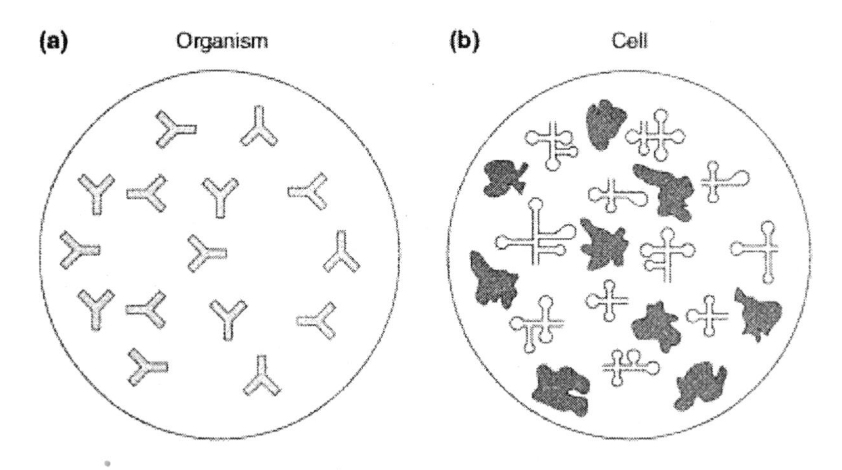

Fig. 12-1. The cell as an elementary immune organism. The left circle *(a)* represents a multicellular organism with Y-shaped antibodies of various specificities. Each antibody has a variable region at the tips of the Y-arms, and a constant region (the rest of the Y-arms and the Y-stem). The right circle *(b)* represents a unicellular organism with a repertoire of "antibody-like" protein molecules (grey) and "antibody-like" RNA molecules (stem-loop structures). These are referred to as "immune receptors" implying that parts of these molecules can interact with intracellular "antigens" (protein or RNA)

On observing such variability we are not surprised, since there is an obvious explanation. This tends to preempt the search for alternative explanations. However, if we consider the possibility that molecular variability might serve more than one purpose, then we can regard the variable *intracellular* mRNA and protein populations in the same way as we regard the variable *extracellular* antibody (immunoglobulin) populations. The latter are the first internal barrier against foreign pathogens ("not-self"), which "select" from among the antibodies (i.e. are recognized by) those with specificity for their coat antigens.

In the case of a virus, coat proteins would be relinquished at the time of entry into a host cell, and the cell's first possible line of intracellular defense

– its first intracellular barrier – would be to recognize the foreign nucleic acid as "not-self," either in its genomic form, or in the form of an early viral RNA transcript. This might implicate dsRNA. Perhaps a segment of a particular host mRNA species, because it happened to have sufficient sequence complementarity, would form a double-stranded segment with viral RNA of sufficient length (at least two helical turns) to trigger an alarm response (see Fig. 6-8). Thus, the variable host mRNA population can be viewed as consisting of an innate repertoire of "RNA antibodies," each preselected over evolutionary time not to react with "self" RNAs, while retaining, and if possible developing, the potential to react with "not-self" RNAs [8].

There is, then, the possibility that one or more resident intracellular molecules would be able to bind molecules from an invading virus with sufficient affinity to tag them as "not-self," thus initiating an adaptive intracellular "immune response". In theory, this diverse "immunological repertoire" could develop either over evolutionary time, or over somatic time. In a unicellular organism, somatic time can be considered as being the period between successive cell divisions, which makes development over evolutionary time far more likely. Whatever the mechanism and timing of the diversification process, there would be a need to eliminate receptors with high affinity for self-antigens. In the case of a clonal population, these intracellular self-antigens should be common to all members of the population.

Unfortunately, given the high replication and mutation rates of viruses relative to those of their hosts, it would seem highly probable that viruses would preadapt to avoid interaction with hostile host macromolecules. What a virus had "learned" (by mutation and selective proliferation) in one host, it would exploit on the next host. However, a high degree of host genome variability (polymorphism) might thwart the virus (see below). Thus, host clonality would be a hazard. Unicellular species that could diversify their genomes (i.e. differentiate and decrease clonality), while retaining their basic phenotypic identities, would be at a selective advantage.

In an elementary unicellular immune system, viruses that, through mutation, acquired the ability to inactivate host apoptotic mechanisms, would preferentially survive (by preventing the "altruistic" self-destruction of their host cell). In the course of the ensuing arms race (see Chapter 10), an intracellular "inflammatory" host response would have evolved to limit viral activities. But in multicellular organisms, death (apoptosis) of the primarily infected cell might limit the opportunity to alert other target cells and cells of the immune system (the lymphoid system), so preventing the initiation of the extracellular host inflammatory response. Thus, further sophistications, including the delaying or prevention of apoptosis, would be expected to have evolved in multicellular organisms.

Polymorphism Creates Host Unpredictability

On average, the haploid maternal and paternal contributions to your diploid genome are likely to differ from each other at least once every 0.5-2.0 kilobases, and general intraspecies differences are even more frequent. Such variations, referred to as "polymorphism," should decrease the extent to which a pathogen from one host could anticipate the genomic characteristics of its next host. "Polymorphism" means literally "many forms" (Greek: *poly* = many; *morph* = form or shape), and the word was used initially to describe anatomical differences among members of a species. However, polymorphism may be present at the genomic level without any correlated anatomical changes (i.e. no change in the conventional phenotype). When the polymorphism affected proteins, it would be likely to affect relative low complexity sequences corresponding to surface protein domains (likely to be hydrophilic and non-globular; Fig. 12-2).

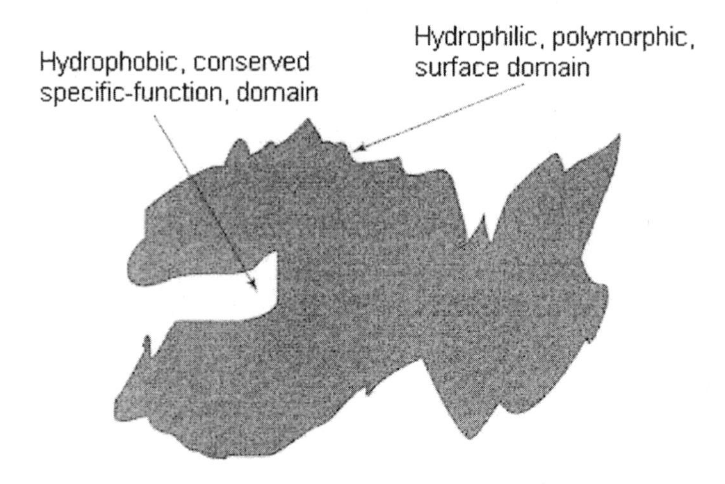

Fig. 12-2. Specific and general functions of a protein as reflected in its structure. Dedicated functions are often associated with conserved, internal, hydrophobic, globular domains. Potential immune receptor functions would be associated with variable (polymorphic), external, hydrophilic, non-globular domains

Thus, these surface domains, usually not critical for the specialized function of the protein, would be available for interaction with complementary molecular patterns presented by intracellular pathogens ("not-self"). These same domains should also have the potential to react with other "self" proteins, sometimes to an extent sufficient to trigger adverse responses in the host (intracellular "autoimmune" pathology). Accordingly, over evolutionary time natural selection would have favored organisms with mutations avoiding this. The intracellular "antibody" repertoire would have been generated, purged and moulded (i.e. fine-tuned) over evolutionary time. But would a single cell be capable of generating an adequate repertoire? In seeking an answer we should consider not just genic information, but the total information in genomes, both genic and non-genic.

Junk DNA

As mentioned in Chapter 1, human beings have a special way of dealing with things that they do not understand. They invent a word for it. They then sometimes confuse word with explanation. For a long time physicians used "psychosomatic" when there was no available explanation for sets of symptoms that did not match a particular disease. Sometimes it was, indeed, "all in the mind," but over the years more and more so-called psychosomatic diseases have been found to have an organic basis. Similarly, as discussed in Chapters 7-9, for a long time changes in DNA sequences that did not appear of adaptive value were dismissed as "neutral" and ascribed to "mutational biases."

The uncomfortable fact that around 98% of our DNA appears non-genic was blunted by a little four letter word – "junk" [9]. We appear to be the products of our genes alone, so therefore surely the rest *must* be non-functional junk? Some organisms, such as the puffer fish, seem to manage with very little non-genic DNA, which therefore cannot be very important. On the other hand, the puffer fish has retained *some* non-genic DNA, so perhaps there is an aspect of its biology that allows it to make do with less? Birds seem under a pressure to rid their genomes of superfluous DNA as shown by the reduction of intron size relative to exons (see Fig. 2-5). So perhaps non-genic DNA and introns are both examples of junk DNA? Perhaps some junk DNA plays a regulatory role? But would 98% of our DNA be needed for this purpose?

What possible other functions could junk DNA have? After all, junk (stuff in your attic that might be useful some day) is not the same as garbage (stuff you want to get rid of). Are there circumstances under which a piece of junk could become useful? When considering usefulness it is very easy, and often correct, to assume that that which is useful has been preserved by natural selection. This can lead to the argument, again often correct, that sequence features that are *conserved* between two species must have been useful. Are there circumstances where the converse might not necessarily apply? Under what circumstances could sequence features that are *not conserved* between two species (i.e. variation) have been useful?

To be functional it is likely that non-genic DNA would have to be transcribed into RNA. Indeed it is. Investigations of the transcriptional activities of entire human chromosomes revealed a "hidden transcriptome," corresponding to a large number of low copy number cytoplasmic RNAs. It is estimated that there is "an order of magnitude" more transcriptionally active DNA than can be accounted for by conventional genes [10–12]. Independent identification of a large number of "**CpG** islands" supports this (see Chapter 15).

Could these low copy number RNAs be dismissed as mere cytoplasmic "junk" – an unavoidable consequence of the existence of genomic "junk"? To understand their role, if any, in the economy of the organism, we need to know, by analogy with known transcriptional processes, whether there are specific promoters, whether there are dedicated RNA polymerases, whether transcription occurs randomly or under specific conditions, whether transcripts are diverse and include appreciable non-repetitive DNA, and whether the transcripts are translated into proteins, or function primarily at the RNA level.

Repetitive Elements

Much non-genic DNA consists of repetitive elements (see Chapter 2). The most prominent of these in humans are the 1,090,000 *Alu* elements. Three of these are shown in Figure 12-3, to which we will return in Chapter 15. It so happened that the discovery of *Alu* elements came at about the same time as the re-emergence of the selfish gene concept (see Chapter 1). The temptation to explain away junk DNA as due to the ability of these "selfish genes" to colonize our genomes then became irresistible.

The power of the selfish gene concept was its ability to provide an explanation for certain social and moral qualities (altruism). Darwin had pondered [13]:

> "How within the limits of the same tribe did a large number of members first become endowed with these social and moral qualities, and how was the standard of excellence raised? ... He who was ready to sacrifice his life, as many a savage has been, rather than betray his comrades, would often leave no offspring to inherit his noble nature ... . Therefore it seems scarcely possible ... that the number of men gifted with such virtues, or that the standard of their excellence, would be increased through natural selection, that is, by the survival of the fittest."

A tribe consists of groups of people that are likely to be related, and so share common genes. A gene, or collection of genes, that, by some chain of cerebral events, encouraged altruistic behaviour, would increase within the tribe if, by sacrificing himself the "savage" had caused many of his relatives (and hence their genes for altruism) to survive.

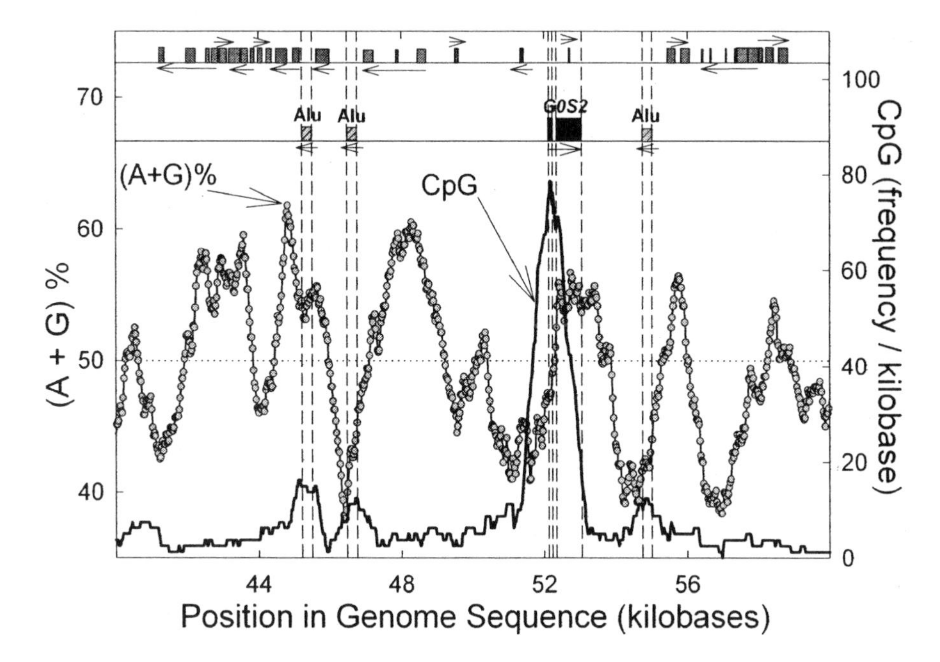

Fig. 12-3. *Alu* and other repetitive elements in part of a 100 kilobase segment of human chromosome one containing the two exon gene, *G0S2*. Horizontal arrows indicate transcription directions. *G0S2* and three *Alu* elements are represented by black and grey (striped) boxes, respectively, on the bottom horizontal line. Their boundaries are demarcated by vertical dashed lines. Other repetitive elements (some of retroviral origin) are represented by grey boxes on the top horizontal line. (**A+G**)% and **CpG** frequency (dinucleotides/kb) were evaluated for 800 base windows moving in steps of 25 bases. **CpG** peaks (macro or mini) are associated with *G0S2* and the *Alu* elements (see Chapter 15). Note that Szybalski's transcription direction rule is obeyed by the *rightward*-transcribing *G0S2* gene (**R** > 50% in top strand) and two of the three *leftward*-transcribing *Alu* elements (**R** < 50% in top strand)

Members of that group within the tribe could increase in numbers, relative to groups whose members did not exhibit altruistic behaviour. Thus, selfish individuals would not survive, but selfish genes (genes governing altruistic behaviour) would. The importance of this was that it brought to light the possibility that, in some circumstances, the focus of natural selection might be seen as at the levels of the gene and the group, rather than as at the level of the individual. However, in the present case there is a better argument than the selfish-gene argument.

Both conventional genes and repetitive elements provide promoters for the transcription of non-genic DNA. Some gene transcripts have been found longer then expected due to a failure of transcriptional termination ("run-on" transcription). Some classes of repetitive element contain promoters from which transcription can initiate and extend beyond the bounds of the element into neighboring genomic regions.

Are such extended transcripts generated randomly in time? In the case of *Alu* elements, transcription (by RNA polymerase III) has been observed to increase at times of cell stress (e.g. viral infection, heat-shock). Indeed, viral infection can simulate the "heat shock response" where there is induction of a set of proteins known as the "heat shock proteins" (see Chapter 13). Thus, it is possible that *Alu* transcription reflects an adaptive response to virus infection (Fig. 12-4), and that natural selection has favored the spread of *Alu* elements for this purpose.

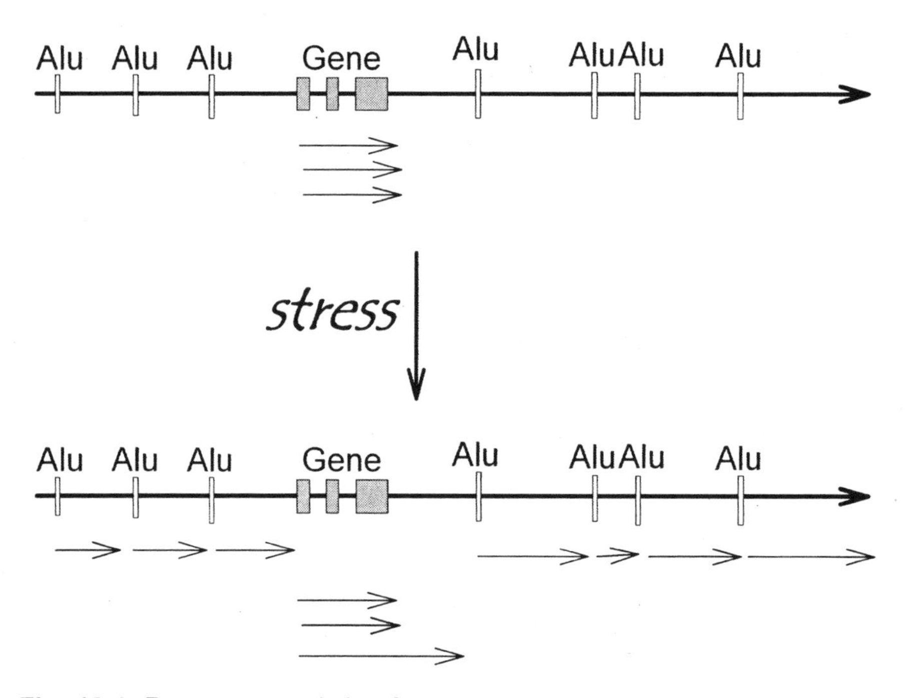

Fig. 12-4. Run on transcription from genes and repetitive elements in cells subjected to the "stress" of virus infection. The direction and extent of transcription is indicated by the thin horizontal arrows. There is evidence that stress may increase run-on transcription of *Alu* elements, but at the time of this writing there is no evidence that stress increases run-on transcription of genes

The Double-Stranded RNA Alarm

Although protein molecules can recognize specific nucleic acids (and the converse), it is convenient here to consider host proteins recognizing foreign proteins and host RNAs recognizing foreign RNAs. In the cytosol RNA molecules adopt characteristic stem-loop configurations (Figs. 5-1, 12-1b), and RNA-RNA interactions must initiate by way of a "kissing" search between bases at the tips of loops. If sequence complementary is found (e.g. **G** pairing with **C**, and **A** pairing with **U**) then two RNA species can pair, partially or completely, to generate a length of double-stranded RNA (dsRNA; Fig. 6-8).

If a virus introduced its own RNA into a cell, would there be sufficient variability among host RNA species for a host "immune receptor" RNA to form a segment of dsRNA with the "not-self" RNA of the virus? Calculations show this to be feasible, especially if the entire genome were available for transcription [14]. Would the dsRNA be able to initiate an adaptive intracellular "inflammatory" response? How would the host cell prevent generation of "self" dsRNAs?

Formation of dsRNA has long been recognized as providing an early cellular alarm response to viral entry. Protein synthesis can be inhibited by very low concentrations of dsRNA. This involves activation of a kinase (an enzyme that adds a phosphate group to proteins) known as "dsRNA-dependent protein kinase" (PKR), and results in inhibition of the initiation of protein synthesis. This is bad news for an invading virus because it wants to synthesize viral proteins, but it is also bad news for the host cell, since the inhibition is not virus-specific, and its ability to synthesize its own proteins is impaired. Evasive viral strategies would include the acceptance of mutations to avoid formation of dsRNA, and direct inhibition of cell components required for the formation of, or the response to, dsRNA.

Virus-infected cells produce "interferons," which can be considered part of the inflammatory response. These induce a general anti-viral state spreading together with various other protein molecules (e.g. "chemokines") from the cell of origin to other cells. Their production was long known to be powerfully stimulated by dsRNA, for which no clear explanation was forthcoming.

There is now growing evidence that, both in animals and plants, another more sequence-specific "inflammatory" response to dsRNA arises as part of an intracellular mechanism for self/not-self discrimination. Furthermore, just as in the adaptive antibody response there is increased production of specific antibody, so, courtesy of enzymes with names such as "RNA-dependent-RNA polymerase" and "dicer," there seems to be increased production of specific "immune receptor" RNAs as part of the specific adaptive response to a virus. These associate with an "RNA-induced silencing complex" (RISC), which mediates sequence-specific strand cleavage of viral nucleic acid (Fig. 12-5). Organisms with mutations in components of this system show increased vulnerability to pathogens [15].

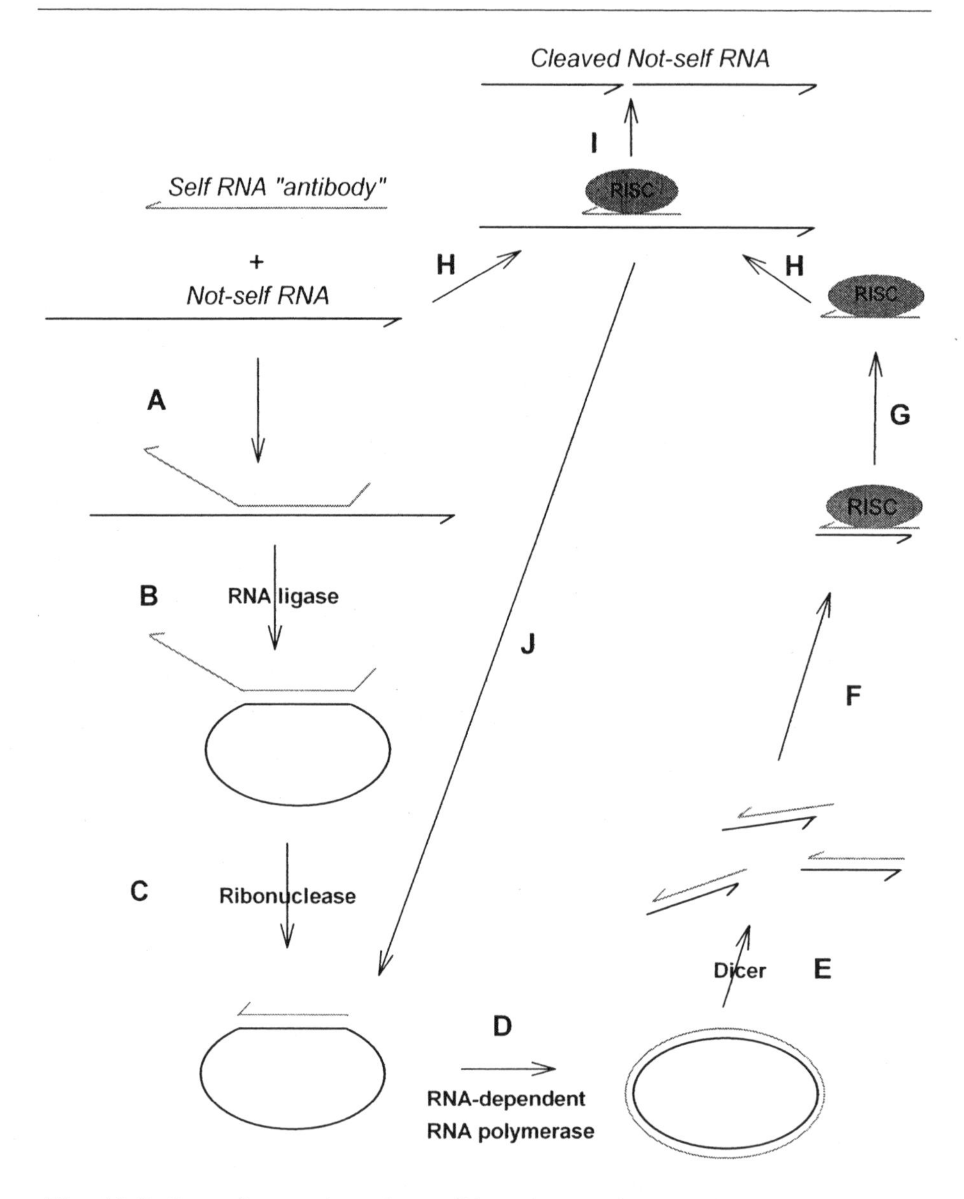

Fig. 12-5. One of a number of possible schemes for an adaptive intracellular self/not-self discriminatory immune system operating at the RNA level. Foreign "not-self" RNA (black line) interacts *(A)* with a member of the diverse "RNA-antibody" (immune receptor) repertoire (grey line) to produce a small, but sufficient (i.e. two helical turns) length of dsRNA. The foreign not-self RNA is circularised by an RNA ligase *(B)*, but a single strand-specific ribonuclease *(C)*, associated with immune receptor RNA, begins to degrade the latter RNA so preventing its circularisation. RNA-dependent RNA polymerase *(D)* uses the paired immune receptor RNA fragment to prime the synthesis of

a longer dsRNA that, when complete, corresponds to the entire not-self RNA. Dicer *(E)*, an enzyme (endonuclease) specific for double-stranded RNA, segments the dsRNA into multiple small (two helical turns) dsRNA fragments. The small dsRNAs are individually bound *(F)* to a RISC protein complex (RNA-induced silencing complex), which *(G)* retains only the strand derived from the original immune receptor RNA (grey line). By virtue of the complementarity between the immune receptor derived RNA (grey line) and the not-self RNA species (black line), the RISC "recognition complex" interacts with *another* member of the not-self RNA species *(H)*. The process may then recycle *(J)* through steps D-H to amplify the number of RISC recognition complexes. Alternatively, an endonuclease *(I)* in the RISC recognition complex cleaves and inactivates the not-self RNA, which will be further degraded by single strand-specific exonucleases (ribonucleases). This area of research is moving very rapidly, and the figure is merely illustrative of some of the principles involved

Purine-Loading to Self-Discriminate

Given the acknowledged role of dsRNA as an intracellular alarm signal, it would seem that viruses, needing to multiply within the host cell, were in some way *obliged* to form dsRNA, possibly as a replicative intermediate. Perhaps viruses, for some fundamental reason, had not been able to evolve to avoid the formation of dsRNA replicative intermediates, and their hosts took advantage of this. However, viruses with dsRNA genomes have adaptations that appear to conceal their genomes from host cell surveillance mechanisms. Furthermore, no obligatory need for a double-stranded replicative intermediate has been demonstrated.

Thus, the hypothesis that dsRNA would be formed by an interaction between host RNA and viral RNA is quite plausible. This requires that the host itself not form dsRNA by interactions between its own RNA molecules. The phenomenon of purine loading (see Chapters 6 and 9) provides a possible mechanism for avoiding the formation of self-self RNA duplexes.

More than twenty contiguous base pairs in a duplex are needed to initiate dsRNA-dependent events (e.g. activation of PKR, or of specific gene silencing). Among the RNA species of a cell ("self" RNAs) there might be two whose members, by chance, happened to have enough base complementarity with each other for formation of a mutual duplex of a length sufficient to trigger alarms. So there would need to have been an evolutionary selection pressure favoring mutations in RNAs that decreased the possibility of their interaction with other self-RNAs in the same cell, a process that might otherwise lead to the negative selection of the host organism. In many cases mutations to a purine would assist this, since purines do not pair with purines. Indeed, interaction with self-RNAs seems to have been avoided by purine-loading the loop regions of these RNAs, thus avoiding the initial loop-loop "kissing" reactions that precede more complete formation of dsRNA. An excess of purines, observed both at RNA and at DNA levels (in RNA-synonymous DNA strands), is found in a wide variety of organisms and their viruses.

Viruses that purine-loaded their genomes to match their hosts (e.g. the AIDS virus, HIV1) would seem to have adopted a sound strategy (from a viral perspective). Viruses that foolishly pyrimidine-loaded would generate pyrimidine-rich mRNAs that should readily base-pair with purine-rich host RNAs to generate corresponding dsRNAs, hence activating intracellular alarms. Among the latter viruses are HTLV1 (see Chapter 8) and EBV (see Chapter 11). However, it should be noted that these viruses are latent (quiescent) in their hosts, and tend only to transcribe under provocation. HIV1 and HTLV1 are both retroviruses, and both have the latency option, but HTLV1 is more profoundly latent, the majority of its human hosts living normal lives.

HTLV1 is localized to certain populations, and is particularly prevalent in Japan. It inserts "seamlessly" (collinearly) into the genome of its hosts and there is no obligatory transcription. In contrast, EBV exists as an independent genomic element ("episome"), and, to regulate its relationship with the host cell, is obliged to transcribe an mRNA encoding the EBNA-1 protein. In stark contrast to most other EBV genes, this gene is purine-loaded. Indeed, it appears to have acquired a long, purine-rich, simple sequence repetitive element solely to facilitate purine-loading (see Chapter 11).

Entropy

Purine-loading is especially high in organisms that live at high temperatures (thermophiles; see Chapter 8) for which there is an "entropic" explanation. Unlike the word "junk," an intellectual crutch that long assured the credulous that 98% of our genomes could safely be ignored, the word "entropy" brings discomfort. Physicists and chemists include entropy among the forces that can drive chemical reactions, but biologists tend not to find the concept so intuitive. Max Lauffer's seminal 1975 text *Entropy-Driven Processes in Biology* has largely been ignored [16]. Biologists use the evolution E-word with great facility, but the other E-word, entropy, is avoided. When he/she emerges, *Homo bioinformaticus* will use both with equal facility. All that is required of the present reader is that he/she accept the fact that entropy can drive molecular interactions, irrespective of any understanding of it as reflecting a universal tendency for ordered systems to break down and become disordered (chaotic).

For example, when you leave a recalcitrant burned saucepan "in soak" overnight before cleaning, you are letting entropy work for you. The burned food is "ordered" to the extent that it is so firmly attached to the bottom of the saucepan that it cannot easily be removed manually. A night's bombardment with water molecules serves to loosen the attachment releasing the burned food, which thus becomes "disordered." The work you have to do the following morning is less to the extent that entropy has been at work. If you had left the saucepan overnight in hot water you would have been spared even more work. The entropy-driven component of a reaction increases as temperature increases.

Exploratory "kissing" interactions between hybridizing single-stranded nucleic acids involve transient base stacking interactions with the exclusion of water molecules that have become ordered around the free bases exposed when nucleic acids are in open single-stranded form. Such reactions have a strong entropy-driven energy component since, although two separate nucleic acid strands become united as a duplex and so are collectively *more* ordered, this is accompanied by the release of many water molecules which become collectively *less* ordered. Overall there is an increase in disorder, and this drives duplex formation.

It seems intuitively obvious that an increase in temperature would increase the internal vibrations and velocities of chemical molecules, so that two interacting molecular species would need to be held together tenaciously at high temperatures by classical bonding forces (ionic bonds, hydrogen bonds and Van der Waal bonds). Thus, DNA duplexes can be "melted" by heating to produce single strands. So it would seem that if you wished molecules to aggregate and stay together you should lower the temperature. However, the release of water molecules that are ordered around macromolecules in solution becomes greater as temperature increases. Accordingly, the entropy-driven component of chemical reactions *increases* as temperature increases. This can be greater than the pressures to break classical bonds.

So, instead of being dispersed by increasing temperature, the aggregation of macromolecules, such as single-stranded nucleic acids forming double-strand duplexes, is favored. The propensity to aggregate increases until the melting temperature is reached, when the strands suddenly separate. The released open single strands immediately adopt stem-loop conformations that are strongly favored entropically at high temperatures. It follows that the need to avoid "kissing" interactions between single-stranded "self" RNA molecules progressively *increases* as temperature increases. Indeed, purine-loading is high in thermophiles (see Chapter 8).

As will be discussed in Chapter 13, the physico-chemical state of the "crowded" cytosol is likely to be highly supportive of reactions with a high entropy-driven component, which include interactions between the codons of mRNAs and anticodons at the tips of loops in tRNAs. This particular interaction results in transient formation of dsRNA segments, which normally do not exceed five base pairs (see Chapter 7). So, protein synthesis, a process of fundamental importance for cell function, is facilitated. But a cytosolic environment that favors mRNA-tRNA "kissing" interactions should *also* favor mRNA-mRNA "kissing" interactions. This would act as a "distraction" impeding mRNA translation, and hence impeding protein synthesis. By generally militating against RNA-RNA interactions, purine-loading should prevent mRNA-mRNA interactions and so free mRNA molecules for interaction with the cytoplasmic components necessary for their translation (e.g. ribosomes and tRNAs).

Thus, purine-loading should facilitate protein synthesis. Because of this, organisms that did not purine-load their RNAs would be negatively selected. The early evolution of such mechanisms to avoid *inadvertent* formation of segments of dsRNAs as a result of interactions between self-RNAs, driven by the need for efficient protein synthesis, would have created an opportunity for the subsequent evolution of mechanisms *utilizing* dsRNA as an intracellular alarm against not-self-RNAs.

In some cases, so important has been the need to accept purine mutations, even the protein-encoding role appears to have been compromised (see Chapter 9). When this compromise is not possible, then the length of the protein can be increased by inserting simple sequence repeats of amino acids with purine-rich codons between functional domains, as in the case of EBV and malaria parasites (see Chapter 11).

The Hidden Transcriptome

The six fundamental steps in the extracellular antibody response, as described above, have at least four analogues in the proposed RNA immunoreceptor system shown in Figure 12-5. Host repertoire generation (step 1) occurs over evolutionary time. Purine-loading militates against reaction with self (step 2). Double-stranded RNA is formed when viral RNA becomes base-paired with a sufficiently long segment of host "immunoreceptor" RNA, and there are enzymes that can amplify the concentrations of the dsRNA complexes (step 4). Double stranded RNA acts as an alarm signal triggering adaptive responses (step 5).

Thus, many disparate observations appear comprehensible in the context of an intracellular RNA immunoreceptor system for self/not-self recognition. Although generation of variability by post-transcriptional RNA editing is feasible [17], in essence the maximum potential repertoire of "RNA antibodies" is limited by genome size. This would explain the existence of a "hidden transcriptome" [10–12]. Repetitive elements (e.g. *Alu* sequences) which change rapidly in evolutionary time, and mobile genetic elements (transposons), can be viewed as germ-line variation-generating devices that create polymorphisms thus making it difficult for viruses to anticipate the genomic characteristics of future hosts. This suggests that regions transcribed as the "hidden transcriptome" should not be evolutionarily conserved. A comparison of libraries of artificial DNA copies of cellular RNAs ("cDNAs") made from humans and mice supports this [18]:

> "It is interesting to note this [human] type of "non-coding" transcript was also found in mouse cDNA collections. ... What was significant was that [the] majority of the examined cDNAs were not evolutionally conserved. In this dataset of mouse genes, identification of 11665 similar transcripts (which would be categorized as 'unclassified' according to our scheme) has also been reported. This suggests that there is little conservation for these 'unclassified' transcripts and/or that there are huge numbers of such transcripts (at least in the order of 100000). Interestingly, ... we have recently examined the promoter activities of randomly isolated genomic DNA fragments on a large scale and observed that there are cryptic promoter activities throughout the genomic DNA (unpublished data). It may be possible that those cryptic promoters may act at low frequency to produce aberrant (or sporadic) transcripts."

To what extent are there analogues of the third (positive selection) and sixth (memory) fundamental steps? One way of creating memory in the extracellular antibody response would be to retain the original eliciting antigen in some cryptic form so that it could "prompt" the immune system from time-to-time, so reminding it of the possibility that the source of the eliciting antigen might one day reappear. While some studies with radioactive antigens have suggested that trace amounts might persist, the main memory strategy is to have a steric complement of an antigen (i.e. a specific antibody protein) and to retain an enhanced capacity to make more of that antibody.

However, in the case of "RNA-antigens," persistence might be a more effective strategy. Thus, at steps D or E in Figure 12-5, an enzyme (a polymerase known as reverse transcriptase) might convert the dsRNA into dsDNA, which might then be inserted into the genome and selectively mutated (or methylated; see Chapter 15) to make it inactive. The sequence would retain sufficient similarity with the eliciting not-self RNA to render it of value, following transcription, in generating "RNA antibodies" at times of future "stress" (Fig. 12-4). If this could occur in germ-line tissue then many sequences with similarities to ancient virus sequences should be present in intergenic DNA, and traces of Lamarckian phenomena (inheritance of acquired characters) might emerge [19]. For example, the HERV elements shown in Figure 12-3 are likely to be remnants of ancient retroviruses.

It is difficult to find an analogue of the third fundamental step, positive selection. There should be selection of individuals in which favorable mutations preventing self dsRNA formation had been collected together by recombination. But this is not the same as the moulding of lymphocyte repertoires that occurs during an individual lifetime (somatic time) as will be discussed in the next chapter. An intracellular example of positive repertoire selection might be a tendency over evolutionary time to accept mutations in some proteins because the mutations confer an ability to aggregate with other self proteins when they mutate to "near-self" (see next chapter). Whether RNAs generated from the hidden transcriptome would operate solely at the RNA level is not known. The possibility that there might also be translation products that might contribute to the intracellular population of protein "immune receptors" will also be considered in the next chapter.

Summary

The full understanding of the genome of any biological species requires an understanding of the genomes of the species with which it has coevolved, which include pathogenic species. Members of a pathogenic species that enter the bodies of members of a host species must be recognized as "not-self" in order that the immunological defences of their hosts be activated. However, hosts whose immunological defences are poised to attack "near-self" versions of not-self, rather than not-self *per se*, should be at a selective advantage. It is likely that immune systems of multicellular organisms are adaptations of the highly evolved immune systems of unicellular organisms, which had already developed the capacity for self/not-self discrimination. From this perspective we can comprehend phenomena such as "junk" DNA, genetic polymorphism and the ubiquity of repetitive elements. That which is evolutionarily conserved is often functional, but that which is functional is not necessarily conserved. Variation may be functional. The "hidden transcriptome," revealed by run-on transcription of genes or repetitive elements constitutes a diverse repertoire of RNA "immune receptors," with the potential to form double-stranded RNA with viral RNA "antigens," so triggering intracellular alarms. Both genic and non-genic DNA would have been screened over evolutionary time (by selection of individuals in which favorable mutations had been collected together by recombination) to decrease the probability of two complementary "self" transcripts interacting to form dsRNA segments of more than 20 bases (about two helical turns). As a result, many RNAs are purine-loaded.

Chapter 13

The Crowded Cytosol

> "If you can talk with crowds and keep your virtue,
> Or walk with Kings – nor lose the common touch,
> Yours is the Earth and everything that's in it."
>
> Rudyard Kipling [1]

While, given the expansion of the world population, it cannot be guaranteed that it will always be so, currently one accepts as a temporary discomfort short periods of crowding, as in a crowded elevator or subway train. However, if, like most intracellular macromolecules, one were both blind and deaf, the need to communicate by touch might make crowding an option of choice. It seems likely that the first cells to evolve soon discovered the advantage of intracellular crowding, which persists to this day [2]. Indeed, the physiological environment of enzymes is very different from the environment we can normally create in the test-tube.

The French microbiologist Antoine Bechamp reported in 1864 that enzymes ("ferments") in yeast would work outside of the cell of origin [3]. More than anything else, this demystified the cellular "protoplasm" suggesting that, in principle, it should be possible to take a cell apart and then reassemble it from its individual components, as if it were a clock. So enzymes were purified and characterized. It even became possible to purchase enzymes "off the shelf" as protein powders. The study of their abilities to convert target molecules (substrates) to other molecules (products) kept biochemists busy for much of the twentieth century, and the fundamental chemistry of life emerged.

For this, the enzymes had to be dissolved in suitable salt solutions. But it is difficult to dissolve proteins at concentrations higher than 10 mg/ml. In contrast cytosolic proteins are collectively at concentrations around 300 mg/ml! Within cells many water molecules are likely to be ordered (made relatively immobile) by crowded macromolecules and so are less able to diffuse freely. This means that many chemical changes within intracellular fluids that involve macromolecules are likely to be entropy-driven (see Chapter 12).

Piles of Coins

The concentration of proteins within extracellular fluids is also very high (e.g. 80 mg/ml). Cells suspended in these fluids can "loose solubility" and aggregate when either the total protein concentration, or the concentrations of certain proteins, exceeds certain limits. Normally red blood cells (erythrocytes) appear as flat red disks suspended in blood plasma. When the concentration of proteins in plasma is increased only slightly the red cells move, as if under the direction of unseen hands, and queue up to form "piles of coins" or "rouleaux". It is possible to watch this and demonstrate the specificity of the aggregation by light microscopy. In mixtures of red cells from different species, cells of the same species preferentially aggregate with each other (Fig. 13-1).

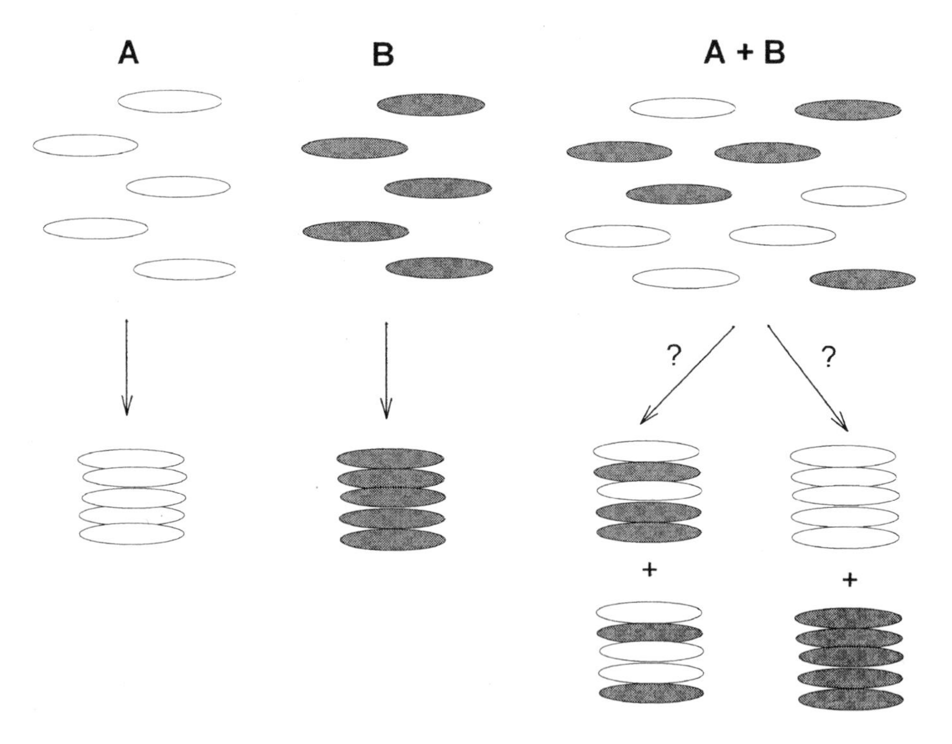

Fig. 13-1. Specificity of rouleaux formation. A rouleaugenic agent (e.g. polymerized blood albumin) is equally active in aggregating red blood cells from animal species *A* or from animal species *B*. The aggregates appear as-rouleaux, or "piles of coins." When the two cell populations are mixed and then treated with the rouleaugenic agent, each of the resulting rouleaux should either contain both *A* and *B* red cells (indicating non-specific aggregation), or all *A* and all *B* red cells (indicating specific aggregation). The latter alternative is found experimentally

Thus, the entropy-driven aggregation shows specificity – "like" aggregating with "like." In mixtures, red cells form homoaggregates (like with like), not heteroaggregates (like with unlike). This reflects a general tendency for shared regularities in structure, and shared molecular vibrations (resonances), to promote self-self interactions between particulate entities, be they whole cells, or discrete macromolecules (see Chapter 2).

It is important to note that the aggregation is a response to an increase in total proteins in the surrounding medium and there is no direct cross-linking of cells by the proteins. The proteins are not acting as cross-linking agents (ligands; Fig. 13-2).

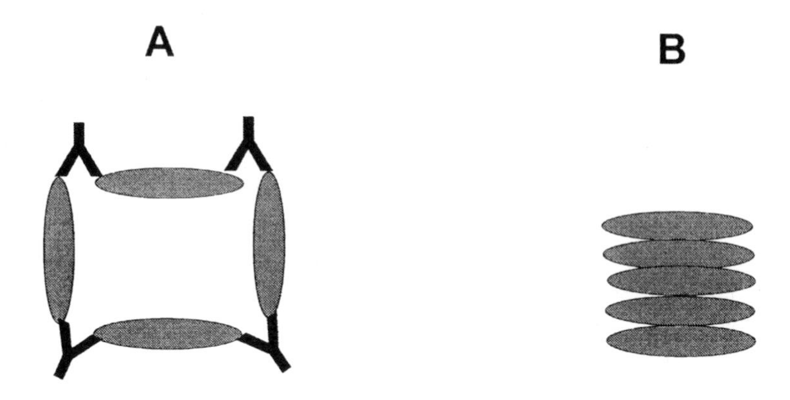

Fig. 13-2. Distinction between *(A)* aggregation of red blood cells by an extracellular cross-linking agent (antibody) and *(B)* aggregation that does not involve a cross-linking agent. In *(A)* the disk-shaped red cells are cross-linked by Y-shaped bivalent antibodies (not drawn to scale). The bonding here is strong and the aggregates are not readily disrupted. In *(B)* the red cells adopt the energetically most favorable (entropy-driven) pile-of-coins conformation (rouleaux). The bonding here is weak and the aggregates are readily disrupted. Within an organism the high protein concentration of the inter-cellular environment should promote the aggregation of similar cells into a common tissue, a process that, initially, might not need cross-linking agents

The increase in protein concentration.can be produced non-specifically by addition of an excess of any one of a variety of proteins. Thus, rouleaux-generation is a *collective* function of proteins. The appearance of rouleaux in a blood sample provides a clinical index of the underlying state of the plasma proteins, not of the red blood cells. However, if red cells themselves are thought of as merely large proteins, then the phenomenon of rouleaux formation can assist our understanding of the phenomenon of protein aggregation as it occurs in concentrated solutions. This can be of help when considering possible mechanisms for intracellular self/not-self discrimination [4].

Homoaggregates

A specific protein can be induced to "self-aggregate" so forming homoaggregates (Greek: *homos* = same). This is brought about either by increasing the concentration of surrounding proteins or by increasing the concentration of the protein itself. In some cases homoaggregate formation reflects an obvious physiological function. Thus, coat protein molecules of tobacco mosaic virus (TMV) naturally aggregate to form the outer coat that protects virus nucleic acid during passage from cell to cell. But if an enzyme aggregates, its activity often diminishes and it may become insoluble (i.e. it precipitates). To this extent, aggregation can be non-physiological and harmful to the cell.

On the other hand, proteins are usually unstable, with life-spans extending from minutes to days, and homoaggregation can assist protein breakdown. This is a stepwise process by which a protein first becomes tagged (marked) as ready for degradation, and then is cleaved into fragments (peptides). Finally the peptides are degraded to the relatively stable amino acid "building blocks" from which new proteins can be assembled. The continuing process of synthesis, breakdown and reassembly of proteins, is referred to as "protein turnover" (see Chapter 2).

Sometimes, however, under special circumstances to be considered later, certain peptide fragments are not degraded. Instead, they are united with peptide-display proteins (major histocompatibility complex proteins; MHC proteins), and taken to the exterior of the cell. Here they are recognized by cytotoxic lymphocytes, which destroy the cell. The underlying principle that emerged in the 1980s was amazingly simple, and I (and probably others) kicked myself for not realizing it earlier. You do not need a whole elephant to diagnose elephant. You do not need a whole protein to diagnose a protein. Immunologists already knew this! They had coined the expression "antigenic determinant" for a part of a protein (antigen) that would suffice for an extracellular immunological recognition event. Yet, no one seemed to have considered the possibility that cells might detach the equivalent of an antigenic determinant from a protein *intracellularly* prior to engaging in a diagnostic recognition event.

In studies with TMV, Lauffer showed that aggregation involves the liberation of water molecules bound to the macromolecules [5]. Thus, while it might appear, from the observed aggregation, that entropy was decreasing, the increase in disorder of the liberated water molecules more than compensated for the increase in order of the macromolecules. System entropy increased. If the aggregation were entropy-driven (endothermic), then it should be promoted by a small increase in temperature. Indeed, aggregation can be induced by increasing the temperature over a narrow range, much lower than would be needed to disrupt the structure of (denature) the protein (Fig.13-3).

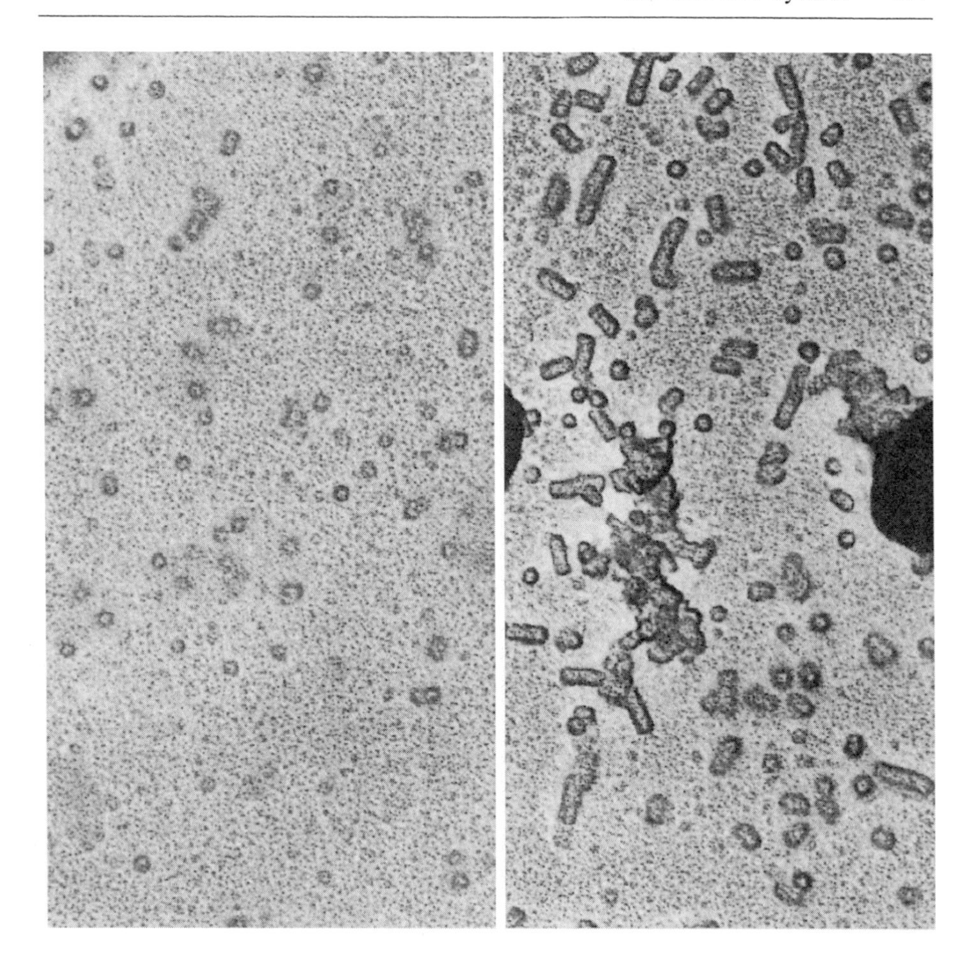

Fig. 13-3. Aggregation of TMV coat protein by increasing the temperature. This is an electron micrograph of a solution of coat protein (0.1 mg/ml) held at either 4°C (left) or 23°C (right). Reprinted from [5]

Collective Pressure

Homoaggregation at high protein concentrations generates specific, but relatively unstable, aggregates (i.e. only weak chemical bonding is involved). If you were to progressively concentrate a mixture of proteins (A, B, C, D) in aqueous solution then, at a certain critical concentration, one of the proteins, say A, would self-aggregate. In this process each molecule of A would loose some of its bound water. In the absence of the surrounding proteins (B, C, D) much *higher* concentrations of A would be required for aggregation to occur. Thus, a group of proteis *collectively* exerts a "pressure" (due to their binding of water) tending to force individual protein species to self-aggregate and give up their bound water. A protein species with the greatest tendency to self-aggregate (a function of factors such as structure, molecular size, and initial concentration) aggregates first. As total protein concentration increases, other protein species aggregate in turn.

The concept of the "crowded cytosol" implies that much intracellular water is bound to proteins, so that there is always a strong *standing* pressure to drive into homoaggregates any macromolecular species that exceeds the solubility limits imposed by the macromolecules surrounding it. Each individual macromolecular species can both contribute to, and be acted upon by, the pressure. Thus, each macromolecular species can have this *collective* function as well as a *specific* function. Both functions can affect phenotype and hence influence selection by evolutionary forces.

Concentration Fine-Tuning

It follows that the concentration of a protein is an important attribute that is not necessarily related, in any simple way, to the role the protein might normally play in the life of a cell. A protein has evolved to carry out a specific primary task. On grounds of economy, it might be supposed that evolutionary forces would have pressed for a maximization of specific activity (e.g. enzyme activity/protein molecule). This would minimize the necessary concentration of the protein. But there is no particular virtue in minimizing the concentration of a protein.

Provided its concentration is not extreme, a protein itself does not burden a cell. Indeed, if a *collective* function (e.g. the ability to exert a pressure to drive other proteins from solution) were an important attribute of a protein, then the steady-state concentration of the protein might tend towards the maximum compatible with the protein remaining in solution without self-aggregation. This would tend to counteract any tendency to maximize specific activity because the number of molecules present would suffice for the necessary level of activity, and there would be no selection pressure for them to improve on a per-molecule basis.

The steady-state concentration of a protein is determined by evolutionary forces acting on parts of the corresponding gene (i.e. base mutations) to affect factors such as mRNA transcription rate, and mRNA and protein stabilities. For example, a mutation that decreases the transcription rate of a gene decreases the concentration of the corresponding mRNA, and hence the concentration of the protein that is made by translating that mRNA. A protein within the cell ends up with a certain specific activity, which can be less than the maximum possible. Over evolutionary time, the concentration of the protein is fine-tuned to the concentrations of its fellow travellers – the other diffusible proteins with which, from generation to generation, it has shared a common cytosol. In this circumstance, whereas normally the protein would be soluble, a mutation in the corresponding gene could result in homoaggregation and insolubility. This might provide an opportunity to register the protein as "not-self."

Heteroaggregates

Why fine-tune? In general, as hinted at in Chapter 2 with the forgery metaphor, fine-tuning creates a narrower frame of reference so broadening the range of events that may be discerned as falling outside that frame. A factor favoring the precise fine-tuning of cytosolic protein concentration would be the need to discriminate self-proteins both from mutated self-proteins and from foreign proteins (such as might be encoded by a virus). We need to discriminate between "self," "near-self," and "not-self." Mutated self-proteins are only slightly changed and so can be considered as "near-self" proteins, rather than "not-self" proteins. But, although "near-self," they need to be registered as "not-self." Virus proteins would be foreign and less likely to correspond with self (i.e. they would be "not-self"). But viruses that could accept mutations making their proteins more like their host's proteins (i.e. approach "near-self") might be at a selective advantage. As we shall see, this advantage would fade if host defenses were attuned for discriminating between self and "near-self" in such a way that the latter would register as "not-self."

As discussed in Chapter 12, a system for intracellular self/not-self discrimination could have evolved when the first unicellular organisms arose and were confronted with the first prototypic viruses. A mutated self-protein might have lost activity either as a direct result of the mutation (e.g. the mutation might have affected the active centre of an enzyme) or because the mutation had decreased solubility, perhaps manifest as homoaggregate formation. However, when mutated, a resulting structural (or vibrational) change might, by chance, have *created* some degree of reactivity with one or more of the many other diffusible protein species within the same cell. Hence, the mutated self- protein might "cross-seed" the aggregation of unrelated proteins. Heteroaggregates might form (Greek: *hetero* = different). There could then be a loss of function not only of the primarily mutated protein, but *also* of the coaggregated proteins. Thus, a mutation in one protein might affect the functions of other specific proteins in unpredictable ways, generating complex mutational phenotypes (pleiotropism). Certain clinical syndromes may be sets of such diverse, often not obviously related, altered phenotypic characters that are observed in disease states.

To the extent that a primary mutation does not result in heteroaggregate formation, then the function of a cell where the mutation has occurred might not be affected, and the cell might persist. Cells with mutations that result in heteroaggregate formation are more likely to be functionally impaired. Unless aggregation were required for its primary function, a protein would have been fine-tuned over evolutionary time so as *not* to interact with the many thousands of other diffusible protein species with which it had been travelling through the generations in the same cytosol. Such interaction might have impaired the function both of the protein and those with which it interacted. So organisms with mutations leading to interactions would have been *negatively* selected.

Heteroaggregate formation could, however, be of adaptive advantage if the primary mutation was in a gene controlling cell proliferation, which might result in a loss of control and hence cancer. In this case, a gene encoding, by chance, a normal protein that would coaggregate with a mutant cancer-causing protein (oncoprotein) would confer a selective advantage, over and above that conferred by virtue of the gene's normal function (i.e. while still retaining its normal function, it would be *positively selected* for encoding an "immune receptor" that would function as a detector of "near self" oncogenic changes). Genes encoding products with such coaggregating functions would tend to make cancer a disease of post-reproductive life. This is a time when selective factors that tend to promote the number and reproductive health of descendents are less important; so cancer prevention would be less evolutionarily advantageous (6).

Protein "Immune Receptors"

From this we see that potential functions of a protein include (i) its primary function (e.g. enzyme activity), (ii) its contribution to the total cytosolic aggregation pressure, (iii) its ability to form heteroaggregates with either mutated self-proteins (e.g. oncoproteins) or foreign proteins that may have been introduced by a virus. In the latter respect, cytosolic proteins can be regarded as "intracellular antibodies," or protein "immune receptors" (see Fig. 12-1).

The genes of a host cell and the genes of an invading virus differ in various ways that might assist discrimination between self and not-self. Within the species-limit, host cell self-genes travelling together through the generations should have had ample opportunity collectively to coevolve and fine-tune to each other. On the other hand, the goal of a virus is to multiply and spread, preferentially within the lifetime of its host. From this it might be thought that the cytosolic concentrations of virus gene products would be less fine-tuned than the cytosolic concentrations of host gene products. However, the high replication and mutation rates of viruses relative to their hosts make it likely that viruses would be no less fine-tuned than their hosts. Indeed, host fine-tuning over evolutionary time would have tended to create a uniform intracellular environment, which would decrease a major virus anxiety – that of anticipating the conditions it would find in its next host. Thus, it is unlikely that the proteins of a "street smart" intracellular pathogen would readily exceed the solubility limits imposed by host proteins in the crowded cytosol.

However, in their role as "immune-receptors," host cytosolic proteins could form a diverse antibody-like environment capable of forming hetero-aggregates with virus proteins, which would accordingly register as not-self (see Chapter 12). This primary self/not-self discrimination event, perhaps in an environment made permissive as part of a response to dsRNA alarms (i.e. cells are alerted to become more conducive to registering not-self), should result in processing of the heteroaggregates by cytoplasmic structures known as proteosomes. Here there is creation of protein fragments (peptides) that are not further degraded to amino acids. Instead, they are displayed at the cell surface by the peptide-display proteins (MHC proteins). This display is recognized by cytotoxic cells of the lymphoid system (T-lymphocytes), which destroy the virus-containing cell. The organism then loses the services of one of its cells. But most tissues can readily replace lost cells by the mitotic division of other cells that are not infected.

How diverse is the intracellular protein immune-receptor repertoire likely to be? At the very least it would include the many diffusible protein species normally present in the cytosol of a differentiated cell type. In addition, when a virus "tripped" the self/not-self discrimination alarm, there might be translation into proteins of some of the RNA products of the "hidden transcriptome" (see Chapter 12) [7]. These might include the products of tissue-specific genes not normally expressed in a host cell of a particular tissue type (i.e. a brain-specific gene might be abnormally expressed in an infected kidney cell).

Phenotypic Plasticity

It would be expected that within, say, a kidney cell, such not-normally-expressed products would be present only at the very low concentrations needed for their role in identifying the virus proteins with which they react (so registering virus proteins as not-self and tripping appropriate alarms). But there would also be the possibility, in the case of proteins normally required at very low concentrations, of an unwelcome effect on the phenotype. For example, very low concentrations of critical regulatory proteins are synthesized at various unique time-points during embryogenesis, so bringing about developmental switches. Thus, developing embryos within pregnant females undergoing virus attack (or an equivalent stress) might produce a developmental switch protein at the wrong time. Offspring would then appear with mutant phenotypes sometimes similar to the mutant phenotypes observed among the offspring of pregnant females exposed to mutagens (e.g. X-rays), which had mutated the DNA of their embryos.

However, if viable, the mutant offspring of a parent that had been under virus attack would not, in turn, be able to pass their mutant characters on to their offspring. In contrast, the mutant offspring of a parent that had been treated with mutagens might be able to pass on their mutant characters if the mutation had affected developing germ-line cells. The mutant forms that resulted from developmental stress would be classified as "phenocopies," rather than "genocopies," since their mutant characters would not be genetically inherited (i.e. there would be no underlying causal genetic change) [8]. In other words, stressed organisms display "phenotypic plasticity" not "genotypic plasticity."

Polymorphism Individualizes

High host polymorphism would make it difficult for viruses to anticipate the "immune receptor" RNA and "immune receptor" protein repertoires of future hosts (see Chapter 12). Furthermore, peptides generated from heteroaggregates of virus and host proteins, would *include* host protein-derived peptides. These self-antigens, as well as, *or instead of*, the antigens of the pathogen, would then serve as targets for attack by cytotoxic T cells [9]. Accordingly, the variability (polymorphism) of intracellular proteins (i.e. differing from individual to individual of the host species) would tend to *individualize* the immune response to intracellular pathogens (and cancer cells). T-lymphocytes from one virus host (or one cancer subject) might not recognize cells of another host infected with the same virus, or afflicted with the same type of cancer. From this perspective, protein polymorphism is not "neutral," as sometimes supposed, but serves to adapt potential host organisms as "moving targets," so militating against pathogen preadaptation.

Thus, the designation by the host of a virus protein as "not-self," might involve both *quantitative* factors (i.e. homoaggregate formation if a virus protein's concentration exceeds a solubility threshold, which the virus might easily avoid by mutation), and *qualitative* factors (i.e. the recruitment of various polymorphic host proteins into protein heteroaggregates, which the virus might not so easily avoid by mutation). However, the T-lymphocytes primed by specific peptide fragments from *self* antigens might, besides multiplying and attacking the virus-infected (or cancer) cell, also react against the same self antigens should they, perchance, be displayed by normal host cells. In cancer patients this could result in immunological diseases of various tissues, other than the primary cancer tissue [10].

An interesting example of such a "paraneoplastic disease" can arise when melanoma tumors are attacked by cytotoxic T-lymphocytes that recognize peptide fragments from melanin pigment, a normal tissue-specific product. The T-lymphocytes can react against both the tumor cells and some normal melanin pigment-forming cells, creating white skin patches (vitiligo) [11]. Under normal circumstances many intracellular proteins (potential self antigens) are not displayed by cells, so there is no deletion (negative selection) of

T-lymphocytes with the potential to react with these proteins when the T-lymphocyte repertoire is being purged and moulded (see below).

Intriguingly, white skin patches are most likely to form where there has been local trauma to the skin [12]. Just as the sound in the wood alerts the sheep, which move away (see Chapter 12), so externally-inflicted stress (not-self) appears to alert skin cells, which are provoked to display fragments of self-antigens. If the concentration of the corresponding specific T-lymphocytes is sufficiently high, then the traumatized, but essentially normal, self-cells are destroyed. By the same token, a minor knock on the head that might normally pass unnoticed, could lead to an immunological disease of the brain if, at that time, cytotoxic T-lymphocytes happened to be rejecting, say, incipient kidney cancer cells that were displaying fragments from a protein that was normally brain-specific (Fig. 13-4). Thus, paradoxically, the first symptom of a kidney cancer might be neurological impairment. Indeed, the T-cell attack might eliminate the provoking cancer, which might then never be detected. Instead the subject would have acquired an autoimmune disease of the brain.

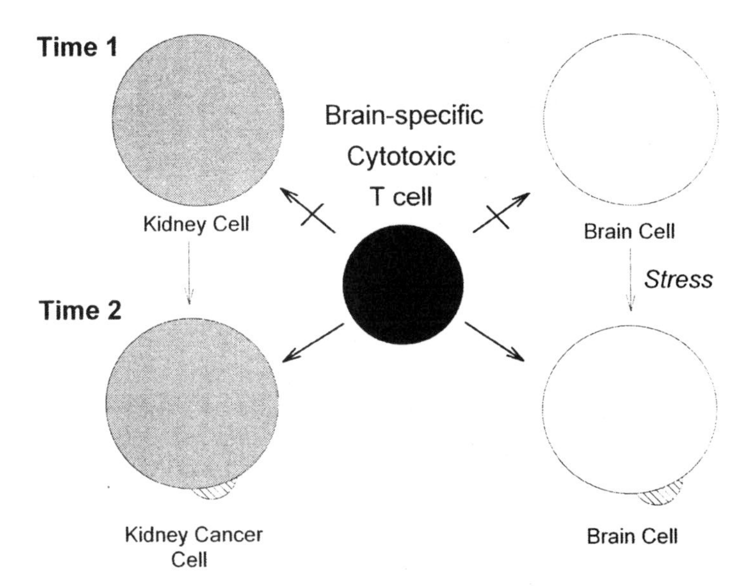

Fig. 13-4. Paraneoplastic disease. Normally a cytotoxic T-cell (central black ball) with a receptor that recognizes a specific MHC-associated brain peptide, will attack neither brain cells, nor cells of other tissues, such as kidney cells (*Time 1*). If the kidney cells later become kidney cancer cells (neoplastic cells), they may then display brain peptides which stimulate the proliferation of specific cytotoxic T-cells (*Time 2*). These can attack both the stimulating cancer cells and *normal* brain cells, particularly if various external stresses provoke the latter to display their own brain-specific peptides.

Death at Home or in Exile

Dividing cells are often seen when tissues are examined under the microscope. For adult organisms of relatively constant size, cell multiplication must be accompanied either by a corresponding number of cell deaths, or by the exile of superfluous cells beyond the body perimeter. Since exiled cells generally die, the options are bleak. Die at home or die in exile! The only exceptions are gametes that can find appropriate partners and so generate new individuals. If the balance (homeostasis) between cell multiplication and destruction is lost, cancer can result.

Death-style options are limited. When there is trauma to a tissue, cells may die by "necrosis," a process that may involve activation of T-lymphocytes and the migration of phagocytic cells (Greek: *phagein* = to eat) from dilated blood vessels. The region may become warm, tender and red (i.e. "inflamed"). Dying cells are ingested by phagocytic cells, which degrade their macromolecules. However, usually cells are eliminated without inflammation. For example, dead skin cells are simply sloughed off into the environment. Each time you undress you discard not only clothes, but also around 400,000 skin cells [13]. Cells that cannot be discarded in this way invoke physiological auto-destructive mechanisms. Without fanfare, the cells self-digest, and their breakdown products are quietly ingested by neighboring cells ("apoptosis"). However, apoptosis is also a possible outcome of an intracellular self/not-self recognition event (see below).

The discarding of cells and/or their secretions into the environment may occur without an organism being aware of it (e.g. sloughing skin cells). However, an organism may be aware of a need to discard, and hence able to consciously control it. Thus, you blow your nose, and cut your nails and hair, at times of your choosing. In your early years, your parents assist this. Even in later years, the result may be more satisfactory if another person is involved (e.g. chiropodist, hairdresser).

At adolescence gamete production begins. In early human communities the discardment of male gametes would usually have involved a sexual partner, who would have been unaware when she was discarding a female gamete. Today, by monitoring the small change in temperature that accompanies ovulation, a human female can know when she is discarding a gamete but, in the absence of medication (e.g. for contraception), she has no conscious control over the timing. In some species, the attention of the male (e.g. visual cues) provokes ovulation.

Selfish Genes and the Menopause

Human females also cannot consciously control the time of the discardment of the uterine cells that have proliferated in anticipation of the arrival of a fertilized ovum. In the absence of pregnancies (or medication), menstrual cycles continue until the menopause. Since natural selection generally favors those who produce most descendents, why is there a menopause? Biologists argue that human females will produce more descendents if, at around the age of 50, they discontinue gametogenesis and expend their energies in attending to the well-being of grandchildren. In other words, individuals with a "selfish" gene that, directly or indirectly, causes arrest of gametogenesis in females, have tended, in the long term, to produce more descendents (who inherit that gene and thus the same tendency for females to discontinue gametogenesis) than individuals who do not have that gene.

This evolutionary trade-off is not so apparent in males, and whether there is a male menopause (e.g. whether being a good grandfather is of greater selective advantageous than continued procreation) is a subject of debate. The discardment of male gametes is usually under conscious control. In modern communities, prior to pair-bonding (legalized as marriage), human males usually discard gametes autonomously at a time of their own choosing. This may involve artificial visual cues (e.g. female images) rather than a partner. Thus, discarded male gametes usually do not prevent menstruation or expand populations. Like other exiled cells, discarded male gametes usually die. Whereas self-discharge of gametes (like menstruation) may have been rare in primitive communities, in modern communities (like other forms of contraception) it is the norm. Yet, over evolutionary time, the shaping of our sexual biology has been mainly influenced by the norms of primitive communities. Biologists argue that "selfish" genes which, by some chain of events, cause us to frown at contraception, must inevitably have increased in human populations. Thus, for most of us the crowded planet is now a more pressing reality than the crowded cytosol.

Molecular Chaperones

Hosts can generate complex multigenic systems for dealing with pathogens. However, pathogens, because of their need to replicate rapidly and disseminate, usually have smaller genomes and so are less able to encode complex systems to counter the increasingly sophisticated host-systems that can arise in an escalating arms race. Large viruses (e.g. with 200 kilobase genomes) have more countermeasures at their disposal than have small viruses (e.g. with 10 kilobase genomes). So the strategy of a large virus must usually be different from that of a small virus.

The armamentarium of the host includes a set of proteins known as "molecular chaperones," which include proteins induced by stresses such as virus infection or heat shock. We came across these "heat-shock proteins" in Chapter 12. Some molecular chaperones play a normal role in cell operations, such as maintaining the structure of self-proteins in order to prevent inadvertent aggregation. This would permit fine-tuning of concentration to approach even closer to the threshold beyond which aggregation would occur. To prevent inadvertent aggregation there would also be a certain margin for error, so that a self-protein would have to "stick its neck out," concentration-wise, before aggregating and triggering alarms that would lead to apoptosis or peptide presentation to cytotoxic T lymphocytes. Figure 13-5 summarizes some factors influencing the normal cytosolic concentration of a diffusible protein [14, 15].

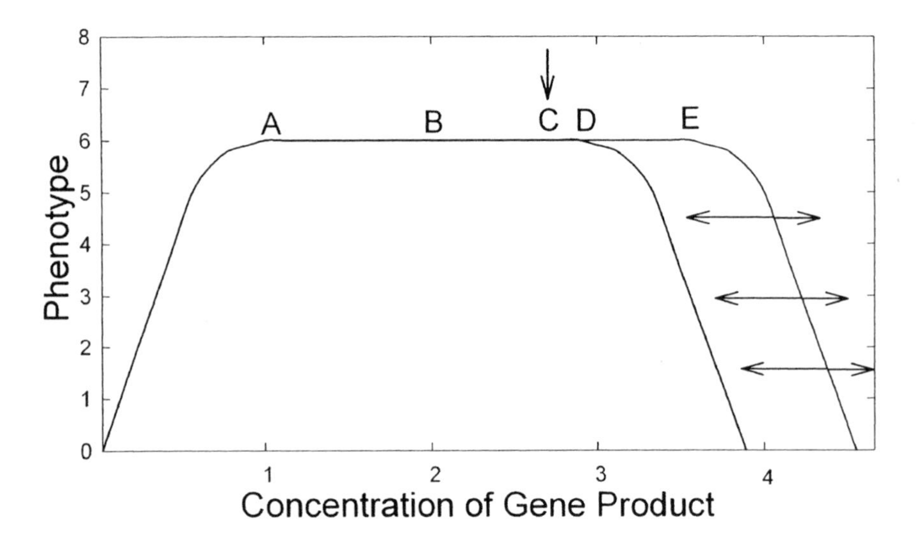

Fig. 13-5. Various factors affecting the concentration of a diffusible protein (e.g. enzyme) within the cytosol of a cell, as revealed by a hypothetical dose-response curve. The dose-response curve shows some quantitative measure of phenotype (e.g. the color of a flower) as it is affected by increasing concentrations of a gene product that generates that phenotype. For example, flower color can depend on the rate of conversion of a colorless substrate to a colored pigment product. This rate would be progressively increased by increasing the concentration of an enzyme (gene product) that catalyzes the conversion. If the concentration of gene product is directly related to the number of active gene copies (as often occurs), then the X-axis can be seen as providing an index of gene dosage (i.e. gene copy number). The phenotypic parameter (measured color) increases with gene product concentration until point *A* when some other factor (e.g. substrate availability) becomes rate-limiting. The curve then plateaus. Increasing the amount of gene product (enzyme) now makes no difference to the colour of the flower (i.e. there is no change in the value of Y).

In a diploid organism, *B* corresponds to the minimum gene dosage required to ensure that the phenotype would be unchanged in a heterozygote. The latter might have only one functional gene copy, and so there would be half the concentration of gene product *(A)* as in the homozygote *(B)*. The phenotype would still correspond to the plateau of the curve. The colour (value on the Y-axis) would be perceived as "dominant," being no less in the heterozygote than in the homozygote (i.e. there would be "haplosufficiency"). *E* corresponds to the concentration at which the gene product would still be soluble (i.e. the color would still be unchanged) *if no other proteins were present*. Above this concentration, the protein would tend to self-aggregate and the phenotypic parameter (Y-axis value) would decrease (i.e. color would fade).

D corresponds to the concentration at which aggregation would occur because of the presence of other cytosolic proteins that would promote such aggregation. The horizontal leftward-pointing arrows symbolize this aggregation pressure exerted collectively by cytosolic proteins, which tends to push the descending limb of the dose-response curve to the left. Another factor, symbolized by the rightward-pointing horizontal arrows, would be the molecular chaperones (e.g. heat-shock proteins) that act to maintain protein conformation and thus decrease aggregation (i.e. tend to push the descending limb to the right).

Thus, it would seem that the concentration of a protein in cells of different members of a species could fluctuate between points *B* and *D*. It is likely, however, that over evolutionary time genes have "fine-tuned" the concentrations of their products to a maximum consistent with avoiding self-aggregation. This point might correspond to *C* (marked by a vertical arrow), which is slightly to the left of *D*, thus providing a margin of safety against inadvertent self-aggregation.

In Figure 13-5 the leftward-pointing horizontal arrows symbolizing the pressure exerted by other cytosolic proteins that tends to reduce the solubility of a given protein. If any particular cytosolic protein mutates in a way that does not affect its concentration (e.g. it may be impaired with regard to maintaining its specific function, but not with regard to maintaining its concentration), then its ability to continue exerting this pressure is unaffected.

Molecular chaperones have the opposite effect to cytosolic proteins in general. The rightward-pointing arrows symbolize the pressure of molecular chaperones to increase the solubility of proteins. A particular type of molecular chaperone has a subset of "client" proteins. Since maintaining the solubility of proteins is a specific function of the molecular chaperones, when they mutate so that this function is compromised, then an important cellular defense against aggregation is removed. Individual protein species that might have just retained their solubility, now more readily cross the insolubility threshold. The proteins no longer have to "stick their necks out" in order to aggregate.

Proteins that are perilously close to the insolubility threshold include mutant proteins which have sustained amino acid changes that affect their conformations, but not their specific functions. As long as they can, with the help of molecular chaperones, maintain their normal conformations, these proteins will maintain their distinctive individual functions. So no effects on the phenotypes of organisms with these mutations will be evident. However, when a molecular chaperone is mutated its client proteins now have to fend for themselves. Many proteins that are close to the insolubility threshold become more prone to aggregate. Mutant phenotypes that were previously hidden (latent) now emerge. They are "conditional mutants" – the condition being that their chaperone is not around to spruce them up. So solubilities (and hence functions) are no longer sustained. Thus, a molecular chaperone can be seen as a mutational "buffer" or "capacitor" that allows organisms to survive certain types of mutation [16].

In the latter case an underlying hidden genetic change can be revealed by chaperone malfunction. This should not be confused with the "phenocopy" phenomenon mentioned above, where an organism undergoing virus attack (or an equivalent stress) at a critical developmental stage might display mutant phenotypes, but there is *no* underlying genetic change [8]. The product of a gene is, because of *environmental* factors, expressed at the wrong time or place, but the gene itself is unmutated.

On the other hand, viral attacks change the levels and modes of expression of heat-shock proteins. If some of these do not remain in chaperone-mode, then some of the observed phenotypes might not be phenocopies, but might reflect the exposure of unbuffered mutations. The expression of these mutations by future offspring would normally also be conditional on failure of chaperone function, but there are suggestions that sometimes mutant expression can switch from conditional to unconditional, so that the expression becomes a permanent characteristic of the line (a mysterious phenomenon known as "genetic assimilation").

All this leads us to distinguish between cytosolic proteins in general, each of which has the potential to react weakly with erring members of a certain *small* subset of its fellow travellers [6], and molecular chaperones – professional interactors – each of which has the potential to react strongly with erring members of a certain *large* subset of its fellow travellers.

Positive Repertoire Selection

Whatever the sophistication of the defence and attack systems of a host and its viruses, the usual initiating event is one of self/not-self discrimination, be it between two RNA species or between two protein species, or be it extracellular or intracellular (see Chapter 12). Accordingly, a virus that could mutate to appear as "self" to its host should have an adaptive advantage. It would exploit the fact that during the development of lymphocytes, each specific for a particular antigen, negative selection of lymphocytes reactive with some "self" antigens had generated "holes" in the repertoire. However, during development (e.g. creation of cytotoxic T-lymphocytes and antibody-forming B-lymphocytes) there is positive selection of those reactive with "near-self" antigens (positive repertoire selection) by mechanisms outlined elsewhere [17–20]. The development of the repertoire of cytotoxic T-lymphocytes, for example, is strongly influenced by the polymorphic MHC proteins of the host. As a virus mutates progressively closer towards the self-antigens of a potential host, the chances that the host will be immunologically prepared become *greater* because its lymphocytes have been positively selected to react with "near-self" MHC proteins. In other words, the virus encounters stiffened host defenses.

In this and the previous chapter we have considered immunology at the intracellular level in a way that is not found in immunology texts. For example, immunologists tend to use the term "altered self," rather than "near self." There is a subtle difference between these usages. "Altered self" implies a difference from self. "Near self" emphasizes how close an entity can come to self yet still be distinguishable from self. Immunologists also tend not to recognize the implications of the crowded cytosol for molecular interactions. But you should be warned that, while the association of peptides with MHC display proteins is well established, at this time the underlying mechanism for the association is not. Then why does a hypothetical mechanism have a place in a text on evolutionary bioinformatics?

One reason, as has been stated, is that the full understanding of the genomes of a species requires an understanding of the genomes of the species with which they have coevolved. These species interactions involve processes of self/not-self discrimination, both extracellular and intracellular. Another reason is that understanding the fundamental role of intracellular protein concentration in self/not-self discrimination can make other evolutionary phenomena intelligible. It is in such terms that we will, in the next chapter, discuss "Muller's paradox" and the mystery of sex chromosome dosage compensation.

Summary

The crowded cytosol is a special environment where weak interactions can be important. Here many macromolecules are close to the limits of their solubility, a condition conducive to weak, but specific, entropy-driven molecular interactions. In addition to being under evolutionary constraint to preserve the functions of their own products, genes encoding specific cytosolic proteins are also under evolutionary constraint, both to support a pressure exerted collectively by proteins to drive other proteins from solution, and to maintain the solubilities of their own proteins in the face of that collective pressure. Thus, genes whose protein products occupy a common cytosol have co-evolved such that product concentrations are fine-tuned to a maximum consistent with avoiding self-aggregation. Cytosolic proteins collectively generate a pressure tending to drive proteins into aggregates. Each individual diffusible protein species both contributes to, and is influenced by, the pressure. Intracellular pathogens must fine-tune the concentrations of their own proteins to the solubility limits so imposed. Aggregates between viral proteins and normal host, antibody-like, "immune receptor" proteins, provide a possible basis for intracellular self/not-self discrimination at the protein level. Molecular chaperones, including heat-shock proteins, modulate this process. In such terms, we can explain Goldschmidt's phenocopy phenomenon, and paraneoplastic diseases. During their development there is positive selection of host lymphocytes reactive with "near-self" antigens (positive repertoire selection). This counters the tendency of pathogens to mutate towards "self". Thus, hosts whose immunological forces are poised to attack "near-self" versions of not-self (a subset of not-self), rather than not-self per se (the entire set of not-self – formidable in range), are at a selective advantage.

Part 6 Sex and Error-Correction

Chapter 14

Rebooting the Genome

> "What makes hybrid male sterility of great current interest is the increasing evidence that the building blocks of this isolating barrier may be radically different from what we had come to believe. ... It is clear that a new paradigm is emerging, which will force us, first, to revised many conclusions ... that had gathered almost unanimous agreement, and, second, to try a completely different experimental approach."
>
> Horacio Naveira and Xulio Maside (1998) [1]

With the notable exceptions of Butler and Miescher, in the nineteenth century the information concept as applied to biological molecules did not extend to information error and the need for its detection and correction (see Chapter 2). Miescher in 1892 thought that: "Sexuality is an arrangement for the correction of these unavoidable stereometric architectural defects in the structure of organized substances" (see Chapter 3). While referring to "left handed coils" being "corrected by right-handed coils," at that time he was unable to relate this to "nuclein," a new substance he had discovered, later known as DNA. However, he appreciated that correction would require some sort of yardstick (i.e. "right hand coils") to permit the fact of error in a molecule of interest (i.e. "left hand coils") to be detected, and then appropriately corrected.

If the various degrees of redundancy found among DNA molecules have a single explanation, it is error-detection and correction. This itself might explain another apparent redundancy, the fact that members of most biological species are either one of two sexes. Yes, this certainly makes our lives more colorful. But, the biological advantage is not obvious. Imagine a world without males in which each woman, on average, was able to produce two offspring per generation asexually. Both of these would be female. In the first generation there would be two women. In the second generation there would be four women. In the third generation eight women, and so on. With a similar limit to offspring number, in a sexual world a woman would be likely to produce one male and one female per generation. Only the latter, on being fertilized by a male, would produce further offspring. In the first generation there would be two individuals, male and female. In the second generation, there would still be two, and so on to the third generation, etc. … . Since the winners in the struggle for existence are organisms that leave the most offspring, sex would seem very disadvantageous [2].

Redundancy

At the outset the two-fold parallel redundancy of naturally occurring DNA duplexes, which contain complementary "top" and "bottom" strands, was recognized as revealing not only how DNA was replicated, but also how it might be corrected. Changes in DNA, manifest as unusual bases (e.g. **U** instead of **T**), or damaged bases (e.g. cross-linked **T**s), would have provided a selection pressure for the evolution of specific enzymes that could recognize and correct errors. Indeed, many such enzymes were discovered [3].

Sometimes, however, a normal base can change into another normal base ("base substitution") so that an unusual base-pair results (e.g. **A** on one strand paired with **C** on the other strand). Again, one can envisage a selection pressure for the evolution of enzymes that *recognize* base mispairing; but after the fact of recognition, how is it known which base to correct (see Chapter 2)? Take the following duplex:

CAGGCTATCGTAA (14.1)
GTCCGATAGCATT

Consider a base change (transition mutation) from **T** to **C** (underlined):

CAGGCTATCGTAA (14.2)
GTCCGACAGCATT

The error might be *corrected* as:

CAGGCTATCGTAA (14.3)
GTCCGATAGCATT

Or *compounded* as:

CAGGCTGTCGTAA (14.4)
GTCCGACAGCATT

There is thus only a 50% chance of actual correction. And there is another actor in the wings. While the "rapid response team" of error-correcting enzymes is alert for base mispairs, and is quick to move into action, one can hear in the distance the whistle and thunder of an approaching DNA polymerase further up the line. If there is any delay, sequence 14.2 may be replicated before it can be corrected. Two new child duplexes (like 14.3 and 14.4) will be produced. One will not contain the error (unmutated duplex). One will contain the error (mutated duplex). In this case there is a 100% chance that there is no correction, at least in the short term. After replication there is no base mispairing in either duplex, so there is no indication, at the duplex level, as to which duplex is the mutant. On partitioning of chromosomes to child cells (segregation), one child cell might malfunction, and on this basis the mutant phenotype might be selected against. To this extent the mutation would be detected and corrected.

Assuming that the error-correcting team will usually arrive before suspicious DNA is replicated, the most obvious way to improve on the 50% correction success rate is to provide the team with another duplex to compare. If sequence 14.2 could be compared with sequence 14.1, the team would "know" that the **A-C** mispairing in 14.2 should be corrected to an **A-T** Watson-Crick base-pair, rather than to a **G-C** Watson-Crick base-pair. Thus, a further level of redundancy is needed, either "in parallel" (i.e. another independent duplex), or "in series" (i.e. repetition further along the line).

Diploidy, in theory, should solve the problem. However, the system is not perfect as indicated by the need to race to correct errors before DNA polymerase can compound them and turn potentially correctable mutations into what may eventually become "accepted mutations." The latter can be detected either experimentally, by those who sequence DNA molecules, or biologically, when there is an observable change in phenotype. In humans, if deleterious mutations affect only somatic tissues and occur after reproductive life, then from the viewpoint of the species there is no major problem (the benefits of infant care by grandparents notwithstanding). On the other hand, mutations in germ-line tissue might result in defects being passed on to future generations. Meiosis appears as a special germ-line process with a low error rate that might both decrease germ-line transmission of mutations, and repair DNA damage. It is the nearest we have to "rebooting" the genome.

Recombination Repair

For recognition and correction of a mutation there must be a template for comparison. Accordingly, you are diploid, and were generated from haploid paternal and maternal gametes that had themselves been subject to meiotic cleansing. You, in turn, can meiotically generate fresh gametes as a rich blend of your parental genomes, due to (i) the random assortment of (i.e.

creation of various combinations of) 23 entire paternal and 23 entire maternal chromosomes (the only constraints being that 23 are allowed in a haploid gamete, and there must be one member of each homologous pair), (ii) the random assortment of (i.e. recombination among) parts of each chromosome pair, and (iii) recombination repair (see Fig. 8-3). However, the effectiveness of the latter correction will depend on the templates. There are three template-critical alternative outcomes:

(i) If your parents were closely related (e.g. incest), then it is likely that you were born with many identical mutations at identical positions in all chromosome pairs (i.e. you are homozygous for these mutations). So the fact of mutation may not be detectable. But the fact that you exist shows that these mutations can be non-lethal when homozygous.

(ii) If you are the result of out-breeding (i.e. your parents were not closely related), then there will be far fewer shared mutations in homologous chromosome pairs, so that many individual mutations can be identified. For example, one of your parents might have donated sequence 14.3 (unmutated) and your other parent might have donated sequence 14.4 (mutated). Thus, you would be heterozygous with respect to the one base pair difference between these sequences. One "good" copy of a gene containing that sequence might suffice (i.e. you would be "haplosufficient" and therefore likely to be viable as an individual; see Fig. 13-5). However, when the two sequences were compared at meiosis as candidates for promotion to gamete status, the fact of heterozygosity might be recognized. Recombination repair might convert one of the sequences to that of the other ("gene conversion;"Fig. 8-3). There would be a loss of heterozygosity and a gain of homozygosity. Without guidance, the conversion would have, on average, a 50% success rate (since which was the "good" copy would not be known).

(iii) Finally, if you are the result of extreme outbreeding the definition as "template" becomes spurious. In one position in the top strand, instead of either **A** (sequence 14.3), or **G** (sequence 14.4 containing an **A** to **G** transition mutation), there might be a **C** or **T** (i.e. a transversion mutation with the purine mutated to a pyrimidine). A few differences of this nature might be tolerated. But many such differences, be they transitions or transversions, will result in a failure of pairing, so that meiosis fails, no gametes are formed and, although you may not be immediately aware of it, you are a sterile hybrid (a "mule;" see Chapter 8).

The second of the above three alternative outcomes, a consequence of parental out-breeding, appears the most advantageous. But if there is only a 50% gene conversion success rate with respect to a base-pair heterozygosity at any *one* position, then the *overall* success rate for multiple heterozygous positions will remain at an average of 50%. Unless there is some guidance as to which parental contribution to your gametes is more advantageous *for each mutant position*, your average gamete will be no better off than you are.

However, 50% of your gametes will, by chance, emerge with more "good" base pairs than you have. And 50% of your spouse's gametes will, likewise, be superior. Thus, if you have several children, some of these will be superior to you and your spouse, with respect to the function which Nature holds supreme, namely the ability to produce many, reproductively fit, children. Accordingly, over the generations, combinations of "good" genes should be selected. This would be emphasized if the positive effects of "good" genes were more than additive (i.e. genetic fitness might increase exponentially as the number of "good" genes increased). A similar argument could apply if the negative effects of "bad" genes were more than additive, so that such genes would tend to be purged collectively from the population.

Strand Guidance

Depending on the type of mutation, sometimes there is guidance as to which copy of a gene is best. For example, there might be a transition micromutation from **C** to **T** in sequence 14.1, resulting in:

5'**CAGGCTATTGTAA**3' (14.5)
3'**GTCCGATAGCATT**5'

Of importance in this case is that the **C** was part of a **CG** dinucleotide sequence in the top strand, which must base pair with a complementary **CG** dinucleotide sequence in the bottom strand (recall that **CG** is read with 5' to 3' polarity, and the two strands are antiparallel; see Fig. 2-4). The **C** in **CG** dinucleotides is sometimes chemically changed (methylated), but the changed base, methylcytosine, is accepted in DNA. (So there are often five, not four, bases occurring naturally in DNA; see Chapter 15). **C** is also sometimes chemically changed (deaminated) to the base **U**. The **U** is recognizably wrong, since **U** does not occur in DNA. But when a methylated **C** is deaminated the resulting base is methyluracil, otherwise known as thymine (**T**), which is, of course, a natural base.

When encountering a base mispair in duplex DNA such that **T** in one strand pairs with **G** on the other, the rapid response team of error-correcting enzymes can check if the latter **G** has a **C** on its 5' side. Better still, it can determine if the **C** is methylated. It then has the information to deduce that there was probably a **CG** dinucleotide on the other strand, so that the error lies in the **T**, not in the **G**.

Nevertheless, the most secure strand guidance would be to have a sequence redundancy greater than diploid (see Chapter 2). A diploid cell, prior to engaging in mitotic, or meiotic, cell division, replicates its DNA in a period referred to as the S-phase of the cell cycle. For a period after this (referred to as G2 phase of the cell cycle) the cell is actually tetraploid with four DNA duplexes (Fig. 14-1).

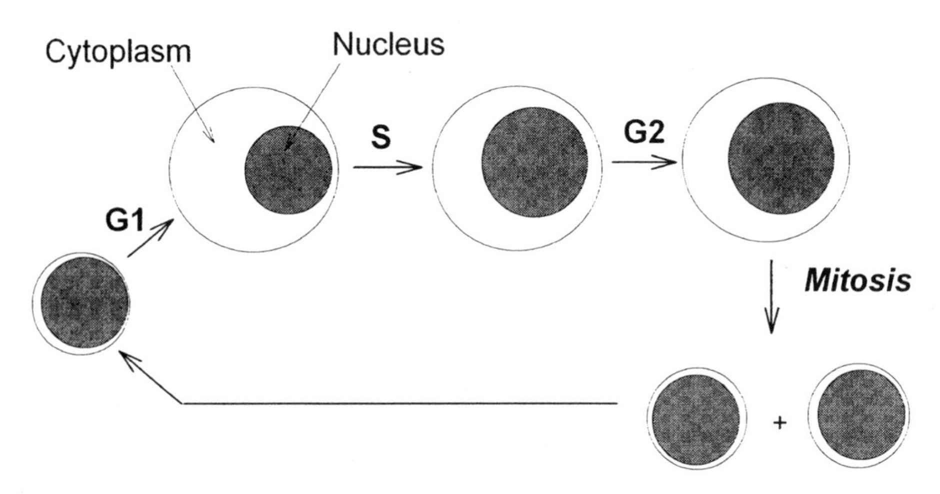

Fig. 14-1. Phases of the cell cycle. After normal cell division (mitosis), child cells (diploid with two copies of each chromosome duplex) may enter the "first gap phase" (G1-phase), in which there is cytoplasmic growth, but little growth of the nucleus where the chromosomes, containing DNA, are located. In S-phase, DNA is replicated and the nuclear volume approximately doubles in size. In the "second gap phase" (G2-phase) the cell is tetraploid with four copies of each chromosome (i.e. four homologous DNA duplexes). Active genes (e.g. "house-keeping genes;" see Chapter 15), and active chromosomes (e.g. the active X chromosome in human females), tend to be replicated early in S-phase. This means that they have shorter G1 phases, and longer G2 phases (i.e. they are longer in the tetraploid state), than inactive genes and chromosomes

After cell division (called mitosis because the thread-like chromosomes become visible; Greek: *mitos* = thread), the cell will return to the diploid state. If a mutation is accepted and remains unrepaired in the period referred to as the G1-phase, it will be faithfully replicated in S-phase, and there will then be two identical mutant DNA duplexes to be repaired in the G2 phase. The relative proportions of mutated and unmutated duplexes will be unchanged. On the other hand, if the mutation occurs after DNA replication (i.e. in the G2 phase) then there will be three DNA duplexes against which the mutant duplex can be compared, so in this case there can be guidance. It is perhaps for this reason that some genes, chromosomes, and cells (e.g. mammalian liver cells and vertebrate oocytes) spend more time in the G2 phase of the cell cycle. There is also a "selfish gene" perspective on this issue, which will not be discussed here [4].

Since mutation is essentially a random event, in a multicellular organism it should strike one duplex in one cell at a time. This means that a diploid cell, with an accepted somatic mutation in one of its two duplexes, could seek help from another diploid cell to verifying which of its duplexes was mutated. This would involve transfer of nucleic acid information between cells, a process that, if it occurred between single cell organisms, would be referred to as sexual. Whether such a process can occur between somatic cells in a multicellular organism is unknown at the time of this writing, but there is suggestive evidence [5].

In meiosis a single cell with two DNA duplexes (of paternal and maternal origin) first replicates to generate four DNA duplexes and then divides twice (producing four cells) with no intervening phase of DNA synthesis. In human male meiosis four haploid gametes result. In human female meiosis one haploid gamete results and three are discarded. Thus, in female meiosis there is a further opportunity for gamete quality control; but if, and how, this would be brought about is unknown [4].

Nevertheless, through the process of sexual reproduction, a given duplex finds itself, from generation to generation, in a succession of bodies each time in the company of a new homologous duplex. A mutant duplex (e.g. sequence 14.2 above) may never be in a situation where it can simultaneously be compared within a cell with three corresponding unmutated duplexes (e.g. sequence 14.1 above), but it can sequentially achieve this result over the generations. If in every generation duplex 14.2 repeatedly encounters duplex 14.1, and never another duplex 14.2, then even with a 50% probability of successful gene conversion on each occasion, a likely outcome will be elimination of the mutation. This process would be assisted if individuals with duplex 14.2 were negatively selected. However, it should be noted that mutations can recur and so may, by chance, restore those that are eliminated.

DNA Damage

An "error" in DNA must be understood in context. Of duplex sequences 14.3 and 14.4 above, one may have good and one may have bad consequences for an organism. On the other hand, sequence polymorphism is a natural phenomenon (see Chapter 12) and, in some circumstances, neither sequence would be inappropriate. DNA damage, however, is always an error. As such it must be "corrected" or "repaired." Strictly speaking these words have the same meaning but, although there is no generally agreed usage, in the literature incorrect base substitutions (micromutations) tend to be "corrected" and damage tends to be "repaired." However, correction by recombination, whether it be of an incorrect base substitution, or of damage, has long been referred to as "recombination repair."

With damage there is likely to be little ambiguity regarding recognition as such, and if the damage affects just one strand then there is likely to be no problem with strand discrimination. In this case, repair could be made on the basis of the complementary sequence in a single duplex (i.e. repair could be effected in a haploid individual). If both strands are damaged (e.g. complementary base-pair damage), then repair must be made on the basis of information present in a homologous duplex that may be either in-series (i.e. sequence duplication) or in-parallel (i.e. diploidy).

One hypothesis of aging is that the selection pressure for efficient damage repair has not been high over evolutionary time because organisms have usually died relatively early from environmental causes (e.g. hunger, predation, disease), rather than from DNA damage. Individuals have had to reproduce before being struck by these apocalypses. So their offspring have tended to inherit traits that enabled them to deal with these, rather than with the ravages of time. Natural selection has mostly favored those with traits that are manifest early in life. Butler gave us a glimpse of this [6]:

> "If heredity and memory are essentially the same, we should expect that no animal would develop new structures of importance after the age at which its species begins ordinarily to continue its race; for we cannot suppose offspring to remember anything that happens to the parent subsequent to the parent's ceasing to contain the offspring within itself. From the average age, therefore, of reproduction, offspring should cease to have any further steady, continuous memory to fall back upon; what memory there is should be full of faults, and as such unreliable."

Thus, in modern populations where apocalypses have greatly decreased, the phenomenon of aging, possibly due to a failure to repair DNA damage, has emerged. The damage would affect the transcription and replication of DNA, with phenotypic consequences that lead to death [3].

Although the disposable soma (that's you and me) may not have highly efficient DNA damage repair systems, such efficient repair is necessary for gamete production in the gonads. Here recombination repair would seem to constitute the last court of appeal, with a sequence in one duplex coming to the rescue of the sequence in the homologous duplex (see Chapter 12). Here again, the templating quality of the donating duplex should be considered. If meiosis worked efficiently in your parents' gonads, you initially inherited undamaged parental DNA sequences. So most of the damages present in your gonads are likely to have been incurred during your lifetime, and are unlikely to be at the same positions in the paternal and maternal haploid contributions to your diploid genome. Thus, the fact that your parents might have been closely related might not be important as far as the repair of DNA damage is concerned. In this respect, parental outbreeding might not necessarily be beneficial. However, DNA damage might summate with the intergenomic differences accompanying extreme outbreeding, so that meiosis would more likely fail (i.e. producing hybrid sterility) than in a genome with less damage.

Note that we have been considering correction and repair at the level of classical Watson-Crick duplexes. Here a **G-T** mismatch would be recognized and corrected. Yet **G-T** pairings occur normally in extruded DNA stem-loop structures (see Chapter 5). Correction of such weak pairings (i.e. formation of a **G-C** or **A-T** pair instead of a **G-T** pair) would increase stem-loop stability, but generate mutations in the corresponding classical duplexes. This might have phenotypic consequences. So it may be presumed that, under biological conditions, error-detecting and correcting enzymes confine their activities to classical duplexes and **G-T** pairings in stem-loops are ignored. Alternatively, such **G-T** pairings are repaired with possible phenotypic consequences, or mutations in the neighborhood that change local stem-loop patterns to avoid **G-T** pairs are accepted [7].

Piano Tuning

From the studies of Wada and others (Chapter 9), individual genes emerge as distinct (**G**+**C**)% entities, each with a distinct potential vibrational frequency. Hence, it seems not unreasonable to regard individual genes much as we regard the individual strings of a musical instrument. A gene with a mutated or damaged base is a gene "out of tune" to the extent that the change may marginally change its vibrational frequency. We should consider the possibility that this has not escaped Nature's notice. Perhaps, there are "tuning" enzymes that can check for changes in resonance, and hence supplement the methods of accurate detection of errors based on direct templated interactions between DNA sequences as considered here. While this might seem fanciful, from time to time there are strange "sightings" in the literature that might well be explained in such terms [8]. The astonishing precision with which entire genomes obey Chargaff's second parity rule (see for example the composition of vaccinia virus in Chapter 4), suggests we still have much to learn about long range interactions in DNA.

Spot Your Reprotype?

In general, outbreeding is beneficial producing "hybrid vigor." But extreme outbreeding could result in disparities in base composition [(**G**+**C**)%] leading to hybrid sterility [9]. From the time of Darwin [10], behavioral studies have suggested how prospective mating partners might have evolved genes inducing courtship displays that would subconsciously allow assessment of the likely vigor of the resulting children should courtship lead to mating. In this context, base compositional differences (i.e. differences in "reprotype;" see Chapter 8) might seem to be below the threshold of detectability by prospective mating partners.

On the other hand, intragenomic conflicts between the pressures on the conventional Darwinian phenotype, and the pressures on the genome phenotype, are sometimes won by the latter (see Chapters 9 and 10). Thus, the genome phenotype can affect the conventional phenotype. In this respect the arrow denoting classical Darwinian natural selection (from environment to phenotype to genotype) is, in principle, reversible. It is thus theoretically possible, although so far without experimental support, that an organism could, as part of its conventional phenotype, display its reprotype – its (**G**+**C**)%. The selective advantage of this (in terms of long term reproductive success) should be considerable, since extremes of inbreeding and outbreeding could be avoided. Thus, hybrid sterility may not be just something that is encountered on the roulette wheel of life. "Your face" may indeed be "your fortune," or rather, your children's fortune.

Haldane's Rule

The advantages of outbreeding argue strongly for the advantages of sexual reproduction, as opposed to asexual reproduction (essentially extreme inbreeding). Up to the hybrid sterility limit, the more disparate the genomes are, the better. As discussed in an earlier book [9], the cycle of reproduction among members of a biological species will proceed continuously, unless interrupted at some stage. The most likely initial stage for such cycle interruption (i.e. the origin of the speciation process) is when disparate haploid parental genomes meet in diploid meiosis and cannot "agree" to pair. Gamete production fails and the organism, a hybrid of paternal and maternal genomes, is sterile (see Fig. 7-4). Turning a diploid into a tetraploid can "cure" hybrid sterility since each chromosome at tetraploid meiosis has a pairing partner (see Chapter 8).

Remarkably, diploid sterility is not gender neutral. As parental genome sequences get progressively disparate, the proportion of sterile children increases. But initially it is the sex that produces two types of gametes (the heterogametic sex) that is sterile. Among humans, males are the heterogametic sex, so boys are sterile and girls are fertile. This regularity was recognized by the English polymath J. B. S. Haldane [11]. It was also recognized, because of its applicability to a wide range of taxonomic groups (e.g. birds, insects, mammals, nematode worms, some plants), that the phenomenon has the potential to tell us much about sex and speciation [12]. However, "Haldane's rule" is really two rules. Haldane's rule for hybrid *sterility* should be clearly distinguished from Haldane's rule for hybrid *inviability*. Diploid inviability is also not gender neutral. It is again the heterogametic sex that is preferentially affected.

A normal human male produces two types of spermatozoa in equal quantities – spermatozoa with an X-chromosome that will confer femaleness on children, and spermatozoa with a Y-chromosome that will confer maleness on children. A new organism is created when one of these spermatozoal types fertilizes a female gamete, which is always an X-bearing gamete. As discussed in Chapter 4, if X-bearing spermatozoa and Y-bearing spermatozoa are in equal quantities, there will be approximately equal chances of a boy (diploid sex chromosome status XY) or a girl (diploid sex chromosome status XX). But if the genes donated to the zygote by the paternal and maternal genomes cannot work together effectively, then there may be a failure in development, so that at some stage zygote-derived cells fail to multiply and differentiate into a viable organism (Fig. 7-4). This "hybrid inviability" usually reflects incompatibilities between gene products and, for reasons discussed elsewhere, preferentially affects the heterogametic sex [13]. This means that boys are either not born, or fail to develop into mature adults. Since only adults produce gametes, *the question of their sterility does not arise*.

If a human female has a tendency towards hybrid sterility (i.e. the two parental genomes she carries are extremely disparate), then there will be decreased production of X-bearing ova. But, although it is likely that there will be fewer children, there is no bias towards boys or girls *by virtue of her sterility*. If a male has a tendency towards hybrid sterility (i.e. the two parental genomes he carries are extremely disparate), then there will be a decreased production of Y-bearing spermatozoa and X-bearing spermatozoa. But again, although it is likely that there will be fewer children, there will be no bias towards boys or girls *by virtue of his sterility*. All that Haldane's rule for hybrid sterility implies is that *the tendency towards complete sterility is facilitated in males*. To understand this, we will first consider sex chromosomes and their role in the anti-recombination process that can lead to speciation.

Sex Chromosomes

In many species males and females are produced in equal quantities. Mendel himself noted that such equal quantities would be produced if a recessive homozygote were crossed with a heterozygote. Thus, if red is dominant to white, when a homozygous white (*WW*, producing one type of gamete, *W* and *W*) is crossed with a heterozygous red (*RW*, producing two types of gamete, *R* and *W*), on average, equal numbers of red and white offspring should be produced (*RW*, *WW*). If sex were similarly determined, this simple scheme would suggest that one sex be homozygous and the other sex be heterozygous for alleles of a particular gene.

As long as only one allele pair was required there would be no reason to regard this process as different from any other genetic process. The chromosome pair containing the gene – let us call them X and Y chromosomes – would be equal in all respects, including size ("homomorphic"). One sex would be homozygous for the recessive allele (the "homogametic" sex) and the other sex would be heterozygous (the "heterogametic" sex). If there were recombination between the chromosome pair in the heterogametic sex, the gene might switch chromosomes, thus converting an X chromosome to a Y chromosome and the corresponding Y chromosome to an X chromosome. The *status quo* would be preserved and equal numbers of differentiated gametes would still be produced.

However, if the complexities of sexual differentiation were to require more than one gene, – say genes X_1 and X_2 on the X chromosome, and the corresponding allelic genes Y_1 and Y_2 on the Y chromosome – the situation would get more complicated. Recombination (crossing-over between chromosomes at meiosis) might then separate the two genes to generate chromosomes (and hence gametes) with genes X_1 and Y_2 together, and genes Y_1 and X_2 together. Sexual differentiation would be impaired (Fig. 14-2).

To prevent this happening either the gene pairs, – X_1 and X_2, and Y_1 and Y_2 – would have to be closely linked on the X and Y chromosomes, respectively (so the chance of a pair being separated by meiotic recombination would be low), and/or one of the chromosomes would have to develop some local mechanism to *prevent* such recombination. If this *anti-recombination activity* could not easily be localized, then the activity might spread to involve other genes on the chromosomes; these genes might themselves play no role in sexual differentiation.

This seems to be the situation that often prevails. The needs of the predominant function, sex determination, overrule, but do not necessarily eliminate, the functions of other genes on the same chromosome. The chromosomes are referred to as the sex chromosomes, even though concerned with many functions not related to sex. However, anti-recombination is the essence of the initiation of speciation (see Chapter 8). The prevention of recombination *between* members of different incipient species, or of an incipient species and the parental species, is a fundamental part of the speciation process ("reproductive isolation"), and extends to all the chromosomes, including the sex chromosomes [9].

On the other hand, sex chromosomes exist *within* a species, and anti-recombination during meiosis is only beneficial to the extent that it prevents recombination between genes specifically concerned with sexual differentiation. When anti-recombination activity spread along sex chromosomes to genes not concerned with sexual differentiation, the latter lost the benefits of recombination.

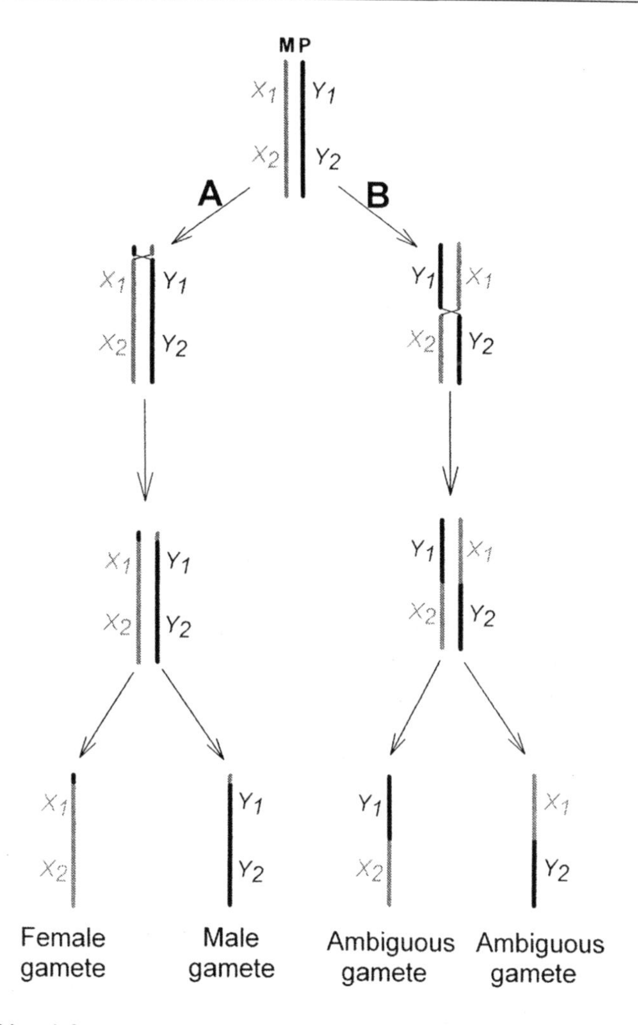

Fig. 14-2. Need for meiotic anti-recombination activity when multiple genes determine sex. In a human male (top), the maternally-donated sex-determining chromosome (M; grey) has a pair of sex-determining genes X_1 and X_2, and the paternally-donated sex-determining chromosome (P; black) has a pair of sex-determining genes Y_1 and Y_2. If path *A* is followed at meiosis, there is an exchange of chromosome segments (recombination) that does not disturb the linkage of each pair. The haploid female gamete (i.e. the gamete conferring the female sexual state) has part of the original paternally-donated chromosome, but X_1 and X_2 remain linked (bottom). The haploid male gamete (i.e. the gamete conferring the male sexual state) has part of the original maternally-donated chromosome, but Y_1 and Y_2 remain linked. If path *B* is followed at meiosis, the exchange of chromosome segments disturbs the linkage, so that each resulting haploid chromosome (at bottom) has both a female sex-determining gene and a male sex-determining gene

To the extent that meiotic recombination allowed recombination repair, then, when such repair was lost, non-lethal mutations remained uncorrected. There was no longer prevention, *within* species, of changes in sex chromosomes, including micromutations (base substitutions, additions, or deletions) and macromutations (segment additions, deletions, transpositions and inversions). These changes are similar to those occurring *between* species as they progressively differentiate their chromosomes [14].

Thus, instead of remaining the same size (homomorphic), the sex chromosomes sometimes came to differ in size (heteromorphic). In some species (e.g. certain fish and amphibia) the sex chromosomes remained homomorphic and appeared to differ only in the genes affecting sexual differentiation. In other species (e.g. mammals, birds) the sex chromosomes became heteromorphic.

Sex and Speciation

Sexual and species differentiations can involve anatomical and physiological changes of similar orders. When classifying organisms into species, two forms initially considered as distinct species have sometimes been found to be merely the sexually differentiated forms of one species [15]. An extreme case of this was Darwin's realization that an apparent member of a parasite species found within the body of a female barnacle was actually the male form of that species ("complemental males") [16]. Bateson and Goldschmidt both drew attention to the possibility of a fundamental similarity between the evolutionary process that divides two groups within a species, such that they become two species, and the evolutionary process that divides two groups within a species, such that they become two sexes. When the chromosomal basis of sexual differentiation became clearer, Bateson in 1922 commented on the chromosomally-borne "ingredients" responsible for this in plants [17]:

> "We have now to admit the further conception that between the male and female sides of the same plant these ingredients may be quite differently apportioned, and that the genetical composition of each may be so distinct that the systematist might, without extravagance, recognize them as distinct specifically [i.e. mistake the different sexes as distinct species]. If then our plant may ... give off two distinct forms, why is not that phenomenon [i.e. sexual differentiation] a true instance of Darwin's origin of species?"

In 1940 Goldschmidt, who had postulated that the initiation of speciation requires a general change in a chromosome's "reaction system" (see Chapters 7 and 8), noted [18]:

> "Another remarkable problem ought to be pointed out ..., though it seems at first sight to be rather remote from the problem of species formation. ... Sexual differences within a species may be of such a nature that, if found distributed among different organisms, they would provide a basis for classification into different species, families, or even higher categories. These differences frequently touch upon practically every single character of the organism, morphological and physiological. Two forms found in nature, which showed sexual differences of such degree ... would never be considered as belonging to the same species ... . In the sexual differences we have, then, two completely different reaction systems in which the sum total of all the differences is determined by a single genetic differential... . The genetics of sex determination ought therefore to furnish information on how a completely different reaction system may be evolved."

Goldschmidt made the distinction between "reaction system" and genes explicit:

> "It is not this or that gene or array of genes which is acting to produce the extreme morphogenetic differences of the sexes, but rather the typical serial pattern within the X-chromosome, or definite parts of it. The chromosome as a whole is the agent, controlling whole reaction systems (as opposed to individual traits). The features which are assumed by many geneticists to prevent a scattering of individual sex genes by crossing over ... actually prevent major changes of the pattern within the chromosome as a whole. Once more I must emphasize that such a conception offers mental difficulties to those steeped in the classical theory of the gene."

As mentioned above, species differentiation and sexual differentiation have an important feature in common. For their maintenance there must be an absence of meiotic recombination, either between one or more chromosome pairs in *both* sexes (species differentiation or "speciation"), or between regions of one chromosome pair in *one* sex (sexual differentiation). For speciation, chromosomes of members of a variant group must not recombine with those of the parental line, otherwise character differences due to multiple genes could blend, and the differentiation would be lost. For sexual differentiation, regions of the sex chromosomes must not recombine because the characters conferring sexual identity could blend, and sexual differentiation would then be lost. The sex chromosomes (e.g. X and Y in humans) must, between themselves, maintain something akin to the reproductive isolation that defines organisms as members of distinct species.

Haldane's rule for hybrid sterility usually concerns crosses between members of different "races" (varieties, breeds, lines) *within* what appears as a single species (defined reproductively) [19]. We are accustomed to recognize as "races," groups within a species, the members of which show some common anatomical differentiation from the members of other groups. However, in principle, members of a group ("race") within a species might begin to differentiate with respect to reproductive potential (no longer retaining full fertility when crossed with parental stock) *before* any anatomical or other physiological differences are evident [20, 21]. When fully differentiated in this way, the members of the group would be reproductively isolated, and so would constitute a distinct species, even if not anatomically or functionally distinguishable from members of the parental group (see Chapter 3).

Thus, Haldane's rule for hybrid sterility has no obligatory anatomical or physiological correlates, and is a phenomenon of incipient speciation *alone.* It seems likely that the preferential sterility of the heterogametic sex can be regarded as a *step,* or *way station,* on the path towards the complete sterility associated with full species differentiation, a process generally accompanied by the anatomical and physiological differentiation of conventional phenotypic characters.

The sex chromosomes tend to progressively differentiate from each other, and in some species meiotic recombination remains possible only in small regions where there is sequence homology, which includes (**G+C**)% compatibility. In this respect the small regions behave like the autosomes, and are referred to as "pseudoautosomal" regions. Other parts of sex chromosomes (the non-pseudoautosomal regions) appear to be kept from meiotic pairing and recombination by a failure of homology [22, 23]. However, recombination repair is still possible when there is *in-series* sequence redundancy (see Chapter 2). Indeed, the human Y chromosome contains extensive palindromic duplications that should facilitate intra-chromosomal non-meiotic gene conversion [24, 25].

If the mechanisms of sexual and species differentiations were similar, then, since the opportunity for recombination between different sex chromosomes (e.g. X and Y) can occur only in the sex with both chromosomes (the heterogametic sex), by *preventing* meiotic recombination that sex could be considered to have taken a step towards speciation. Whereas, for species differentiation, the homogametic sex (e.g. human females) would have to differentiate *both* sex chromosomes *and* autosomes, the heterogametic sex (e.g. human males) would have only the autosomes left to differentiate. By virtue of this head-start, among the progeny of crosses between an incipient species and its parental stock, *the heterogametic sex would be preferentially sterile* (i.e. Haldane's rule for hybrid sterility) [26].

The possibility of general changes in the base composition [(**G+C**)%] of chromosomes, perhaps the modern equivalent of Goldschmidt's changes in "reaction pattern," is consistent with studies of Haldane's rule in the fruit fly, where, like humans, the male is the heterogametic sex. In 1998, Naveira and Maside concluded [1]:

> "The total number of sterility factors [on chromosomes] must probably be numbered *at least* in the hundreds. The individual effect on fertility of any [one] of these factors is virtually undetectable, but can be accumulated to others. So, hybrid male sterility results from the [experimental] co-introgression [combining] of a minimum number of randomly dispersed factors (polygenic combination). The different factors linked to the X, on the other hand, and to the autosomes, on the other, are *interchangeable* [i.e. are equally effective] … . Recent experiments on the nature of these polygenes suggest that *the coding potential of their DNA may be irrelevant.*"

In the latter respect it was further noted:

> "The effect detected after inserting non-coding DNA suggests that the coding potential of the introgressions … might be … irrelevant for hybrid male fertility. *It might be only a question of foreign DNA amount, … .*"

In the general case, later-developing *secondary* mechanisms of isolation, reinforcing and/or substituting for a primary micromutation-dependent mechanism, would include chromosomal macromutations (likely to increase hybrid sterility), single or collective *genic* incompatibilities (likely to produce hybrid inviability) [13], and mating incompatibilities (also likely to be of *genic* origin; see Fig. 7-4).

Sex Chromosome Dosage Compensation

The overwhelming influence of the need to correct errors appears to explain how DNA came to "speak in palindromes" (see Chapter 5), and how many species came to have two sexes. The need for two sexes, in turn, required that at least two entire chromosomes be somewhat constrained in their error-correction activities – a constraint that often led to the degeneration of one of them (e.g. the Y-chromosome in human and fruit fly) [27]. So males found themselves with, essentially, only one set of X-chromosome genes, and hence with the potential for only one dose of X-chromosome gene products, the majority of which played no role in sexual differentiation.

In contrast, females with two X-chromosomes had the potential for two doses of X-chromosome gene products. Thus, although the sexes might be equal with respect to the doses of autosomal gene products, females might have an extra dose of X-chromosome gene products. Many of these products would be proteins corresponding to genes whose expression might not be subject to short-term fluctuations in response to environmental influences or developmental programs (i.e. the genes might be constitutively expressed) [28]. Would this potential difference between the sexes in intracellular protein concentration be accepted (dosage acceptance), or corrected (dosage compensation)?

Many species engage in some form of sex chromosome dosage compensation. In humans this takes the form of a random inactivation of one of the X-chromosomes in women. Since one X-chromosome was inherited from her father, and the other was inherited from her mother, this means that in about half her cells a woman expresses her paternal X-chromosome and in the other half she expresses her maternal X chromosome. Since it is women who have to do the accommodating, it has been wryly remarked that God must be a man! However, in fruit flies it is the male who accommodates, by doubling the activity of his solitary X chromosome.

Muller discovered sex chromosome dosage compensation in fruit flies in the late 1920s, and wondered how it could have evolved [29]. He had already deduced that, in homozygotes, gene, and hence gene-product, dosage is usually well along the plateau of the dose-response curve (see point B in Fig. 13-5). Males would be expected to have half the dosage of X-chromosome gene products, but this dosage would still be on the plateau of the dose-response curve (see point A in Fig. 13-5). Thus, function would still be maximum. Muller deduced that in this circumstance there could be no selection pressure, based on differential gene product function, for dosage compensation to have evolved. Blinkered by the genic paradigm, his attempt to explain this paradox in terms of "exceedingly minute differences" in gene dosage is described in an earlier book [9].

A possible resolution of "Muller's paradox" can be derived from the need to fine-tune, not the functions, but the *concentrations* of individual proteins within cells, in order to provide a uniform collective aggregation pressure (see Chapter 13). Over a series of generations, a Y-chromosome and its descendants define, and so exist in, a succession of male cells (M). This path may be simply expressed as:

$$M \rightarrow M \rightarrow M \rightarrow M \rightarrow M \rightarrow M \rightarrow M \rightarrow M \rightarrow \ldots \qquad (14.6)$$

Over evolutionary time, factors such as the transcription rates of genes on the Y-chromosome and the stabilities of their products (mRNAs and proteins) could have become fine-tuned to the needs of this relatively stable intracellular environment. A relatively constant concentration of each gene product could have become established (Fig. 14-3). On the other hand, an X-chromosome and its descendants alternate between male (M) and female cells (F). A typical path might be:

$$M \to F \to M \to F \to F \to M \to F \to M \to \ldots \quad (14.7)$$

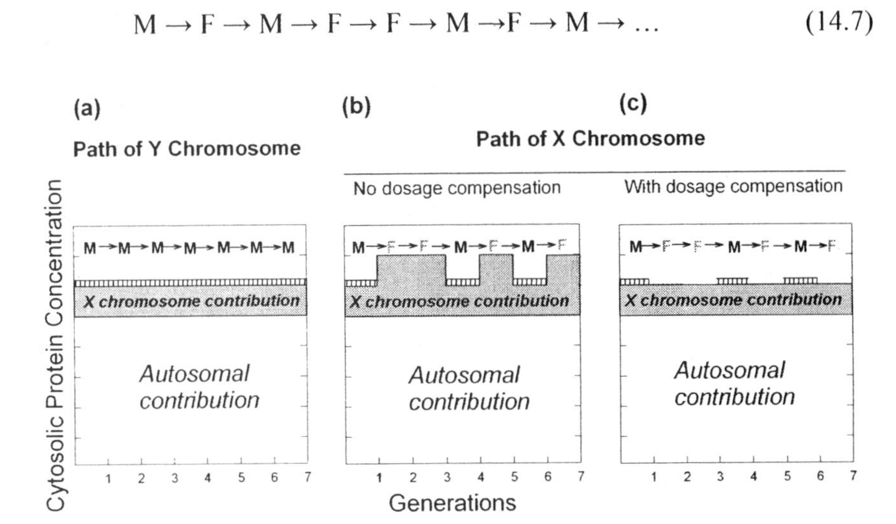

Fig. 14-3. Fine-tuning of protein concentration is not possible without X-chromosome dosage compensation. Passage of Y- or X-chromosomes through the generations occurs either in male (M) or female (F) cells. On average an X-chromosome exists for one third of its time in a male cell. Potential contributions of chromosome gene-products to the cytosolic protein concentration are shown for autosomes (white), X-chromosomes (grey), and Y-chromosomes (thin strip with vertical stripes). *(a)* The presence of the Y-chromosome defines an organism as male, and thus the Y-chromosome always exists in a male cell, in the presence of one X-chromosome. Fluctuation of total protein concentration is minimal. *(b)* The contribution of the Y-chromosome is very small so that in female generations, defined by the presence of two X-chromosomes, there would be a great fluctuation in total protein concentration, in the absence of dosage compensation. *(c)* Halving the contribution of the X-chromosome (dosage compensation) in female generations keeps the cytosolic protein concentration essentially independent of the sex of the host cell

Over evolutionary time it would be difficult to fine-tune, both for cells containing a second X-chromosome, and for cells containing a solitary X-chromosome. By inactivating one X whenever two are present, the intracellular environment would be stabilized and the collective fine-tuning of protein concentrations could continue. This would facilitate collective protein functions, such as intracellular discrimination between self and not-self.

How is an entire chromosome inactivated? While beyond the scope of the present work, it should be noted that an important factor in the actual inactivation process is the RNA product of a gene, *Xist* ("X inactivation specific transcript"), which is the only gene that remains active on the inactive X chromosome in human females. *Xist* RNA coats the inactive X-chromosome, but not the active X-chromosome. *Xist* is an example of a gene without a protein product. Its immediate transcription product is a long RNA molecule that contains introns. The processing of the RNA involves excision of the introns to generate the mature *Xist* RNA, which executes the gene's functions. The allelic copy of *Xist* on the active X chromosome ("sense") is kept inactive by virtue of a counter-transcript ("antisense") from a gene known as *Tsix*. The RNA products of *Xist* and *Tsix*, presumably recognize each other by "kissing" interactions that result in formation of a duplex in the region of complementarity. Inactivation of one X chromosome is accompanied by extensive DNA methylation, as will be discussed in Chapter 15.

Summary

The need for genomes to detect and correct errors has been a major force in evolution, driving DNA to "speak in palindromes" and splitting members of a species into two sexes. Errors associated with DNA mutations or DNA damage may be imperfectly corrected in our bodies ("soma"), but should be perfectly corrected in the germ line. To this end, meiotic recombination repair, in which maternal and paternal genomes are compared, is the last court of appeal. For this "rebooting" of the genome, parental genomes must be neither too similar, nor too disparate. If too similar (inbreeding), differences will not emerge at meiosis in their child's gonad. If too disparate, an exploratory speciation process (anti-recombination) may initiate, manifest as a healthy, but sterile, child (hybrid sterility). For the initiation of speciation, human females have to complete three steps, (i) differentiation of their sex chromosomes (X and X), (ii) differentiation of non-sex chromosomes (22 autosomal pairs), and (iii) activation of "check-points" which respond to such differentiations by disrupting meiosis. In contrast human males, being already advanced in the first step due to differentiation of their sex chromosomes (X and Y), have essentially to complete only the latter two steps. Thus, the first sign of speciation (incipient speciation), manifest as incipient hybrid sterility, is production of sterile males (Haldane's rule for hybrid sterility). Anti-recombination activity prevents repair of Y-chromosomes, which consequently degenerate, thus loosing many X-chromosome equivalent genes. Human males have potentially only one dose of many X-chromosome gene products, whereas females have potentially two doses. Dosage compensation in human females, leaving only one X-chromosome active, buffers fluctuations in intracellular protein concentrations between male and female generations. This permits a gene, independently of the sex which may harbor it, to fine-tune the concentration of its protein product to the concentrations of other proteins with which it has been travelling through the generations. In this way, collective protein functions, perhaps including intracellular self/not-self discrimination, are facilitated.

Chapter 15

The Fifth Letter

> "We have had some absurd attempts ... to apply mathematics to biology, but ... my hope is still that I may live to see mathematics applied to biology properly. The most promising place for beginning, I believe, is the mechanism of pattern."
>
> W. Bateson to G. H. Hardy (1924) [1].

In the biological sciences few generalizations are absolute and we have already noted strange bases in RNA in addition to the usual – **A**, **C**, **G** and **U** (Fig. 5-1). For many purposes DNA can be considered solely in terms of its four major bases – **A**, **C**, **G** and **T**. However, in written languages single letters are sometimes qualified with accents. We should not be surprised to find that there are similar qualifications in the DNA language. The most evident of these is methylcytosine, where the base **C** acquires a chemical grouping (methyl) [2]. Thus, in many organisms DNA has five letters – **A**, **C**, Me-**C**, **G** and **T.** Apart from the pattern of the four regular bases, there is a pattern of methylation at intervals along a DNA sequence. A brief consideration of the fifth letter is needed to conclude our discussion of evolutionary bioinformatics.

Post-Synthetic Modifications

DNA is synthesized as a string of the four base letters, but subsequently some of the **C**'s are modified so that they become Me-**C**. In similar fashion, someone writing in French might review a sentence and place an acute accent over certain e's. The enzymes that bring this about in DNA (methyl transferases) can distinguish between **C**'s on the basis of their 3' (i.e. downstream) nearest neighbors. Thus, **C**'s in the **CG** dinucleotide can be post-synthetically modified, but **C**'s in the dinucleotides **CA**, **CC**, and **CT** are usually not subject to modification. The nature of the 5' (i.e. upstream) base is usually irrelevant. Note that we are referring here to a base *sequence* (i.e. an ordered pair of bases). This is to be distinguished from a sequence's total content of two bases (its base composition). To avoid ambiguity, the dinucleotide **CG** is sometimes written as "**CpG**", with the "p" referring to the phosphate in the chain of phosphate and deoxyribose residues – the "medium" upon which the base sequence is "written" (see Fig. 2-4).

CG is a dinucleotide which, when considered at the level of a DNA duplex, has "palindromic" characteristics to the extent that **CG** on the top strand is matched by **CG** on the bottom strand (see sequence 14.1 in Chapter 14). The dinucleotide is a member of the set of four self-complementary dinucleotides (see Table 4-1a). In some circumstances methylation of a **C** in the **CG** of one strand will only occur if the **C** in the **CG** in the complementary strand is already methylated. Thus, prior to replication of a DNA duplex (see Fig. 2-3) the **CG**'s in the parental strands may already be symmetrically methylated. Immediately after replication each new duplex will then contain one parental strand with its original me-**C**'s, and one child strand without me-**C**'s. The methylating enzymes then methylate the child strands so that the symmetrical methylation is restored (Fig. 15-1). To this extent, the pattern of methylation in a DNA sequence is inherited by a succession of cells – a form of inheritance sometimes referred to as "epigenetic" [3]. This word, however, has had a variety of historical usages, and must be understood in context.

Restriction Enzymes

Many bacteria use post-synthetic modification of specific sequences to mark their DNA as "self." The DNA of a virus that infects bacteria (bacteriophage) is recognized as "not-self" because of the absence of the modification. Enzymes of bacterial origin are thus able to distinguish self from not-self. The enzymes cut the viral DNA duplex at or near to the specific recognition sequence (i.e. they are often sequence-specific endonucleases). This restricts the growth of the virus in a most definitive way, by cleaving its DNA into fragments. Thus, the "restriction enzymes," like the modification enzymes, recognize specific base patterns, often palindrome-like (see Chapter 4), and generate "restriction fragments." For example, a restriction enzyme named "HpaII" in the bacterium *Haemophilus parainfluenzae* recognizes the four base sequence **CCGG** in foreign duplex DNA. This sequence will already have been recognized in the bacterium's DNA by its own modifying enzymes and appropriately modified to declare it as self, and therefore uncuttable. The following is unmodified, and therefore declared "foreign" (not-self) and cuttable:

5' **NNNNNNNNNN<u>CCGG</u>NNNNNNNNNNNNNNN** 3'
3' **NNNNNNNNNN<u>GGCC</u>NNNNNNNNNNNNNNN** 5' (15.1)

The DNA is cut asymmetrically to generate fragments:

$$\begin{matrix} 5'\mathbf{NNNNNNNNNN}\underline{\mathbf{C}}3' & 5'\underline{\mathbf{CGG}}\mathbf{NNNNNNNNNNNNNN}\ 3' \\ 3'\mathbf{NNNNNNNNNN}\underline{\mathbf{GGC}}5' & 3'\underline{\mathbf{C}}\mathbf{NNNNNNNNNNNNNN}\ 5' \end{matrix} \quad (15.2)$$

Since the cutting enzyme (nuclease) attacks the nucleic acid internally, it is an "endonucleases," rather than an "exonuclease" (enzymes which degrade nucleic acids from the ends; Latin, *exo* = outer; *endo* = inner). Some restriction enzymes attack a foreign DNA in such a way that sets of restriction fragments of uniform size are generated. The sets can be separated as individual bands in a gel, which can be stained to allow the bands to be visualized. The pattern of bands is generally unique for a particular DNA sequence, so "restriction digests" of a DNA sample allow "restriction-site mapping,"of the DNA. From such maps, and the sequences of individual, overlapping, fragments, large genome sequences can be assembled.

As you might expect, bacteriophages do not sit around passively letting prospective hosts demolish their DNA. Bacteriophage defence strategies include the encoding of "antirestriction enzymes" that can inactivate host restriction enzymes – yes, this seems to be another example of a biological arms race (see Chapter 10). Whether their hosts have yet evolved "anti-anti-restriction enzymes" remains to be seen!

Bacterial restriction enzymes are co-inherited with bacterial modification enzymes specific for the same sequence, and the corresponding genes are co-localized in the genome as a "restriction-modification gene complex." This constitutes a prototypic package for intracellular self/not-self discrimination. Co-localization (linkage)of the genes would be expected, since this decreases the possibility of their separation by recombination, and facilitates coordinate regulation. A mutation in the restriction enzyme gene could impair a bacteria's ability to repel a foreign virus. On the other hand, a mutation in the modification enzyme gene might result in a failure of modification (a lack of site-specific DNA methylation). The mutant bacterium's DNA would then become vulnerable to its own restriction enzymes. This attack on self-DNA might not occur immediately, since the DNA would already have been modified at the time the mutation occurred, and the existing modifying enzymes might still be operative, although subject to degradation and turnover (see Chapter 2). After one round of replication the DNA duplexes would still retain one parental methylated strand, and this might suffice to protect the DNA. But after two rounds of replication, some child cells would have completely unmodified DNA duplexes. By that time, the pre-existing modification enzymes would probably have been so degraded, or so diluted by cell growth, that their concentrations would have been insufficient to maintain DNA modifications.

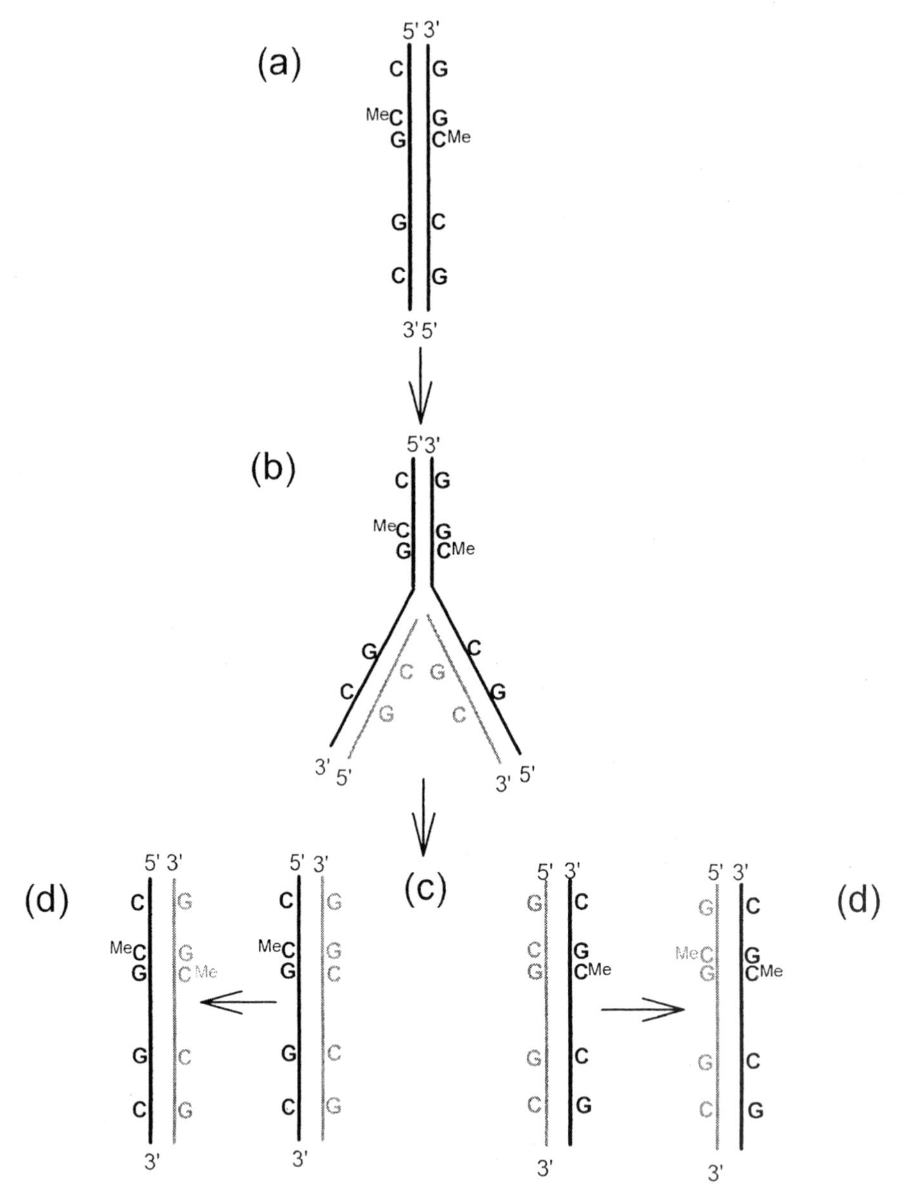

Fig. 15-1. Post-synthetic modification of DNA. In a parental DNA duplex in which only complementary **C-G** Watson-Crick base-pairs are shown, Me-**C** is present in one complementary **CG** dinucleotide pair *(a)*. The two strands of the parental duplex (black) are replicated *(b)* to produce child strands (grey). Initially, the child strands in the two freshly replicated duplexes *(c)* do not have methylated **CG** dinucleotides. These hemimethylated nascent duplexes are subsequently methylated *(d)*. Thus the symmetrical parental methylation pattern is inherited by child cells

CpG Islands

Regulation of the transcription of a eukaryotic gene is often brought about by the binding of various protein "transcription factors" near to the 5' end of the gene, where there is a "promoter" site to which RNA polymerase molecules bind. Transcription factors generally provide for short-term, on-off, control of transcription – either activating or inactivating transcription. A more definitive stamp of inactivation involves the methylation of **CG** dinucleotides in the promoter region. This can "lock-in" the inactivation for many cell generations.

A dramatic example of this somatic lock-in process involves the inactivation of one of the female X-chromosomes, which has been designated for inactivation by *Xist* gene RNAs acting on the chromosome from which they were derived (see Chapter 14). In this case there is a general spread of methylation to the promoter regions of all genes of the chromosome, except, of course, the *Xist* gene itself. In the remaining active X-chromosome the allelic *Xist* gene is itself inactive, a process requiring another gene (*Tsix*) and methylation of the *Xist* gene promoter. "Unwanted" genetic elements, which may intrude into DNA sequences (e.g. retroviruses, or gene duplicates), are sometimes inactivated by **CG** methylation and/or mutation (e.g. "repeat-induced point mutation") [4].

However, methylation of **C** means that if the **C** subsequently becomes subject to deamination, there will be a mutation to **T**, which is a normal DNA base. This **T** will now exist in duplex DNA attempting to engage in Watson-Crick pairing with the previous pairing partner of the **C** – namely a **G**. The **T-G** mispairing sometimes creates a problem for the repair enzymes, which may not know which is the correct base. So a region where **CG** dinucleotides are methylated (e.g. sequence 14.5 in Chapter 14) is especially vulnerable to mutation.

When such a mutation occurs in the DNA of somatic (non-germ line) cells, it is not passed on to children. When it occurs in the DNA of germ line cells, gametes are affected and so it can be passed on to the children, who can then pass it on to their children, etc.. As a consequence, sequences that are inactive (and hence methylated) in the germ line (e.g. genes that are inactive in the gonad, but are later required for a function specific to a particular somatic tissue, such as the production of insulin by pancreatic cells), have tended to lose **CG** dinucleotides. The sequences then maintain the inactive germ-line state by means other than **CG** methylation (i.e. they are now found to be deficient in **CG** dinucleotides and inactive, and it can be presumed that at one point in time they sustained the inactive state by methylation).

On the other hand, sequences that are active in the germ line (e.g. "housekeeping" genes that are required for some general cell function) do not have methylated **CG** dinucleotides and so are less subject to mutation. These sequences have retained their **CG** dinucleotides, which may be recognized as "**CpG** islands." The human *G0S2* gene provides an example of this. In Figure 12-3 the gene is shown with a major peak of **CG** dinucleotides. Careful inspection also reveals minor peaks, some associated with the *Alu* elements. This suggests some *Alu* expression in the germ line and, thus, a "housekeeping" function for *Alu* elements. Depending on the cut-off criteria by which **CG** peaks are defined as "islands," in human chromosomes there are far more **CpG** islands than can be accounted for either by conventional genes, such as *G0S2* (macro-islands), or by *Alu* elements (micro-islands) [5]. This points again to a "hidden transcriptome," perhaps with a "housekeeping" role in intracellular defense (see Chapter 12).

It was noted in Chapter 4 that 2-tuples such as **CG**, like n-tuples in general, have frequencies that are characteristic of the species. And it was noted in Chapter 9 that there can be mutational biases away from certain bases and towards others. In many eukaryotes, the tendency of Me-**CG** to mutate to **TG** (see Chapter 14) is an example of a random mutational bias that arises intrinsically from the underlying chemistry of DNA – in this case the instability of an amino group in **C**, which readily deaminates to become **U** (see Chapter 6 for the role of this in mitochondria). As a result there is suppression of **CG** frequencies and an elevation of **TG** frequencies. If uncorrected, **TG** on one strand will persist with the complement **CA** on the opposite strand (see Table 4-1b). Thus, a species with **CG** suppression sometimes shows an elevation of the frequencies of **TG** and **CA**. When there is **CG** suppression, then the frequencies of the 3-tuples (trinucleotides) containing **CG** (**CGN** and **NCG**, where **N** is any of the four usual bases) should also be suppressed. This is shown in Figure 4-2b, where a segment of human chromosome 19 is shown to have decreased frequencies of many **CG**-containing 3-tuples.

CG suppression is particularly evident in primates, but is not evident in the gut bacterium *E. coli*. However, *E. coli* protects its DNA against its own restriction endonucleases by post-synthetic modifications that include methylation of **CG**. Relative to the DNA of the somatic cells of its primate hosts, *E. coli* DNA has more **CG** dinucleotides that, in general, are less methylated. Hosts that could recognize this might have a selective advantage. Indeed, DNA with these characteristics (as might be released from dying bacteria) can be taken as a not-self signal by a host [6]. Thus, species differences in **CG** frequencies are of considerable interest with respect to immunological defenses.

When comparing **CG** dinucleotide frequencies in different species it is usual to "correct" for base composition. By virtue of its base composition alone, a high (**G**+**C**)% genome would be expected to have many **CG** dinucleotides (i.e. if the sequence were shuffled, more **C**'s would be found with a downstream **G** on their 3' side than in the case of a low (**G**+**C**)% genome).

Thus, **CG** frequency can be considered as having base *composition*-dependent and base *order*-dependent components (see Chapter 5). Knowing base composition, an "expected" **CG** frequency can be calculated for a given DNA (i.e. by multiplying the independent probabilities of **C** and **G**). Thus, **CG** frequency can be presented as a ratio of "observed" **CG**'s (actual number of **CG** dinucleotides counted) to "expected" **CG**'s (see Fig. 15-2). In some species Nature has been adjusting base order to enhance **CG** frequency in most DNA segments, and in other species Nature has been adjusting base order to suppress **CG** frequency in most DNA segments. In other species some DNA segments show enhancement and others show suppression [5].

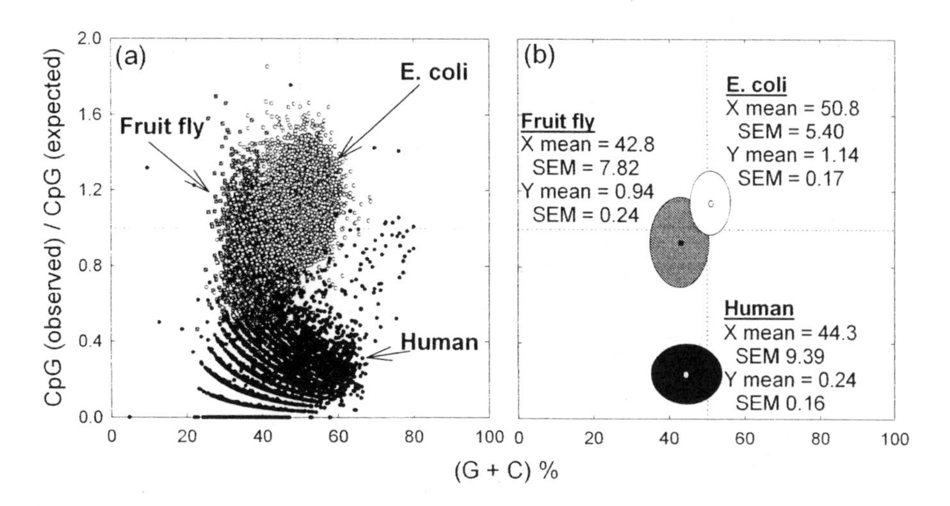

Fig. 15-2. **CpG** suppression. 5000 randomly selected 500 base segments from three genomes (the bacterium *E. coli*, the fruit fly *Drosophila melanogaster*, and the primate *Homo sapiens*) were each assessed both for the ratio of observed to expected frequencies of the **CG** dinucleotide (scoring 1.0 if there is neither suppression nor enhancement), and for base composition, (**G**+**C**)%. For each organism in *(a)* the data points form distinct clusters, the fruit fly points (dark grey squares) being partly hidden by *E. coli* points (white circles) and human points (black circles). The scattered points are summarized in *(b)* as a single point (defined by a mean X-axis value and a mean Y-axis value) surrounded by an oval that provides a visual measure of the scatter about the central point (SEMs; standard errors of the mean with respect to the mean X-axis and mean Y-axis values). Note that most *E. coli* segments show slight enhancement, whereas most segments from its human host show profound suppression. Some segments from the fruit fly genome show enhancement and some show suppression. Differences in **CG** dinucleotide frequencies generate corresponding differences in oligonucleotide frequencies (see Fig. 4-4b). Data for preparing this figure were kindly provided by Daiya Takai and Peter Jones

Methylation Differences in Twins

Normally a fertilized ovum develops into one individual. But sometimes it divides to produce two daughter cells that each develop into an individual. Two individuals that are essentially genetically identical are then born. Such monozygotic twins are used in attempts to distinguish the relative roles of inherited and environmental factors. Phenotypic differences between monozygotic twins become more evident as they age and, accompanying this, there are increasing differences in patterns of methylation and gene expression. Whether these differences are random (i.e. the two individuals drift apart as the time from their divergence from a single egg increases), or are the result of differing inputs from the environment, can be determined by comparing twins that have been raised together, with twins that were separated into different environments at an early age. Current evidence suggests that environment has a significant input [7]. Indeed, there is growing evidence for transgenerational transfer of epigenetic states (see below).

Imprinting

A major theme of this book has been conflict. It would seem that the genomes of male and female gametes, having united to form a diploid zygote, would then have a common interest and would cooperate unconditionally. Nature (i.e. the "hand" of evolution) should not have "intentionally" designed a conflict between them at the level of their embryo. The better the genomes cooperated, surely the more likely would be the zygote to complete its development and, in turn, produce healthy haploid gametes for the next generation? The phenomenon of imprinting that occurs in flowering plants and mammals tells us that this is not necessarily so [8].

By checking the methylation-status of the promoter region of a gene we can determine whether it is likely to be active or inactive. In the case of most genes, if they are needed for development then normally both the paternally-derived copy (paternal allele) and the maternally-derived copy (maternal allele) will be *unmethylated* (i.e. active). If the genes are not needed for development, then normally both copies (alleles) will be *methylated* (i.e. inactive).

In the case of some genes, however, one allele is always turned on (unmethylated) and one allele of always turned off (methylated). This determination (the "decision" to methylate or not) was made in the gamete prior to fertilization (i.e. it was a determination made *from the perspective of each individual parent*). The differential methylation of certain allelic genes in the haploid genomes of male and female gametes is referred to as "gametic imprinting."

One might be inclined to dismiss this as trivial. Perhaps the developing embryo can manage quite well with one dose of a gene (i.e. it is haplosufficient), and does not need the other allele to be expressed. However, when the nature of the genes themselves was examined, it was found that genes that might promote embryonic development resulting in a high birth weight tended to be unmethylated (i.e. active) in the genome contributed by the father, but they tended to be methylated (i.e. inactive) in the genome contributed by the mother [9].

An ingenious explanation for this was that, in the long term, paternal and maternal interests in their children, biologically, are not the same. For a human mother, each pregnancy is a nine-month commitment to one child, during which time she can produce no further children. During her reproductive life there is a ceiling on the number of children she can produce. If her "selfish genes" wish to maximize their representation in future generations they must accept mutations that ensure that the birth-weight of the first child is not so great as to exhaust the mother's resources, so that she is less able to produce future children. Mutations that, by some chain of events, ensure during development of the maternal gamete that there will be future inactivation of genes whose overexpression in the embryo might increase birth-weight, would tend to have been accepted into human DNA.

For a human father there is no such ceiling on the number of children he can produce. Over evolutionary time, males who can fertilize the most females have tended to pass on more genes to future generations. Biologically, they have no necessary interest in those females' abilities to produce further children, which, indeed, might be fathered by other males (i.e. biologically, it is argued that male monogamy is less "natural" than male polygamy). Thus, responding to an evolutionary pressure to maximize the representation of a father's selfish genes in future generations, humans must have accepted mutations in genes that, by some chain of events, ensure that in male gametes a state of activation is imposed on genes that will promote the development of a large embryo (thus giving the newborn a better start in life).

Because imprinting is imposed at the gamete level and inherited in the cells of the offspring, some of which will become the gametes of the next generation, patterns of methylation are usually erased (cancelled) and re-established each generation. But patterns of methylation may sometimes be inherited unchanged over many generations (transgenerational inheritance). Indeed, the story of the fifth letter can be dated, like many stories in genetics, back to Bateson. In 1915 he and Caroline Pellew investigated the "rogue" phenomenon in peas [10]. From time to time lines of culinary peas "throw rogues." The phenotype of these "weed-like" individual pea plants does not follow the usual Mendelian pattern of inheritance. "Rogues" when crossed with normal plants always yield "rogue" offspring, and the non-rogue phenotype is permanently lost. We can now understand this in terms of the sudden imposition of an inerasable pattern of methylation [11]. There is also growing evidence for transgenerational inheritance in mammals (people feel uncomfortable calling it Lamarckian). For example, there are data suggesting that nutritional deprivation in childhood of grandfathers increases the life-span of their grandsons, but not of their granddaughters [12].

Summary

Methylcytosine, the fifth base letter in DNA, is generated by modifying a **C** in a **CG** dinucleotide that has already been formed during DNA synthesis as one of the usual string of four base letters – **A**, **C**, **G** and **T**. The pattern of methylation is retained when cells replicate and may persist for many generations both of cells (epigenetic inheritance) and of organisms (transgenerational inheritance). However, often methylation patterns are imposed and cancelled during one generational cycle. In a bacterium, methylation is an example of a post-synthetic change that identifies its DNA as "self" and prevents the DNA being cleaved by the bacterium's own restriction endonucleases. The methylation of a eukaryotic cell's own DNA serves to "lock in" the inactivation of either a single gene, or a set of genes in the form of an entire chromosome (e.g. the inactivation of one X-chromosome in human females). This involves methylation of **CG** dinucleotides in regions where RNA polymerase binds to DNA (promoters). Genes that are *active* in the germ-line (mainly "house-keeping" genes) have less methylation of their **CG** dinucleotides, so that if the **C** undergoes spontaneous deamination, the base **U** is formed, which is readily recognized as inappropriate for DNA and an appropriate correction made. Thus, an unmethylated **CG** dinucleotide will tend to persist. Accordingly, **CG** dinucleotides are preserved in eukaryotes as "**CpG** islands" (macro and mini), which accompany the promoters of house-keeping genes and the putative "genes" that generate the "hidden transcriptome."However, genes that are *inactive* in the germ line (mainly tissue-specific genes) have methylated **CG** dinucleotides. If the **C** undergoes spontaneous deamination, the base **T** is formed. This mutation is not readily recognized as inappropriate and so there is less likely to be an appropriate correction. Since most mutations are disadvantageous, organisms with the mutation may die or produce fewer descendents. Accordingly, the genes evolve to maintain the inactive state with fewer **CG** dinucleotides. Hence, the dinucleotide **CG** has come to be under-represented in many eukaryotic genomes. As such **CG** can serve as a host (self) marker permitting recognition of the DNAs of microbial invaders that have different patterns of methylation.

Epilogue

To Perceive is Not To Select

> "His nature was too large, too ready to conceive regions beyond his own experience, to rest at once in the easy explanation, 'madness,' whenever a consciousness showed some fullness and conviction where his own was blank."
>
> George Eliot, *Daniel Deronda* (1876) [1]

Librarians, booksellers and publishers like books that can be conveniently classified (dare we say speciated?). Books on science go on one shelf. Books on science politics go on another [2, 3]. Most students encounter science first, and read books that give no inkling of the underlying politics. This has been described as "... an enormous problem of which students, or people who look only at standard textbooks and sanitized histories of science, are usually unaware. ... You get the impression that the history of science is a totally progressive, orderly, logical development of ideas" [4].

To some extent the science literature has tended to become a *pat-on-the-back* literature. To pass the peer-review gate, authors are inclined to paint a rosy picture, avoid controversy, and positively emphasize the work of possible gatekeepers. An exception is where controversy is exploited in a *holier-than-thou* literature that attempts to slip flawed arguments by the reader, while criticizing easy targets (such as people who are no longer alive), and admonishing the reader to avoid unsavoury authors who try to slip flawed arguments by. Students will inevitably learn about science politics in their later years if they become engaged in one of the most exciting aspects of science – the construction of hypotheses based on prevailing knowledge, and the making of discoveries that extend or refute that knowledge.

Percepts

All scientific knowledge rests on hypotheses. Immanuel Kant argued that percepts without concepts are blind [5]. This point was reiterated by physiologist John Scott Haldane when writing in 1891 about his uncle, the original "JBS," John Burdon Sanderson, a mentor to Romanes [6]:

> [He] "would say ... that he is very tolerant about theories -- [but] that what really tells is facts. But then what are facts that are essential? It's the theory that determines that. I would simply disregard as trivial and misleading heaps of things which he considers essential, and vice-versa. And even the simplest 'facts' are expressed - perceived - through theory."

We perceive facts in the context of theory. Theories themselves, until the evidence becomes overwhelming (e.g. the theory that the earth is not flat), are unsubstantiated theories, or mere hypotheses. In an ideal world, competing hypotheses would be dispassionately analyzed and a decision that one is more correct than another made with much diffidence (see Appendix 3). Darwin in 1868 set the standard [7]: "It is a relief to have some feasible explanation of the various facts, which can be given up as soon as any better hypothesis is found."

Torch Passed

But we do not live in an ideal world. In an ideal world, the ideas advanced by Gregor Mendel in 1865 would have been seized upon by scientists worldwide as a basis for further experimentation. By 1870 the ideas would have entered university-level biology curricula and students at Cambridge, such as Romanes and Bateson, would have read Mendel along with their Darwin. Yet, as has been told many times, had Mendel never lived progress in genetics would not have been much affected [8]. At least, in the years before the independent discovery of Mendel's laws in 1900, his work was not castigated:

> "For historians there remains the baffling enigma of how such distinguished biologists as ...W. Bateson ... could rest satisfied with such a crassly inadequate theory. ... The irony with which we must now read W. Bateson's dismissal of Darwin is almost painful."

This remark in 1983 by Richard Dawkins, deservedly one of the most influential scientists of our times, is representative of the multiplicity of attacks on William Bateson that occurred both before and after his death in 1926 [9].

It cannot be said that progress in genetics would not have been affected had Romanes and Bateson never lived. Their many contributions have been acknowledged and extended. However, Bateson's detractors repeatly proclaimed that, through his refusal to accept the conventional genic wisdom, he had delayed progress in genetics. I have argued both here and in my earlier books that, to the contrary, Bateson was light-years ahead of his contemporaries and of many who came after. It is his detractors who may have delayed progress [10]. Nevertheless, the Batesonian torch was passed through the twentieth century by Richard Goldschmidt, by Gregory Bateson, by Michael White, and, with qualifications, by paleontologist Stephen Jay Gould.

In an article entitled "The Uses of Heresy," Gould in 1982 introduced a reprint of Goldschmidt's classic text *The Material Basis of Evolution*. Here Gould described "the counterattack" of the neo-Darwinians that included his Harvard colleague Ernst Mayr's despair at "Goldschmidt's total neglect of" the "overwhelming and convincing evidence" against his ideas [11]. In Gould's opinion Goldschmidt "suffered the worst fate of all: to be ridiculed *and* unread" [Gould's italics], although his "general vision" was held to be "uncannily correct (or at least highly fruitful at the moment)," and "interesting and coherent, even if unacceptable today."

In 1980 Gould himself had came close to embracing the Batesonian-Goldschmidtian argument in an article in the journal *Paleobiology* entitled: "Is a new and general theory of evolution emerging?"[12]. But, after a two decade struggle with both the evolution establishment and cancer, Gould recanted, while still maintaining "a hierarchical theory of selection." In his *The Structure of Evolutionary Thought*, published shortly before his death in 2002, he wrote [13]:

> "I do not, in fact and retrospect (but not in understatement), regard this 1980 paper as among the strongest ... that I have ever written ... . I then read the literature on speciation as beginning to favor sympatric alternatives to allopatric orthodoxies at substantial relative frequency, and I predicted that views on this subject would change substantially, particularly towards favoring mechanisms that would be regarded as rapid even in microevolutionary time. I now believe that I was wrong in this prediction."

Generally courteous when responding to those who did not share his evolutionary views, Gould, like Romanes a century earlier, characterized his many opponents as "ultraDarwinian" fundamentalists, to be contrasted with his few "pluralist" supporters. There are remarkable parallels between Gould and Romanes [14]. Both were of the establishment and centre-stage. Steeped in the substance and history of evolutionary science, their writings were welcomed by the leading journals of their days. Both wrote prolifically for the general public, as well as for scientists. Both faced attacks from the very top of their profession – Romanes from Huxley, Wallace and Thiselton-Dyer – Gould from Dawkins, Mayr and Maynard Smith. Sadly, both were stricken with cancer in their forties. A difference was that, with modern therapies, Gould survived another two decades, whereas Romanes died at age forty-six. Furthermore, Romanes never recanted and, I suspect, never would have, even if given extra time.

Gould was strongly tested in 1995 by John Maynard Smith, the "Dean of British ultra-Darwinians" [15]:

> "Gould occupies a rather curious position, particularly on his side of the Atlantic. Because of the excellence of his essays, he has come to be seen by non-biologists as the pre-eminent evolutionary theorist. In contrast, the evolutionary biologists with whom I have discussed his work tend to see him as a man whose ideas are so confused as to be hardly worth bothering with, but as one who should not be publicly criticized because he is at least on our side against the creationists. All this would not matter, were it not that he is giving non-biologists a largely false picture of the state of evolutionary theory."

In the USA Mayr protested that Gould and his allies "quite conspicuously misrepresent the views of [biology's] leading spokesmen." Other evolutionists were less restrained [16]:

> "Evolutionary biology is relevant to a large number of fields – medicine, neuroscience, psychology, psychiatry, cognitive science, molecular biology, etc. – that sometimes have an impact on human welfare. Many scientists in these fields look to Gould, as America's most famous evolutionist, for reliable guidance on his field, and so the cumulative effect of Gould's 'steady misrepresentation' has been to prevent the great majority of leading scientists in these disciplines from learning about, or profiting from, the rapid series of advances made in evolutionary biology over the last thirty years."

It seems not to have occurred to these authors that if Gould had, in fact, provided "reliable guidance," then it will be *their* "steady misrepresentation" that will have negatively impacted human welfare. While some biohistorians have expressed dissatisfaction with the prevailing orthodoxy [17, 18], following Gould's demise it has been left largely to scientists such as Patrick Bateson (a relative of William and Gregory) and myself, to bear the torch onwards into the twenty first century [10, 19]. The task is now a little easier because, as has been shown here, we can begin to flesh out the Batesonian-Goldschmidtian abstractions in both informatic and chemical terms.

Voting with Facts

The ground was well prepared when Watson and Crick presented their model for DNA structure in 1953. The comments of Muller and Wyatt noted in Chapter 2 show that the possibilities of a helix and base-pairing were "in the air." The time was ripe. The model soon won wide acceptance. However, many new hypotheses are not so readily accepted.

> "Another man was convinced that he had the mathematical key to the universe which would supersede Newton, and regarded all known physicists as conspiring to stifle his discovery and keep the universe locked; another, that he had the metaphysical key, with just that hair's-breadth of difference from the old wards which would make it fit exactly. Scattered here and there in every direction you might find a terrible person, with more or less power of speech, and with an eye either glittering or preternaturally dull, on the lookout for the man who must hear him; and in most cases he had volumes which it was difficult to get printed, or if printed to get read."

These words of Victorian novelist George Eliot, who knew Romanes, and almost certainly knew of Butler, were written three decades before Einstein superseded Newton with his special theory of relativity [1].

Acceptance of scientific ideas can be seen as a struggle between two voting democracies – the democracy of the people, and the democracy of the facts. A scientific theory comes to be "accepted" when a major proportion of the "experts" in the field agree that this should be so. This is a true democracy of "the people" in the sense that, since all democracies contain "non-experts", namely those who are insufficiently qualified to have a vote (e.g. children), then their vote is assigned implicitly to those who are deemed to be qualified.

However, usually a theory appears to be consistent with certain facts and inconsistent with others. In the mind of the theoretician the facts *themselves* vote. Theoreticians embrace theories that are consistent with the most facts. Indeed, theories are usually elaborated in the first place to explain facts that seem discordant (i.e. are paradoxical). In many cases it is likely that a theoretician first finds an explanation for a few facts. As the theory develops, more and more facts fall into line (i.e. vote positively). Predictions are then made, and further facts are found that add to the positive vote. Thus, even before attempting publication or seeking experimental support, the theoretician is confronted with an array of voting facts. Even if the votes of the experts end up not supporting his/her theory, the theoretician can still affirm the probable correctness of the theory to the most significant critic – *his/her doubting self* – based on the voting of the facts. Thus, a disconnection can arise between the theoretician and his/her peers.

As described in Chapter 1, Romanes was unable to follow Butler's conceptual leap that heredity was due to stored information (memory). Today, this seems rather obvious. But, much in awe of Darwin, in 1881 Romanes, aged thirty-three and already a Fellow of the Royal Society, negatively reviewed Butler's *Unconscious Memory* and scoffed at the exposure of the flaws in Darwin's theory [20]:

> "Now this view ... is interesting if advanced merely as an illustration; but to imagine that it maintains any truth of profound significance, or that it can possibly be fraught with any benefit to science, is simply absurd. The most cursory thought is sufficient to show ... ."

One reason for this, of course, is that each person on this planet is somatically different from every other person, and occupies a unique position in time and space. Each person has a unique perspective. Einstein worked in a Swiss patent office when time zones were being established and there was an enormous interest in accurate time-keeping. Switzerland was the centre of the clock-making industry. It is likely that he reviewed many submissions on the synchronization of clocks, a major principle underlying relativity theory [21]. This focused his mind in a way denied to his peers in their academic ivory towers. It is to their credit that they did not bar publication of his work in 1905 and, within a decade, most came to recognize its worth.

Yet, in all probability, Einstein would still have believed his theory correct, even if his peers had not accepted it. Butler was also far from the ivory tower. By the time of his death in 1902, the Victorian establishment had scarcely moved an inch towards acknowledging his contribution. This annoyed, but did not overly disconcert [22]:

> "If I deserve to be remembered, it will be not so much for anything I have written, of for any new way of looking at old facts which I may have suggested, as for having shown that a man of no special ability, with no literary connections, not particularly laborious, fairly, but not supremely, accurate as far as he goes, and not travelling far either for his facts or from them, may yet, by being perfectly square, sticking to his point, not letting his temper run away with him, and biding his time, be a match for the most powerful literary and scientific coterie that England has ever known."

Judicial wisdom may not be displayed even by people of scientific eminence who, earlier in their careers, have themselves failed to convince sceptical peers. Several years after Einstein's seminal paper on special relativity, Max Planck and other distinguished physicists, while embracing relativity, incorrectly cautioned that Einstein "may sometimes have missed the target in his speculations, as, for example, in his hypothesis of light quanta" [4].

Planck's own ideas on the second law of thermodynamics had not been appreciated by Helmholtz, whose ideas, in turn, had also not been recognized. Indeed, Helmholtz appreciated that "new ideas need the more time for gaining general assent the more really original they are" [23], but he may here have been thinking more of the originality of his *own* ideas than those of others! The work of the seventeenth century polymath Robert Hooke was condemned to oblivion because of "the implacable enmity of Newton .... But deeper than this is perhaps ... that Hooke's ideas ... were so radical as to be inassimilable, even unintelligible, to the accepted thinking of his time" [24]. In this way it is argued that peers are obliged to discard "premature" claims, to avoid being overwhelmed by false leads [25].

So if a theoretician's peers do not accept, what is an impartial observer to do? In the words of Butler in 1890 [20]:

> "We want to know who is doing his best to help us, and who is only trying to make us help him, or bolster up the system in which his interests are invested. ... When we find a man concealing worse than nullity of meaning under sentences that sound plausibly enough, we should distrust him much as we should a fellow-traveller whom we caught trying to steal our watch."

Similarly, a century later Dawkins noted [26]:

> "It's tough on the reader. No doubt there exist thoughts so profound that most of us will not understand the language in which they are expressed. And no doubt there is also language designed to be unintelligible in order to conceal an absence of honest thought. But how are we to tell the difference? What if it really takes an expert eye to detect whether the emperor has clothes?"

Butler had a solution [20]:

> "There is nothing that will throw more light upon these points than the way in which a man behaves towards those who have worked in the same field with himself ... than his style. A man's style, as Buffon long since said, is the man himself. By style, of course, I do not mean grammar or rhetoric, but that style of which Buffon again said that it is like happiness, and *vient de la douceur de l'âme*. ... We often cannot judge of the truth or falsehood of facts for ourselves, but we most of us know enough of human nature to be able to tell a good witness from a bad one."

More specifically, four questions need to be asked. Is the theoretician sane? Is the theoretician objective? Is the theoretician honest? Does the theoretician have a credible track record in the field? A negative answer to these questions will not disprove the theory. A positive answer, however, should serve to counter a negative vote from alleged experts.

For Bateson and Goldschmidt, the answers are quite positive. Their demeanors, relationships and writings attest to their sanity, objectivity and honesty, the latter being most evident towards the person they would most likely deceive, *themselves*. Like Romanes before them [10], having been immersed in evolutionary science for many decades, they were both very well prepared. Yet, like Romanes before them, they were denied the benefit of the doubt and were subjected to personal attacks. George Eliot wrote [1]:

> "Like Copernicus and Galileo, he was immovably convinced in the face of hissing incredulity; but so is the contriver of perpetual motion. We cannot fairly try the spirits by this sort of test. If we want to avoid giving the dose of hemlock or the sentence of banishment in the wrong case, nothing will do but a capacity to understand the subject-matter on which the immovable man is convinced, and fellowship with human travail, both near and afar, to hinder us from scanning any deep experience lightly."

People are not impartial. As suggested here and elsewhere [3], people act from a variety of motives other than that of seeking the truth. Ignoring the admonitions of Wall Street, they will not diversify and hedge their bets. And sometimes, despite the best intentions, they just do not seem to find time to do their homework. Writing in 1932 to the US geneticist Thomas Hunt Morgan, Ronald Fisher noted [27]:

> "One of the chief reasons why, in spite of raising so much dust, we are not making in this generation more rapid progress, is that we do not really give ourselves time to assimilate one another's ideas, so that all the difficult points, the things really worth thinking about, have to be thought out independently, with great variations in efficiency and success, some hundreds of times."

Marketing

To select is not to preserve, and without preservation there can be no future selection. But selection, whether before or after preservation, depends on perception, and perception can be imperfect. To perceive is not to select. Too often blandishments, in Madison Avenue mode, can convince that round pegs fit square holes, and square round. The conceptually challenging, abstract, arguments of Romanes, Bateson and Goldschmidt denied Darwinian dogma. For this they were dismissed by generations of biologists bewitched by genes and all that their marketers promised. With overwhelming rhetoric to the contrary, we were slow to appreciate the deep truths these scientists were attempting to convey [10, 14].

So, in the twenty first century, scientific understanding continues to advance at a pedestrian pace with much waste of time and energy, and of public and private funds. As indicated in the Prologue, commercial interests continue to exert an undue influence on what science is supported and published. In Canada, bioethicist Margaret Somerville comments [28]:

> "Conditions that are attached to government funding can affect the purposes and values upheld, especially when those conditions require academic-industrial partnerships for research to be eligible for funding, as in the case of the Canadian government's $300 ... million investment in a series of genomics research centres (Genome Canada). Structuring funding in this way leaves out the funding of research that will not result in marketable products, and excludes those researchers who undertake it."

It should not be imagined that the situation will easily change. In 1931, the author of *Brave New World*, a grandson of Thomas Huxley, wrote to Fisher [29]:

> "The really depressing thing about a situation such as you describe is that, the evil being of slow maturation and coming to no obvious crisis, there will never be anything in the nature of a panic. And as recent events only too clearly show, it is only in moments of panic that anything gets done. Foresight is one thing: but acting on foresight and getting large bodies of men and women to accept such action when they are in cold blood – these are very different matters."

Thus, the western democracies, driving to ever increasing prosperity with individuals liberated to explore and indulge their perceived interests and desires, appear somewhat like headless monsters. Such a situation appears infinitely to be preferred over a totalitarian alternative, *provided* there can be appropriate feedbacks from periphery to center so that things do not spiral out of control. This requires informed individuals with the courage to challenge the conventional wisdom, open-minded information gatekeepers, a free press, and an attentive and responsive executive.

Joining the Dots

The "intelligence failure" of September 11th 2001 occurred while this book was in preparation. It turned out that, prior to the event, there had been substantial clues, but the "Butlers" and "Mendels" in government agencies had been overlooked. The "authorities" had failed to sift and weigh. They had failed to "join the dots" [30, 31]. Likewise, as argued here, the biomedical research enterprise, while "concealing difficulties and overlooking ambiguities" has sometimes failed similarly. Sadly, there has been little recognition of this. Indeed, the very opposite, as the outcry of the complacent against Gould reveals.

Hailing the sequencing of the human genome as again demonstrating the vigor of free-market systems, commentators can contrast this with the evil example of Trofim Lysenko who in the 1930s won the support of Joseph Stalin, and sent his scientific competitors to their deaths, so setting back Soviet genetics by decades [32]. Yet few are aware of the emergent "new Lysenkoism" in the West that, while sparing their bodies, has forced academic suicide (i.e. loss of research funding, tenure, etc.,) on those who could see beyond their noses and refused to embrace the prevailing orthodoxy [2, 3]. Is "suicide" too strong a word? Well, "It is suicidal" was how Marshall Nirenberg's plan (ultimately successful) to decipher the genetic code was greeted by a colleague at the US National Institutes of Health in 1959 [33].

While Romanes, a friend of the editor of *Nature*, had had little trouble getting his anonymous editorials and signed articles accepted [34], in the early decades of the twentieth century William Bateson battled the peer-review gatekeepers who controlled access to publication. He had to content himself with the less visible *Reports to the Evolution Committee of the Royal Society* [35], or publication at his own expense. Butler had to self-publish most of his works on evolution, and negative reviews by the Victorian establishment ensured that he would suffer repeated financial loss. In his early days, Fisher submitted a paper on the probable error of the correlation coefficient to the journal *Biometrika*. The Editor, the biometrician Karl Pearson, rejected it since he was "compelled to exclude all that I think is erroneous on my own judgement, because I cannot afford controversy" [36].

Chargaff himself was subject to the "hissing incredulity" of the funding agencies. As an innovative and vigorous experimentalist (he lived to be 96), it is likely that our understanding of the base compositional regularities he described would by now have been much further advanced had he not been forced into retirement [2]. And, just as it is the victors in war who write the history books, so it can be the successful marketers and their disciples who write the history of science. As late as 1996 the monthly article on biohistory in the prestigious journal *Genetics* was portraying William Bateson as "irredeemably confused about evolution" and as "one of the enemies battled against during the modern synthesis"[37, 38]. As a service to the genetics community, the history articles were collected together in book form [39]. A bold reviewer had the temerity to remark: "Some of the historical content needs to be treated with caution while it cannot be claimed that all articles give an unbiased account of their subject" [40].

Cogito Ergo ...?

It is my hope that, having completed this book, you will have captured my enthusiasm for evolutionary bioinformatics and my belief that studies in this area will greatly advance our progress towards solutions to some major problems of humankind, such as AIDS, malaria and cancer. But this Epilogue implies caution. If you intend to engage in this area, be prepared for difficulties. If you have a penchant for joining the dots, and find that you can see beyond your nose, then be prepared for a hidden life with quick passage to an academic tomb inscribed *cogito ergo non sum.*

Yet, there is a ray of hope suggesting that times may now be different from when Aldous Huxley penned those depressing words above [29]. Through the Internet amazing new lines of communication have opened to all, and developments in search-engine design promise a sifting and manipulation of information that would have been unimaginable to our fathers. The headless monster (Adam Smith would have called it "the invisible hand") of our biomedical/industrial enterprise, has generated more data than in can handle. This is accumulating in open databases worldwide. Inspired programmers are writing "open source" software packages that can be downloaded with a click of a mouse. The playing field is being levelled in strange ways.

Canadian chemist John Polanyi points out that although now, more than ever before, information is freely available to all, it "only *appears* to be freely available to all. In fact, it is available only to those who understand it's meaning, and appreciate its worth" [3]. The present book, hopefully, will help you understand its meaning and worth. Voltaire in 1770 wrote "*On dit que Dieu est toujours pour les gros bataillons*" [41]. Today, *les petits bataillons* may be similarly favored. Even so, the difficulties will be great, and you may try, and fail, as did George Eliot's sad figure, Casaubon, in her novel *Middlemarch*. But, unlike Casaubon, through trying you may come to understand, and hence become able to communicate, the wisdom of others, so that the works of tomorrow's Mendels, Herings, Butlers, Romanes, Batesons, Goldschmidts and Chargaffs will not go unrecognized. Let George Eliot close [42]:

> "Her full nature ... spent itself in channels which had no great name on the earth. But the effect of her being on those around her was incalculably diffusive: for the growing good of the world is partly dependent on unhistoric acts; and that things are not so ill with you and me as they might have been, is half owing to the number who live faithfully a hidden life, and rest in unvisited tombs."

Appendix 1

What the Graph Says

> "Since the probability of two bombs being on the same plane is very low, the wise statistician always packs a bomb in his/her luggage."
>
> Anonymous (before the modern era of high airport security)

The pursuit of evolutionary bioinformatics requires the handling of large amounts of sequence data, sometimes from multiple biological species. This is greatly assisted by plotting the data points on a graph. Invariably the points are widely scattered, and the volume of data – sometimes many thousands of points – can seem overwhelming. Fortunately, there are software packages (like Microsoft Excel) that will plot the points and draw the statistically best lines through them.

The resulting graphs can sometimes be understood immediately without resort to numerical analysis. But confidence increases when visual interpretations are numerically confirmed. Thus, throughout this book graphical displays include values for slopes of lines, squares of the correlation coefficients (r^2), and probabilities (*P*). These capture key features of what might otherwise have seemed a hopeless jumble of points. Out of a cloud meaning emerges, sometimes discerned only as very faint, but still statistically significant, signals.

Often lines are linear (rectilinear), but sometimes they are curved (curvilinear). Often rectilinear lines are tilted, but sometimes they are horizontal. Sometimes no linear interpretation is evident, but clusters of points can be demarcated from others by drawing circles around them. Without resort to a statistics text, the essence of what the various values tell us can be shown by constructing dummy graphs and then applying a standard software package.

Horizontal Lines

Table 17-1a shows a spreadsheet with X-axis values increasing from 1 to 20. Four sets of Y-axis values were preselected as detailed below.

(a)

X-axis	Y-axis			
	Column 1	Column 2	Column 3	Column 4
1	6.0	6.0	6.0	6.0
2	5.9	5.0	6.9	6.0
3	6.1	7.0	8.1	9.0
4	5.8	4.0	8.8	7.0
5	6.2	8.0	10.2	12.0
6	6.0	6.0	11.0	11.0
7	5.9	5.0	11.9	11.0
8	5.8	4.0	12.8	11.0
9	6.1	7.0	14.1	15.0
10	6.2	8.0	15.2	17.0
11	6.1	7.0	16.1	17.0
12	6.2	8.0	17.2	19.0
13	5.9	5.0	17.9	17.0
14	5.8	4.0	18.8	17.0
15	6.0	6.0	20.0	20.0
16	6.1	7.0	21.1	22.0
17	5.9	5.0	21.9	21.0
18	6.0	6.0	23.0	23.0
19	6.0	6.0	24.0	24.0
20	6.0	6.0	25.0	25.0

(b)

Mean	6.0	6.0	15.5	15.5
Standard error	0.133 [SEM]	1.333 [SEM]	0.133 [SEE]	1.333 [SEE]
r^2	<0.0001	<0.0001	0.999	0.951
Slope = a	0.003	0.003	1.003	1.003
P **(slope probability)**	0.954	0.954	<0.0001	<0.0001
y_0	6.0	6.0	5.0	5.0

Table 17-1. A spreadsheet of dummy data values *(a)* intended to generate graphical plots that are either horizontal (columns 1 and 2), or sloping (columns 3 and 4). It is also intended that, around the line that best fits the points (as determined by a standard computer program), there be either a small scatter of points (columns 1 and 3), or a large scatter of points (columns 2 and 4). Various statistical parameters are shown in *(b)* below the related columns. Scatters around regression lines are referred to as standard errors of the estimate (SEE). However, when a line is horizontal the SEE is the same as the standard error of the mean (SEM) of the data points.

In Table 17-1 Y-axis values for four graph plots (plots of different sets of Y values against the X values) are displayed in columns 1 to 4. Numbers were placed in the first two Y columns with the intention of plotting two horizontal lines that would cross the Y-axis at a value of 6 (i.e. Y_0, the value for Y when the X value was zero, would be 6).

For column 1, the numbers were allowed to deviate very little from the mean of 6. Thus, in the first row, 6 was entered. Then, in the second row, 5.9 was entered. Then, to compensate for this, so that an average of 6.0 was sustained, in the third row 6.1 was entered. This was continued to generate 20 values for Y with minima no lower than 5.8 and maxima no higher than 6.2.

For column 2, the numbers were allowed to deviate further from the mean of 6. Thus in the second row 5 was entered. To compensate for this, so that an average of 6 was sustained, in the third row 7 was entered. This was continued to generate 20 values for Y with minima no lower than 4, and maxima no higher than 8.

The computer was then asked to plot the two sets of Y values against the set of X values. The resulting graph (Fig. 17-1) shows two horizontal lines ("regression lines"), which are so close that they cannot be distinguished by eye. Values from column 1 (open circles) sit close to the lines. Values from column 2 (grey circles) are more widely scattered. The statistical read-out from the computer is shown below the corresponding columns in Table 17-1. As expected, the averages (arithmetical means) of the two sets of twenty Y values are both 6. A measure of the degree of scatter about the regression lines is provided by the characteristic known as the standard error (SEM in these cases). As expected, the values in column 1 are less scattered about the regression line than the values in column 2 (0.133 versus 1.333).

The squares of the correlation coefficient (r^2; sometimes called the coefficient of determination) are essentially zero in both cases. This means that no part of the variation between the twenty data values for Y (the "dependent variable") can be attributed to the changes in the twenty data values for X (the "independent variable"). Both sets of numbers vary (in column 2 more than in column 1). But, whatever caused that variation, it was not related to X (as far as we can tell from the numbers at our disposal).

Being a near horizontal line, the slope value (change in Y value per unit X) is very close to zero (a = 0.003 in each case). That this slope value is *not* significantly different from zero is shown by the associated probability values (*P*), which normally range over a scale from zero to one. The *P* values are both 0.954 showing that, for the number of data points available, it is *most* probable that there is *no* significant difference from a slope value of zero. By convention, *P* values must be less than 0.05 to establish likely significance. Finally the values for the intercept on the Y-axis (Y_0 values) are both 6, as expected.

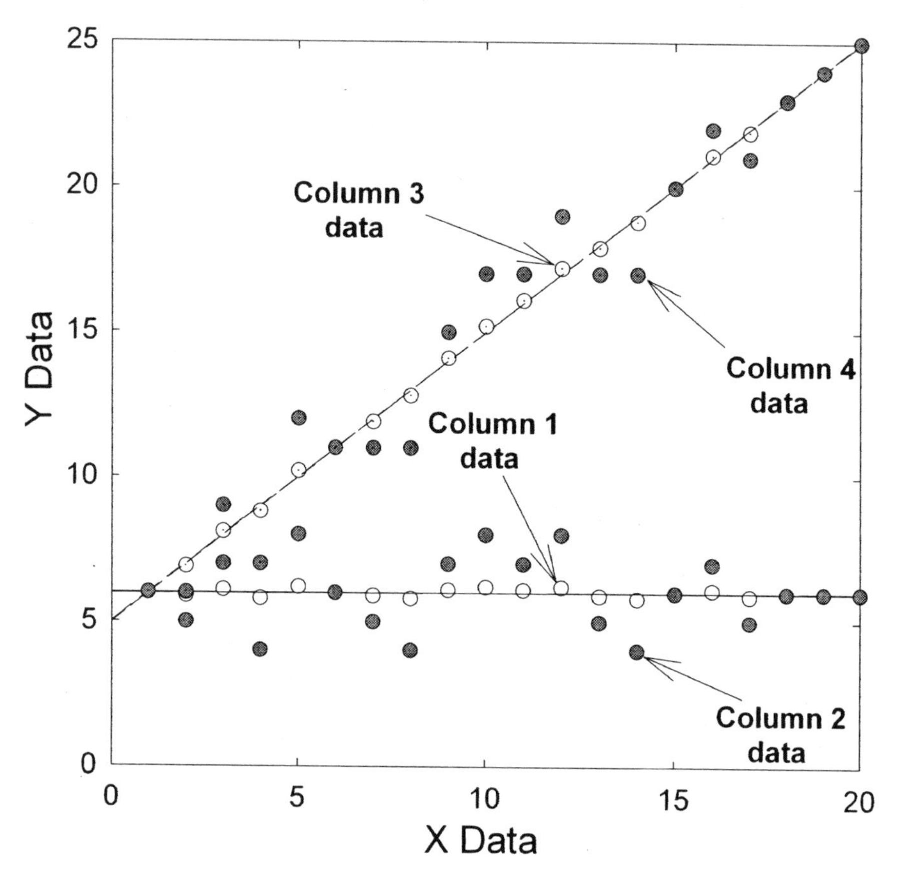

Fig. 17-1. Computer-generated lines that best fit the plots of the data shown in Table 17-1. Open circles (column 1 data); grey circles (column 2 data); open circles with central dot (column 3 data); grey circles with central dot (column 4 data). There are two horizontal lines for columns 1 and 2 and two sloping lines for columns 3 and 4, but the line pairs are so close that they cannot be distinguished (i.e. they overlap)

Tilted Lines

It was next intended to tilt the two horizontal lines so creating two new lines differing from the originals in slope and in intercept at the Y-axis (Y_0). Accordingly, values were generated for columns 3 and 4 by adding increasing numbers to the values in columns 1 and 2, respectively (Table 17-1). Thus, zero (0) was added to the numbers in the first row (columns 1 and 2), so the numbers both remained at 6 (columns 3 and 4). Then, 1 was added to the numbers in the second row (columns 1 and 2) to generate values of 6.9 and 6.0 for columns 3 and 4, respectively. Then, 2 was added to the numbers in the third row to generate values of 8.1 and 9.0, respectively. Then, 3 was added to the numbers in the fourth row to generate values of 8.8 and 7.0, respectively. This was continued to fill columns 3 and 4, each with twenty numbers.

The resulting computer plots (Fig. 17-1) show two new sloping lines (which are again indistinguishable from each other by eye). They retain the same scattering of points as in the horizontal plots from which they were derived. As expected, the mean values for the sets of numbers in columns 3 and 4 are greater than the corresponding sets in columns 1 and 2 (Table 17-1b). But, since the scatters about the regression lines are unchanged, the standard errors (SEE in these cases) are unchanged. The standard errors for columns 1 and 3 are identical. The standard errors for columns 2 and 4 are identical.

On the other hand, the r^2 values, which like *P* values range from zero to one, are 0.999 and 0.951, respectively. Thus, in the case of the plot with less scatter of the points around the regression line (Y values from column 3), much of the variation among these values can be explained on the basis of the X data values (i.e. the r^2 value of 0.999 means that essentially 100% of the variation among the twenty Y values correlates with the underlying variation among the twenty X values). In the case of the plot with more scatter of the points around the regression line (Y values from column 4), much of the variation between these data values can again be explained on the basis of the X data values. However, in this case the r^2 value is 0.951 showing that, while 95% of the variation among the twenty Y values correlates with the underlying variation in the twenty X values, 5% of the variation might relate to another factor, or other factors.

The slope values (both 1.003) are now no longer close to zero, as in the case of the horizontal plots generated from the data in columns 1 and 2. Furthermore, the associated *P* values (<0.0001) affirm that the slope values are likely to be significantly different from the horizontal (i.e. there is only a very low probability that the slope values are *not* significantly different from zero). The Y_0 values show that both slopes extend back (extrapolate) to cross the Y axis when Y is equal to 5. Thus, all four lines intercept (have a common value) when X is 1 and Y is 6.

Curved Lines

In Figure 17-1 the data points are economically fitted to straight lines, as is the case with many graphs in this book. Each slope can then be described mathematically by a single value (often designated as "a" in what is known as a "first order" relationships). However, sometimes the lines curve (see figures in Chapters 9 and 11). In these cases, description of the slopes requires further values (e.g. "a" and "b" in what are known as a "second order" relationship). For dissection of more complicated relationships, procedures such as "principle component analysis" and "best subsets regression" are available in software packages (e.g. Matlab, Minitab). Table 8-1 is an example of the latter.

No Lines

Sometimes data points form a cluster and no linear relationship can be discerned. In this case the points can be represented as a single central point with some measure of their scatter around that point. The central point is determined by the two values for the means of the X-axis and Y-axis values. This point is surrounded by an oval, the limits of which are determined by the scatters of X and Y values along the corresponding axes. For this the corresponding standard errors of the means (SEMs) provide a measure (e.g. see Figure 15-2). There are computer programs that can examine a large cluster of points and determine whether it can be divided into sub-clusters. Thus, although discriminated sets of points were plotted in Figure 15-2a, we could have plotted all the points without discrimination and then could have asked a computer program to attempt to independently cluster them.

A Caution

Huxley in 1864 noted "three classes of witnesses – liars, damned liars, and experts." While retaining its truth, in modern time this has further mutated to three classes of lies – "lies, damned lies, and statistics." There are many reasons to be cautious about statistics. My favorite statistician story is that of the statistician who, after the Second World War had ravaged Holland, noted a strong correlation between the subsequently rising birthrate and the return of storks to nest on the housetops. However elegant the mathematics, if applied without an awareness of "the big picture," statistics can be very misleading.

Of concern to the student of bioinformatics is that sequences are now available in GenBank because, at some point in time, people *decided* to sequence them. Sequencers being of the species *Homo sapiens*, they tended to prefer DNA either from *Homo sapiens*, or from species of economic importance to *Homo sapiens*, which includes species that are pathogenic for *Homo sapiens* and his dependent species (domesticated animals and cultivated plants). Thus, the sequences currently in databases are far from being a random sample. We must keep this in mind when selecting sequences for our studies, and when interpreting results.

Appendix 2

Scoring Information Potential

"A man who knows the price of everything and the value of nothing"
Oscar Wilde, *Lady Windermere's Fan*

Information scientists measure information as binary digits or "bits" (see Chapter 2). The value of each unit in a sequence segment depends on the number of types of possible unit that are present in the segment (e.g. base composition), not on the way the units are arranged (e.g. base order). According to their compositions, different segments can be scored for their respective *potentials* to carry order-dependent "primary" information. The same segments retain their potentials to carry composition-dependent "secondary" information. In spoken information the text can be regarded as primary, and the accent or dialect can be regarded as secondary.

Two Units

If, in a hypothetical nucleic acid sequence that can contain only two types of base, the bases are distributed randomly but with equal probability (i.e. at a particular position the uncertainty as to which base is coming next is maximum), then each base adds 1 bit of potential base order-dependent information. Thus, $2 = 2^1$. The left 2 refers to the number of bases. The right 2 refers to the type of decisions (binary yes/no decisions; see Chapter 2). The 1 bit on the right side of the equation is referred to both as the "exponent" of the 2 on the right side of the equation, and as the "log," or "logarithm," of the 2 on the left side of the equation.

If only one of the two available bases is present in a segment ("window" in the sequence) there is no uncertainty as to which base is coming next, and each base adds 0 bits of potential base order-dependent information ($1 = 2^0$). So variation along a nucleic acid sequence, which contains only one or two types of base, can be assessed in terms of its local content of potential base order-dependent information in sequence windows on a log scale scoring from 0 to 1 for each base (i.e. the number of bits in a window is equal to the logarithm to the base 2 of the number of available base types in that window). A ten base window consisting of only one base (e.g. **AAAAAAAAAA**) would score zero (10 x 0). A ten base window consisting of equal quantities of the two bases (e.g. **ATTATAATTA**) would score 10 (10 x 1). This score would not change if the order of the bases were varied.

Four Units

Along similar lines, if the four bases of a natural nucleic acid sequence are distributed randomly, but with equal probability, along a sequence (i.e. at a particular position the uncertainty as to which of the four bases is coming next is maximum), then each base adds 2 binary digits (bits) of potential base order-dependent information ($4 = 2 \times 2 = 2^2$). If only one base is present in a sequence segment, there is no uncertainty and each base adds 0 bits of potential base order-dependent information ($1 = 2^0$). Again, the left side of the equation refers to the number of bases, and the exponent of 2 on the right side refers to the number of binary (yes-no) decisions required to distinguish between those bases at a single position in the sequence segment.

Here, potential base order-dependent information content (the exponent) is on a scale from 0 to 2. A nucleic acid sequence containing one to four bases can be assessed in terms of its local content of potential base order-dependent information on a log scale scoring from 0 to 2 for each base (i.e. again, the number of bits is equal to the logarithm to the base 2 of the number of bases).

Twenty Units

A protein, with a higher ordered structure even more elaborate than that of a nucleic acid, is hardly susceptible to analysis in these terms. Nevertheless, similar reasoning can be applied to protein sequences with one to twenty amino acids. Here the scale is from 0 ($1 = 2^0$) to 4.322 ($20 = 2^{4.322}$). Following the binary decision-making principle, one begins by dividing the pool of twenty amino acids into two *equal* groups. A hypothetical informed respondent, on being asked whether a certain pool contains the next amino acid, replies either positively or negatively, and one has reduced the number of possibilities to ten.

This ten is then divided into two *equal* groups of five, for a second round of interrogation. Third, fourth and, sometimes, fifth, interrogations follow. Since five amino acids do not divide into two equal groups, the number of interrogations is not unitary, but has to occur, on average, 4.322 times (i.e. on average, 4.322 binary (yes-no) decisions are needed to specify which of the twenty amino acids occupies the next position in a protein sequence). The number of bits is equal to the logarithm to the base 2 of the number of types of amino acid in a window.

Meaning

The above calculations display a *quantitative* aspect of information. Given the number of each type of base in a DNA segment (its base *composition*) we can determine the segment's potential to carry different base order-dependent messages. Its potential is low if the segment contains only one of the four bases. Its potential is maximized if the segment contains equal proportions of the four bases (see Chapter 11). However, of all the potential base order-dependent messages only those of a small subset are likely to convey meaning (i.e. they are messages conveying information that may or may not be helpful to the receiver). We need to distinguish sense, which informs the receiver, from non-sense, which does not inform the receiver. Meaning requires the existence of a *code* through which some messages can convey meaning. The same code may be applied to other messages, but a meaning may not emerge. Sometimes meaning requires recognition of base composition alone (i.e. the code maps different meanings to different base compositions; see Chapter 8). However, often critical to meaning is base *order*. This is a *qualitative* aspect of potential information.

Take, for example, various three-unit message sequences composed of two unit types – say 9 and 1. At each position in each sequence there is either a 9 or a 1. There are $2 \times 2 \times 2 = 2^3 = 8$ alternatives (i.e. the length of the sequence is equal to the logarithm of the number of possible messages). These eight messages are 999, <u>991</u>, <u>919</u>, <u>911</u>, <u>199</u>, <u>191</u>, <u>119</u>, and 111. The first and last contain only one unit type. Of the remaining six (underlined), three have two nines and a one, and three have one nine and two ones. Thus, there are *four* composition-dependent messages – three nines (999), two nines and a one (991, 919, 199), one nine and two ones (911, 191, 119), and three ones (111).

Which of the eight messages has order-dependent meaning depends on the existence of appropriate codes. Perhaps each message has a distinct meaning. Perhaps none has. 911 is a widely recognized "universal" alarm call. For all I know, 191 may be the "pin number" you enter when using your credit card. Thus, it is possible to distinguish *general* codes that are not individual-specific (i.e. they are observer-independent), and *local* codes that are individual specific (i.e. they are observer-dependent).

However, if you passed on your pin number to your children, and they passed on the same pin number to theirs, etc., then your pin number could become more general. All organisms in our planetary biosphere are considered to have evolved from a common ancestor. This seems to predict that all biological codes should be general, as is the case with the codes we have discovered so far (with a few minor wrinkles). So coding studies carried out on the bacterium *E. coli* are helpful in understanding coding in *Homo sapiens*.

A DNA sequence may be natural or artificial. A natural DNA sequence has *evolved* and so has distinguished itself from the large theoretical subset of DNA sequences of the same length and base composition. Thus, the entire sequence is likely to have base order-dependent biological meaning if the information channel through which it has "flowed" through the generations has a limited carrying capacity. If carrying capacity is largely unlimited then the unlikely notion that genomes can carry many long meaningless messages for more than a few generations (see Chapter 12) remains on the table.

Imagine that carrying capacity in an information channel is so limited that only three units out of a set of nines and ones can be accommodated. So we can send a three-unit alarm signal (911), which can be interpreted as "primary" order-dependent information and the appropriate code applied. The same three units can also be interpreted as "secondary" composition-dependent information and an appropriate code applied. In this case, 911 is one of the set with one nine and two ones (911, 191, 119). Each of these would have the capacity to convey the same composition-dependent message (whatever that might be).

Thus, if appropriate codes exist, 911 can simultaneously convey *two* messages. The order-dependent code (for "primary information") allows the recipient to distinguish between a total of eight possible meanings. The composition-dependent code (for "secondary information") allows the recipient to distinguish between a total of four possible meanings. The recipient would have to know from the context whether the encoder intended the message to be read as primary or secondary information, or both. However, given that only three units can be accommodated, what if the encoder wanted to send both an alarm signal (911) and a composition-dependent signal that to encode required three ones (111)? There would then be a *conflict*. The encoder could either send 911 or 111, but not both. Furthermore, 911 would give the wrong information if decoded as secondary information. 111 would give the wrong information if decoded as primary information. The encoder would have to weigh the relative importance of the two messages and decide which to send, and in which context. This, in a nutshell, is the topic of this book.

Appendix 3

No Line?

> "The ancient wisdoms are modern nonsenses. Live in your own time, use what we know and as you grow up, perhaps the human race will finally grow up with you and put aside childish things."
>
> Salman Rushdie. *Step Across the Line* (2003) [1]

In the struggle for literary existence this book may not survive to provoke the comments of future historians. In the event that it does, it is probable that, but for this appendix, they would have wryly noted that, while at the turn of the twentieth century religious extremism was the centre of attention – the self-immolation of terrorists and the so-called "intelligent design" version of creationism – this book, like many other contemporary evolution texts, disregarded the topic. Indeed, evolutionists (or rather, evolutionists that get published) tend to draw a politically-correct line between science and religion, arguing for separate, non-overlapping, domains, or "magisteria" [2, 3]. But certain approaches used by scientists can be applied to issues deemed religious, while remaining within bounds deemed scientific.

Setting the Stage

Apart from the genetic information stored in our DNA, we also contain non-genetic "mental" information (see Chapter 1). The latter is stored in human heads in some, at the time of this writing, undefined form, and may be externalized to databases, often in digital form. Human memory being limited and humans being ephemeral, each generation depends on those proceeding to have selected and externalized some of their non-genetic information to databases, and to have preserved it there in forms that can be accessed. Databases *per se* have not, as far as we know, acquired the ability to manipulate information in the way humans do – a function to which we give the name "consciousness." Human consciousness keeps alive (i.e. holds in trust) what would otherwise be dead information. As Butler pointed out, it is like money in your pocket, dead until you decide to spend it. And human consciousness itself is kept alive by passage of DNA from generation to generation. The dependence on previous generations is of much consequence.

It is an inescapable fact that your grandparents appeared on this planet before your parents, and your parents before you. Thus, your ancestors had the opportunity, which they often availed themselves of, to "set the stage." On arriving you found that much around you, and the accompanying admonitions, made sense. They told you not to touch the candle flame. Ooops! Won't do that again! And like many of your six billion fellow planetary residents, you came to ask how we came to be here and, the not necessarily related question, how we should conduct ourselves during a stay that, with the passage of years, can appear increasingly brief.

Having been wisely guided on less substantial issues, many of the six billion are inclined not to doubt the wisdom of ancestral answers to these two substantial ones. Salman Rushdie – whose rebellion against religious authority (reminiscent of Samuel Butler's a century earlier) [4], led a head of state with parliamentary support to condemn him to death (issue a fatwa) [1] – put it this way:

> "You will be told that belief in 'your' stories [i.e. your people's stories] and adherence to the rituals of worship that have grown around them, must become a vital part of your life in this crowded world. They will be called the heart of your culture, even of your individual identity. It is possible that they may at some point come to feel inescapable, not in the way that truth is inescapable, but in the way that a jail is. They may at some point cease to feel like the texts in which human beings have tried to solve a great mystery, and feel, instead, like the pretexts for other properly anointed human beings to order you around. And its true that human history is full of the public oppression wrought by the charioteers of the gods. In the opinion of religious people, however, the private comfort that religion brings more than compensates for the evil done in its name."

William Bateson put it no less emphatically in 1889 in a letter to his future wife [5]:

> "For me and for most other people in this year of grace I believe the practice of religion to be an outward and visible sign of inward and spiritual duplicity. Of course there are a very few men who feel things heavenly as vividly as things earthly, but they are very rare. For me to be married in a church would be acting a lie and though I love the old services as I do, the old buildings, as some of the fairest things left to us in an age of pollution and

> shoddy[ness], yet my feeling is that it is ours as a trust somehow from our forefathers, in which we have no part lot, and I should feel just as false if I went to Church and took credit for sanctity as I should if passed a false cheque."

So, the stage may have been so well set that you may have been persuaded that the two questions are indeed related – namely, that the origin of planetary life came replete with instructions ("commandments") as to how that life should be lived.

Rules of Thumb

Science deals with observations such as the sun rising in the east and setting in the west, and that the surface of the earth appears flat. As has been shown in these pages, through hypothesis, experiment, and fresh observations, scientists arrive at schemes of relationships that correspond to reality (i.e. are "true") to the extent that they increase our understanding of, and hence potentially our command over, our environment. However, in addition to this general approach (i.e. *the* scientific method), there are three "rules of thumb" that scientists – indeed, thinking people in general – have found to be valuable, *albeit not infallible*, adjuncts.

As mentioned in Chapter 3, the first, attributed to William of Occam, is that, given a number of alternative hypotheses to explain a set of observations (facts, data, phenomena), one should first go by (i.e. have most faith in) the simplest consistent with the observations, and only turn to the more complex when the simplest has been tested and found wanting.

The second is a variant of the first and is sometimes attributed to Albert Einstein. It is that one should indeed prefer the simple over the complex, but not the *too* simple. For example, the hypothesis that everything is a balance between the forces of good and evil can be made to explain anything. The sun rises in the east because the forces of good are supreme, and sets in the west when the forces of evil get the upper hand. You were wide awake at the beginning of a lecture because of the forces of good. You were fast asleep at the end because the forces of evil had prevailed.

Finally, there is what can be called "the feather principle." Confronted with a number of competing hypotheses (of which one is "none of the above"), be most dubious of the hypothesis that is pressed by those who have much to gain materially should the hypothesis come to be generally accepted. In other words, suspect those who may be seeking, either consciously or unconsciously, to "feather their own nests" by advocating a particular hypothesis. As discussed in Chapter 1 and the Epilogue, scientists themselves can have vested interests in particular viewpoints that may play the feathering role to the extent that their power and authority in a hierarchical system is sustained. This may sometimes be seen in its negative form – a failure to admit the existence of a particular, rival, hypothesis. Thus, when confronted with two rival texts authored by A and B, where A, albeit negatively, cites the previous work of B, but B, through ignorance or disdain, does not cite the previous work of A, be most dubious of B.

Even if we accept the assertion that the scientific method *per se* cannot be applied to religious issues [2, 3], nevertheless we can still apply the three rules of thumb. The hypothesis of one, quasi-anthropomorphic, intelligent creator, accords well with Occam's principle. Quite simply, an entity exists that understands, and can ordain, everything. You have a problem? Be assured that things are that way because that entity decided it should be so. Next question please!

Of course, this argument contradicts the second rule of thumb. The argument is just too simple. It explains everything, yet it explains nothing. Furthermore, applying the feather principle, we can note that the material benefits accruing to its advocates have often been substantial. Indeed, for millennia thinking men and women who lack material resources have opted for the cloistered life where, in return for dispensations of what Karl Marx would have called "opium" to the populace, they have been left in peace to pursue whatever they might find agreeable.

Mendel took this path. The necessity for a governing Prince to simulate religiosity if he wished to continue to enjoy the benefits of his office was stressed by Machiavelli [6], and modern politicians disregard this at their peril [1]. However, the creationists do quite well in terms of the negative feather principle – their opponents tend not to cite their texts, whereas creationists tend not to reciprocate by omitting citation.

Probabilities

On balance, creationism fails by the rule-of-thumb principle. But what can science offer instead? Can what science offers also resist rule-of-thumb scrutiny? First we should recognize that, as noted by Thomas Huxley [7], it is difficult for a scientist to be a 100% atheist. A scientist, to remain credible as a scientist, must assign a probability to each member of a set of competing hypotheses. A hypothesis deemed to be most unlikely, such as that of the existence of a supreme creator of the universe, can be assigned a very low probability, but it cannot be excluded.

For a scientist, a declaration that one is say, a Darwinist, merely means that, on balance, one adopts (i.e. has most faith in) Darwin's view of the power of natural selection. Similarly, one may declare oneself to be an atheist, meaning that, on balance, one prefers (i.e. one believes in) this hypothesis over agnostic and religious alternatives. This means that scientists go into combat with one hand tied behind their backs when arguing the merits of a case with a non-scientist who is not so encumbered. In Rushdie's words [1]:

> "One of the beauties of learning is that it admits its provisionality, its imperfections. This scholarly scrupulousness, this willingness to admit that even the most well supported of theories is still a theory, is now being exploited by the unscrupulous. But that we do not know everything does not mean that we know nothing."

Given all this, can a scientist speak objectively about religion in general, and about creationism in particular? By now, scientists have a good track-record of providing examples where apparently objective human perceptions have been shown false. Early man probably considered the earth to be flat. Scientific observation of astronomical objects led to the realization that we – all six billion of us – are standing on the surface of an approximately spherical earth. The hypothesis was proven by circumnavigation of the planet by early explorers. So, unless there is something we have overlooked, the surface of the earth taken as a whole (i.e. disregarding small segments) is not flat. More precisely, the probability that the planet's surface is flat may be considered very low. Perhaps, likewise, scientists can show that some religious postulates may be insecure.

No Beginning?

A fundamental observation that underlies much creationist thinking is that everything around us appears to have a beginning and an end. To create, after all, means making something begin. If there is no beginning there can be no creationism. I have sketched out here and in my previous book [8] – with a degree of detail that I believe even the most recalcitrant of bickering evolutionists will be able to live with – a likely path from complex organic molecules to complex living organisms. Others have made the case for the derivation of complex organic molecules from inorganic precursor molecules. Let's examine this further.

The building up of inorganic molecules from even simpler units has been, and remains, an area of intense study based on the premise that there is some fundamental entity ("strings" in modern parlance) from which everything is constituted [9]. The important point is that there is a unified chain of relationships down from highly complex organisms to a fundamental substratum which, quite simply, *exists*.

Thus, as a point of departure, most scientists accept the concept that there is something fundamental, of a relatively low order of complexity, that exists through all time. Their concern is to find what that fundamental is. They feel no need to invoke an abstract creator who might have made that fundamental.

In short, scientists can accept something as *being*. Accordingly, there is no beginning or, necessarily, end, for that something.

So scientists can accept the concept of being. They accept the concept of existing through time. But so can creationists. In their case the entity that is constant through time is the creator who, by "intelligent design," has put it all together. Whereas the constant entity for the scientists is relatively simple, the constant entity for the creationists would appear to be, given its key attribute as creator, highly complex. The two curves (A and B) shown in Figure 19-1 indicate the roles of this entity, termed "God", first to create the fundamental units out of which everything derives, and then, *perhaps*, to prescribe (influence) the path this derivation will take. The fact that the latter curve (B) is not obligatory for the God concept is overlooked by some scientists who hold that our ability to explain the path without resorting to a supernatural power destroys the concept [10, 11].

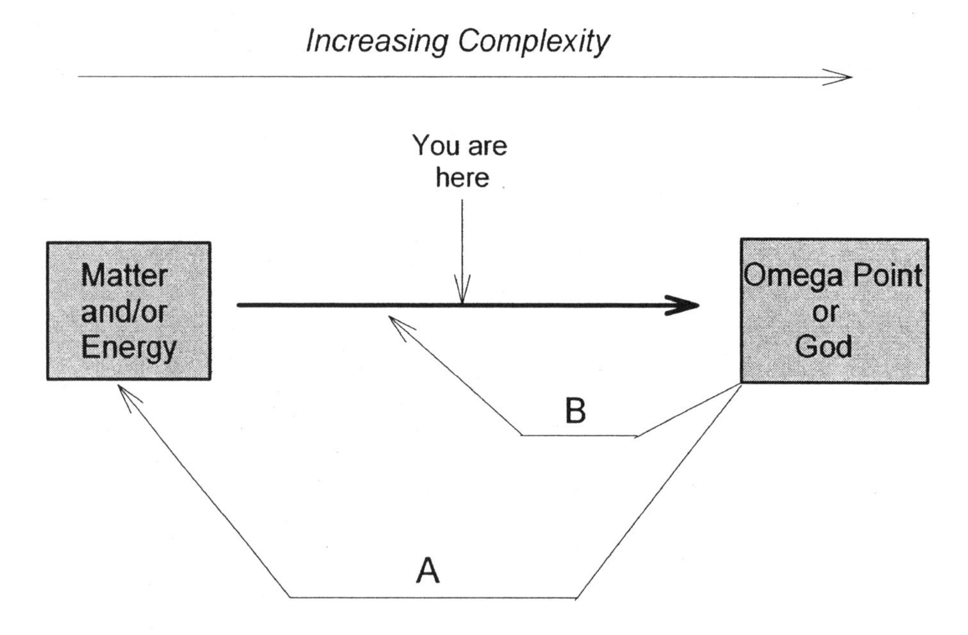

Fig. 19-1. Two extremes of the complexity scale (boxed) that exist through time. Intermediate states appear ephemeral. The curved arrows indicate the possibilities of feedbacks either at the beginning (*A*) or during (*B*) the process of increasing complexity (the thick horizontal arrow). For details see text

One scientist has embraced both simple and complex constant entities. Disregarding the writings of mathematicians and physicists on what may be the true nature of time (as, through ignorance, do I), the priest and anthropologist Pierre Teilhard de Chardin stood back from the phenomena before him, and saw just one process – matter becoming aware of itself [12]. This required time that would stretch endlessly backwards and forwards from the present, so that what we can conceive as happening, either has already happened (if it is possible to happen at all), or is part of some recursive cycle, or seems intrinsically without end (like the number of decimal places in the ratio of the circumference of a circle to its radius).

In Chardin's view the process is ongoing and tends to produce forms and relationships of increasing complexity that can be extrapolated to an ultimate "omega point" where the degree of complexity would approach a maximum, tending to be constant in time. All this can be summarized as two boxes and an arrow (Fig. 19-1). The difference between Chardin's omega point and the creationists' God, is that the latter is often held capable of influencing the path from simple to complex (curved arrows), whereas the former does not. There is the possibility, doubtless explored in the science fiction literature, of a distant omega point having arisen independently of us, which would choose somehow to influence our evolutionary path, but Chardin does not countenance this.

Can Intermediate States Exist?

Under the scheme shown in Figure 19-1, we six billion humans are located somewhere in space and time along the arrow leading from low to high complexity. The following question can now be asked: If the two extremes can, quite simply, *exist*, is it not possible that an intermediate state, such as that in which we currently find ourselves, could also, quite simply, *exist*, or, at least, have the potential to exist? To this, most of the six billion would answer with a resounding "no." Everything that they see – including themselves, their friends, and their relatives – appears ephemeral, impermanent. Everything has a beginning and an end. However, many presumptively ephemeral, impermanent, things can be rendered less ephemeral, and perhaps permanent, *if we so will it*.

For example, for millennia people have agreed to live together on the banks of the River Thames in a geographic region named London. Just as the individual molecules of water in the river are for ever changing, so are the people who make up London's population. But London remains. Those who *will* this far outnumber, and/or so far have defeated, those who might will otherwise. So London has the potential to persist through time to constitute part of that which will exist at Chardin's omega point.

Similarly, through each person – indeed, through each organism on the planet – there is a constant flux of molecules. By recycling, replacement and repair (see Chapter 2) today's new molecules substitute for yesterday's. We, as individuals – a collective of molecules – persist. And, hypothetically, there is no limit to the duration of this persistence, so that we also have the potential to constitute part of that which will exist at Chardin's omega point.

Cells taken from cats, dogs or humans thrive equally well when cultured away from the body, and can appear potentially immortal. Yet cats and dogs seldom live beyond twenty years – the time it takes humans to reach their physical prime. Our understanding of the biology of this is increasing (Chapter 14). To some scientists it does not seem unrealistic to suppose that, in centuries hence (some hyperoptimists/hyperpessimists – depending on how you view it – think sooner), humans, perhaps having opted finally to liberate themselves from religious mythology, will choose to "permanise" themselves as entities that, quite simply, *exist*.

By this is meant that biological knowledge is approaching a state such that all diseases could be either prevented or cured. Considering the process of biological aging as a disease, it also might be slowed, or even halted. This, in itself, would serve partly to dispel religious mythology, much of which seeks to ameliorate the fear of death. We can envisage a scenario such that, as perceptual dust and cobwebs clear away, and births and deaths and the associated rituals become infrequent events, people will increasingly come to recognize the existential nature of their being. There will still be churches and priests. But the latter will bring comfort through truth, rather than through mythology. They will describe religions as relics of ancient ways of thinking, celebrate the present, and help people to look afresh at the two fundamental questions. Our evolution as biological entities will slow, giving way to the evolution of the information each generation holds in trust.

Contradiction Shall Reign!

So, where will the priests ("officients") say we came from? It seems that we came from something fundamental, from which inorganic matter is composed, and which, quite simply, *exists*. How shall we conduct ourselves? Again, quite simply, in such ways as should further an agreeable existence. And we will bicker endlessly, without fear of recrimination, on all topics, including that of what constitutes an agreeable existence. In short, we will be free. In Rushdie's words [1]: "Freedom is that space in which contradiction can reign, it is a never ending debate." Or, as Samuel Butler put it a century earlier [13]: "Whenever we push truth hard she runs to earth in contradiction in terms, that is to say, in falsehood. An essential contradiction in terms meets us at the end of every enquiry."

We also will debate possible omega points towards which we might collectively strive, and some may come to agree with Thomas Huxley [14] that:

> "The purpose of our being in existence, the highest object that human beings can set before themselves, is not the pursuit of

> some chimera as the annihilation of the unknown; but it is simply the unwearied endeavour to remove its boundaries a little further from our sphere of action."

Butler of course said it more simply, and with his usual twist of humor [13].

> "If I thought by learning more and more I should ever arrive at the knowledge of absolute truth, I should leave off studying. But I believe I am pretty safe."

In other words, we will strive to further the evolution of the sort of information that will help our understanding.

How does all this accord with the three rules of thumb? Quite well with the first two. In not invoking elusive supernatural powers, one arrives at a scenario of moderate simplicity. And those who advocate this view, including those trained in science, while having much to gain intellectually from its wide acceptance, usually do not gain materially to such an extent as the ardent advocates of religious viewpoints. Thus, we should not be overly suspicious of the motives of those advocating the view (third rule of thumb).

We should neither regard such individuals as shallow, nor impute an absence of sensitivity, or an innate inability to be moved at the wonder of it all. There is no evidence that atheists are less sympathetic to the needs of others than the religiously inclined; nor are they moved less by great music, great art, the quiet splendor of a sunset, and the silver path to the moon at sea. The lives of Samuel Butler and William Bateson attest to this. The difference is that atheists consider the latter to be works of individual genius, or of natural forces, not of a God working through man or through Nature. When things go wrong they see that "the fault, dear Brutus, lies not in the stars, but in ourselves" [15]. There may be no line between us and "the stars."

Acknowledgements

I am indebted to Jim Gerlach and David Siderovski for much patient advice in the 1980s as I struggled to understand computers. My bioinformatic studies were enthusiastically assisted by a host of computer-literate students in the Department of Biochemistry at Queen's University in Canada. Special thanks to Isabelle Barrette, Sheldon Bell, Labonny Biswas, Yiu Cheung Chow, Anthony Cristillo, Kha Dang, Previn Dutt, Scott Heximer, Gregory Hill, Janet Ho, Perry Lao, Justin LeBlanc, Shang-Jung Lee, Feng-Hsu Lin, Christopher Madill, James Mortimer, Robert Rasile, Jonathan Rayment, Scott Smith, Theresa St. Denis, and Hui Yi Xue.

Insight into the differing perceptions of the "info" people and the "bio" people came in the 1990s when Janice Glasgow of the Department of Computing and Information Science at Queen's University invited me to join her in initiating a graduate course in Bioinformatics, and Paul Young of the Department of Biology invited me to contribute to his undergraduate Genomics course. His text, *Exploring Genomes*, was published in 2002. Andrew Kropinski of the Department of Microbiology and Immunology advised on bacteriophage genomes. David Murray of the Department of Psychology provided valuable biohistorical information and, among many, the works of biohistorian William Provine (Cornell University) were of great help in my getting "up to speed" in that area. Peter Sibbald advised on bits and bytes. Charlotte Forsdyke copy-edited. Polly Forsdyke advised on German usages, and translated the French and German versions of Delboeuf's 1877 paper into English. Sara Forsdyke advised on Latin and Greek usages. Ruth Forsdyke believed in me, sometimes more than I did myself, and suggested the term "reprotype." My wife Patricia was and is a constant source of inspiration and encouragement.

Particular thanks are due to Sheldon Bell, Christopher Madill and James Mortimer for writing programs that greatly facilitated our bioinformatic analyses. Michael Zuker of the Rensselaer Polytechnic Institute, kindly made available his nucleic acid folding programs. Dorothy Lang gave me information on mirror repeats prior to publication. The Genetics Computer Group (GCG) suite of programs ("the Wisconsin package") kindly acquiesced to modification by my primitive UNIX scripts.

Joseph Burns, Deborah Doherty and Marcia Kidston of Springer (Norwall) smoothed the passage from manuscript to final copy. Original data for figures were generously provided by Kenneth Wolfe and Paul Sharp (Figure 7-3), Honghui Wan and John Wootton of the National Center for Biotechnology Information (Figure 11-1), and by Daiya Takai and Peter Jones of the University of Southern California (Figure 15-2). Most other data were publicly available from sources such as GenBank and the ExInt Database (Fig. 10-2). The photograph of Erwin Chargaff was from the collection of the National Library of Medicine, Washington. The photograph of Samuel Butler is from a self-portrait held by St. John's College, Cambridge, with permission of the Master and Fellows. The photograph of Friedrich Miescher was from the University of Basel. The photograph of Gregory Bateson was from the Imogen Cunningham Trust, California. The photograph of Richard Goldschmidt is from the personal papers that Alan G. Cock (deceased 2005) entrusted to me in 2004. Figure 2-1 was kindly adapted by Richard Sinden of Texas A & M University from his book *DNA Structure and Function*. Elsevier Science gave permission to use this, and also Figure 5-5 from an article by Michael Zuker and his colleagues in *Methods in Enzymology*, and Figure 13-3 from Max Lauffer's *Entropy Driven Processes in Biology*. Figure 2-5 is adapted from an article by Austin and Marianne Hughes with permission of the authors and Nature Publishing Group.

Permissions to reproduce or adapt materials from my own works were provided by Springer (Figures 4-2, 4-3, 4-4, 7-2, 8-5, 9-10, 10-4, 10-5), Cold Spring Harbor Laboratory Press (Figure 6-2), McGill-Queen's University Press (Figures 7-4, 7-5 and 14-3), Elsevier Science (Tables, 8-1, 10-1; Figures 5-3, 5-4, 6-1, 6-7, 10-6, 10-7, 11-2, 11-4, 11-5, 11-6, 11-7, 12-1, 12-2, 12-3, 13-5), Adis International Limited (Table 4-3; Figures 9-1 to 9-8), Oxford University Press (Figure 10-3), and World Scientific Publishing Company (Figures 8-4, 13-1 and 13-2). For space reasons many important studies are not directly cited. However, the studies that are cited often contain references to those studies.

My web pages are hosted by Queen's University (for internet locations see the beginning of the reference section of this book). These display full-texts of various scientific papers from the nineteenth century onwards, and much supplementary material. Here the reader will find regular updates on new work that appeared after the book went to press. To further supplement the book, I have written short biographies of W. Bateson, E. Chargaff, J. B. S. Haldane, G. J. Mendel, H. J. Muller and G. J. Romanes, and an article on "Functional Constraint and Molecular Evolution." Nature Publishing Group placed these in both on-line and paper editions of the *Nature Encyclopedia of Life Sciences* (2002) and/or the *Nature Encyclopedia of the Human Genome* (2003). In 2005 these became the *Encyclopedia of Life Sciences*, published by John Wiley & Sons, and available on-line at http://www.els.net/.

Finally, there is a rather unusual acknowledgement of an anticipatory rather than retrospective nature. First, a quotation from *Unconscious Memory* concerning Samuel Butler's attempts to make the Victorian evolutionists recognize both the ideas of Hering and himself, and their own intellectual debts to Georges Louis Leclerc de Buffon and Charles Darwin's grandfather, Erasmus Darwin:

> "My own belief is that people paid no attention to what I said, as believing it simply incredible, and that when they come to know that it is true, they will think as I do concerning it. ... My indignation has been mainly roused ... by the wrongs ... inflicted on dead men, on whose behalf I now fight, as I trust that some one – whom I thank in anticipation – may one day fight on mine."

The reader will recognize that, a century after his death, I have here fought on Butler's behalf. I respectfully accept his thanks, while noting his caution *To Critics and Others* that the future might come to "see in me both more and less than I intended". But the battle is not yet won. So, in turn, I thank in anticipation that some one who may, one day, continue our work.

References and Notes[1]

Prologue – To Select is Not To Preserve

1. Romano T (2002) Making Medicine Scientific. John Burdon Sanderson and the Culture of Victorian Science. John Hopkins University Press, Baltimore, p 105 [Sanderson was the great uncle of the twentieth century's "JBS", John Burdon Sanderson Haldane, and may have partly inspired Lydgate in George Eliot's *Middlemarch*.]

2. Brenner S (2002) The tale of the human genome. Nature 416:793–794

3. Forsdyke DR (2001) Did Celera invent the Internet? The Lancet 357:1204

4. Wu X, Li Y, Crise B, Burgess SM (2003) Transcription start regions in the human genome are favoured targets of MLV integration. Science 300:1749–1751 [The authors note: "The two children that developed leukaemia in the MLV trials both had an integration near the oncogene *LMO2*. We observed a preference for actively transcribed genes with MLV; however, even assuming no bias for individual genes, the data are troubling. In the X-linked severe combined immune deficiency syndrome clinical trials, more than 5×10^6 cells with MLV integrations were injected into each child. Assuming that 20% of integrations are near transcription start sites, there will be … an

---

[1] Full text versions of some of these references, together with much supplementary information, may be found in my web-pages, which are likely to be updated after the publication of this book. The pages may be accessed at:

http://post.queensu.ca/~forsdyke/homepage.htm
http://crystal.biochem.queensu.ca/forsdyke/homepage.html

Early versions of the latter web-pages are archived at:

http://www.archive.org
https://qspace.library.queensu.ca/html/1974/136/homepage.htm

The date of a written work sometimes does not coincide with that of its publication. When this problem arises I give the date of the written work after the author's name, and the date of publication with the publisher's name.

average of 55 integrations into the 5' region of the *LMO2* locus per treatment."]

5. Brenner S (1991) Summary and concluding remarks. In: Osawa S, Honjo T (eds) Evolution of Life: Fossils, Molecules and Culture. Springer-Verlag, Berlin, pp 453–456
6. Little P (1995) Navigational progress. Nature 277:286–287
7. Hood L, Rowen L, Koop BF (1995) Human and mouse T-cell receptor loci. Genomics, evolution, diversity and serendipity. Annals of the New York Academy of Sciences 758: 390–412
8. Baldi P (2001) The Shattered Self: The End of Natural Evolution. MIT Press, Cambridge, MA
9. Wolfram S (2002) A New Kind of Science. Wolfram Media, Champaign, IL
10. Badash L (1972) The complacency of nineteenth century science. Isis 63:48–58
11. Butler S (1914) The Humour of Homer and Other Essays. Kennerley, New York, p 255
12. Gould SJ (2002) The Structure of Evolutionary Theory. Harvard University Press, Cambridge, MA, pp 36–37
13. Forsdyke DR, Mortimer JR (2000) Chargaff's legacy. Gene 261:127–137
14. Morris R (2001) The Evolutionists. Norton, New York
15. Forsdyke DR.(2005) Web-Pages. http://post.queenu.ca/~forsdyke/homepage.htm
16. Forsdyke DR (2001) The Origin of Species, Revisited. McGill-Queen's University Press, Montreal
17. Dunn LC (1965) A Short History of Genetics. McGraw-Hill, New York, p xxii
18. Dawkins R (1986) The Blind Watchmaker. Longman, Harlow, p ix [Dawkins' preface reads: "This book is written in the conviction that our own existence once presented the greatest of all mysteries, but that it is a mystery no longer because it is solved. Darwin and Wallace solved it, though we shall continue to add footnotes to their solution for a while yet." This appears to rest on the assumption that matter and energy – whatever they are – interchangeably exist, and there is nothing more fundamental. So it seems our task is to determine the mechanisms by which living beings arose from this raw material; see also Appendix 3]

Chapter 1 – Memory: A Phenomenon of Arrangement

1. Pope A (1711) Essay on Criticism. Macmillan, London (1896)
2. Darwin C (1859) The Origin of Species by Means of Natural Selection, or the Preservation of Favoured Races in the Struggle for Life. John Murray, London, p 490
3. Bateson W (1899) Hybridization and cross-breeding as a method of scientific investigation. In: Bateson B (ed) William Bateson, F.R.S. Naturalist. His Essays and Addresses. Cambridge University Press (1928) pp 161–171
4. Darwin C (1872) Letter to Moggridge. In: Darwin F, Seward AC (eds) More Letters of Charles Darwin. John Murray, London (1903) Vol 1, p 337
5. Gelbart WM (1998) Databases in genome research. Science 282:659–661
6. Delbrück M (1971) Aristotle-totle-totle. In: Monod J, Borek F (eds) Of Microbes and Life. Columbia University Press, New York, pp 50-55
7. Lyell C (1863) Geological Evidences on the Antiquity of Man with Remarks on Theories of the Origin of Species by Variation. Childs, Philadelphia, p 467
8. Mendel G (1865) Versuche uber Pflanzen Hybriden. Verhandlung des naturforschenden Vereines in Brunn 4:3–47
9. Nägeli C von (1884) Mechanisch-physiologische Theorie der Abstammungslehre. Munich, p 73 [Translated in Darwin F (1908) British Association for the Advancement of Science. Presidential Address.]
10. Darwin C (1868) Provisional hypothesis of pangenesis. In: The Variation of Animals and Plants under Domestication. John Murray, London, Chapter 27
11. Dawkins R (1976) The Selfish Gene. Oxford University Press, Oxford
12. Huxley TH (1869) The genealogy of animals. In: Darwiniana. Collected Essays. Macmillan, London (1893)
13. Roux W (1881) Der Kampf der Theile im Organismus. Liepzig
14. Hering E (1870) Über das Gedachtniss als eine allgemeine Function der organisirten Materie. Karl Gerold's Sohn, Vienna [On memory as a universal function of organized matter. A lecture delivered at the anniversary meeting of the Imperial Academy of Sciences at Vienna, translated in Butler S (1880) *Unconscious Memory*. David Bogue, London, pp 97–133.]
15. Butler S (1985) The Notebooks of Samuel Butler. Jones HF (ed) Hogarth Press, London, pp 58–59, 70–71
16. Bateson W (1909) Heredity and variation in modern lights. In: Bateson B (ed) William Bateson, F R S, Naturalist. His Essays and Addresses. Cambridge University Press, Cambridge (1928) pp 215–232

17. Butler S (1923) Life and Habit, 3rd edition. Jonathan Cape, New York, pp 109, 152–160, 215, 240–250 [The first edition, dated 1878, appeared in December 1877. G. H. Lewes was a Victorian philosopher/biologist and member of the Physiological Society (London), who is best known as the partner of authoress George Eliot.]

18. Butler S (1872) Erewhon or Over the Range. Penguin Books, Harmondsworth (1935), pp 197–198

19. Romanes GJ (1884) Mental Evolution in Animals, with a Posthumous Essay on Instinct by Charles Darwin. Appleton, New York, p 131

20. Butler S (1884) Selections from Previous Works, with Remarks on Mr. G. J. Romanes' "Mental Evolution in Animals," and a Psalm of Montreal. Traubner, London, pp 228–254

21. Butler S (1880) Unconscious Memory. David Bogue, London, pp. 252, 269–273 [Tekke Turcomans were members of a nomadic tribe occupying an area somewhere to the north of what was then Persia.]

22. Butler S (1920) Luck or Cunning as the Main Means of Organic Modification. 2nd Edition. Jonathon Cape, London, pp 259–260 [The first edition was in 1887.]

23. Butler S (1914) The Humour of Homer and Other Essays. Kennerley, New York, pp 209–313

24. Provine WB (1971) The Origins of Theoretical Population Genetics. University of Chicago Press, Chicago

25. Olby R (1974) The Path to the Double Helix. University of Washington Press, Seattle

26. Crookes W (1866) On the application of disinfectants in arresting the spread of the cattle plague. In: Appendix to Third Report of the Commissioners Appointed to Inquire into the Origin and Nature, etc. of the Cattle Plague. Houses of Parliament, London, pp 187–201

27. Romano TM (2002) Making Medicine Scientific. John Burdon Sanderson and the Culture of Victorian Science. John Hopkins University Press, Baltimore, pp 63

28. Harris H (1999) Birth of the Cell. Yale University Press, New Haven, CT

29. Galton F (1872) On blood-relationship. Proceedings of the Royal Society 20:394–402 [Although not named as such, there is also here a distinction between the "phenotype" corresponding to "patent" elements, and "genotype" corresponding to both "patent" and "latent" elements.]

30. Weismann A (1904) The Evolution Theory, Vol 1. Edward Arnold, London, p 411

31. Vries H de (1910) Intracellular Pangenesis. Gager CS (translater) Open Court, Chicago [Translated from Vries H de (1889) *Intracellulare Pangenesis.* Fischer, Jena.]

32. Olby R, Posner E (1967) An early reference on genetic coding. Nature 215:556

33. Romanes GJ (1893) An Examination of Weismannism. Open Court, Chicago, pp 182-183 [Romanes pointed out that Weismann's "germ-plasm" and Galton's equivalent (called "stirp" from the Latin *stirpes* = root), had in common the properties of stability and continuity. However, unlike today's DNA, "germ-plasm" as originally conceived by Weismann could only vary by the mixing with another germ-plasm through sexual reproduction. Galton held open the possibility that "stirp" might also vary by other means and, to "a very faint degree" (perhaps in deference to his cousin's theory of pangenesis"), Lamarckian principles were admitted.]

34. Bateson W (1894) Materials for the Study of Variation Treated with Especial Regard for Discontinuity in the Origin of Species. Macmillan, London, p 33

35. Bateson W (1908) The methods and scope of genetics. In: Bateson B (ed) William Bateson, F R S, Naturalist. His Essays and Addresses. Cambridge University Press, Cambridge (1928), p 317–333

36. Bateson W (1913) Problems of Genetics. Yale University Press, New Haven, p 86

37. Bateson W (1914) Presidential address to the British Association, Australia. In: Bateson B (ed) William Bateson, F R S, Naturalist. His Essays and Addresses. Edited by Bateson B, Cambridge University Press, Cambridge (1928) pp 276–316

38. Schrödinger E (1944) What is Life? The Physical Aspect of the Living Cell. Cambridge University Press, Cambridge

39. Portugal FH, Cohen JS (1977) A Century of DNA. MIT Press, Cambridge, MA

40. Loewenstein WR (1999) The Touchstone of Life. Molecular Information, Cell Communication, and the Foundations of Life. Oxford University Press, Oxford

41. Kay LE (2000) Who Wrote the Book of Life? A History of the Genetic Code. Stanford University Press, Stanford

42. Weismann A (1904) The Evolution Theory, Vol 2. Edward Arnold, London, p 63

43. Celarius (alias Butler S) (1863) Darwin among the machines. The Press, Christchurch, New Zealand [Reprinted in Butler S (1914) *The First Year in Canterbury Settlement with Other Early Essays*. Fifield, London.]

44. Kellogg VL (1907) Darwinism Today. Holt, New York, pp 274–290

45. Butler S (1924) Evolution, Old and New. 4th edition. Jonathan Cape, New York, p 35 [The original edition was in 1879.]

46. Haeckel E (1909) Charles Darwin as an Anthropologist. In: Darwin and Modern Science. Cambridge University Press, Cambridge, pp 137–151

47. Jablonski NG, Chaplin G (2000) The evolution of human skin coloration. Journal of Human Evolution 39:57–106

Chapter 2 – Chargaff's First Parity Rule

1. Darwin C (1871) The Descent of Man, and Selection in Relation to Sex. Appleton, New York, pp 57–59 [The survival of favored words in individual nervous systems was considered in R. Semon's *Mnemic Psychology* (1923; Allen and Unwin, London): "There is thus – and perhaps this is at the bottom of all Mneme, a competition between what has been and what *is* and continues; there is the victory of the present, which accepts from the past only as much as it can integrate with its substance and turn to its uses."]

2. Watson JD, Crick FHC (1953) Molecular structure of nucleic acids. A structure for deoxyribose nucleic acid. Nature 171:737–738

3. Chargaff E (1951) Structure and function of nucleic acids as cell constituents. Federation Proceedings 10:654–659

4. Wyatt GR (1952) Specificity in the composition of nucleic acids. Experimental Cell Research, Supplement 2:201–217

5. Wyatt GR, Cohen SS (1953) The bases of the nucleic acids of some bacterial and animal viruses. Biochemical Journal 55:774–782

6. Israelachvili J, Wennerstrom H (1996) Role of hydration and water structure in biological and colloidal interactions. Nature 379:219-224

7. Muller HJ (1941) Resumé and perspectives of the symposium on genes and chromosomes. Cold Spring Harbor Laboratory Symposium on Quantitative Biology 9: 290–308

8. Sinden RR (1994) DNA Structure and Function. Academic Press, San Diego

9. Watson JD, Crick FHC (1953) Genetical implications of the structure of deoxyribonucleic acid. Nature 171:964-967

10. Kornberg A (1989) For the Love of Enzymes: the Odyssey of a Biochemist. Harvard University Press, Cambridge, MA

11. Muller HJ (1936) The needs of physics in the attack on the fundamental problems of genetics. Scientific Monthly 44:210–214

12. Crick F (1970) Molecular biology in the year 2000. Nature 228:613–615

13. Gatlin LL (1972) Information Theory and Living Systems. Columbia University Press, New York

14. Burton DW, Bickham JW, Genoways HH (1989) Flow-cytometric analyses of nuclear DNA contents in four families of neotropical bats. Evolution 43:756–765

15. Hughes AL, Hughes MK (1995) Small genomes for better flyers. Nature 377:391

16. Waring MJ, Britten RJ (1966) Nucleotide sequence repetition: a rapidly reassociating fraction of mouse DNA. Science 154:791–794

Chapter 3 – Information Levels and Barriers

1. Bateson G (1967) Cybernetic explanations. American Behavioural Scientist 10:29–32

2. Romanes GJ (1891) Aristotle as a naturalist. Contemporary Review 59:275–289

3. Delbrück M (1971) Aristotle-totle-totle. In: Monod J. Borek F (eds) Of Microbes and Life. Columbia University Press, New York, pp 50–55

4. Lyell C (1863) The Geological Evidences of the Antiquity of Man with Remarks on Theories of the Origin of Species by Variation. Childs, Philadelphia, p 467

5. Shaw GB (1913) Pygmalion. In: Bernard Shaw. Complete Plays with Prefaces. Volume I. Dodd, Mead, New York (1963)

6. Galton F (1876) A theory of heredity. Journal of the Anthropological Institute 5:329–348

7. Butler S (1923) Life and Habit. 3rd Edition. Jonathan Cape, London, pp 140–147, 163–165 ["*Locus poenitentiae*" is Latin for a "place of repentance." Legally the term signifies a, perhaps brief, temporal safe-haven where there is an opportunity to change one's mind. Thus, between the last bid and the fall of the auctioneer's hammer there is a *locus poenitentiae* for a decision, which cannot be stretched out for ever.]

8. Olby R, Posner E (1967) An early reference on genetic coding. Nature 215:556

9. Jenkin F (1867) The origin of species. The North British Review 46:277–318

10. Alberts B, Bray D, Lewis J, Raff M, Roberts K, Watson JD (1994) Molecular Biology of the Cell. 3rd Edition. Garland Publishing, New York, p 340

11. Lehninger AL, Nelson DL, Cox MM (1993) Principles of Biochemistry. 2nd Edition. Worth Publishers, New York, p 789

12. Goldschmidt R (1956) Portraits from Memory. Recollections of a Zoologist. University of Washington Press, Seattle [Goldschmidt refers to Richard Semon who wrote two books on the theory of the "mneme." Building on the work of Butler, Semon proposed that memory was "the consequence of an accumulation of mnemetic engrams produced by environmental action," which can compete for neural space. Richard Dawkins was unaware of this when he suggested the word "meme," which could be "thought of as relating to memory" and is defined as "a self-replicating element of culture, passed on by imitation;" see *A Devil's Chaplain* by Dawkins (2003) Houghton Mifflin, Boston.]

13. Bateson W (1894) Materials for the Study of Variation Treated with Especial Regard to Discontinuities in the Origin of Species. Macmillan, London, pp 16, 69 [While there are many references to Galton, there is no reference to Jenkin, and Romanes is cited only with respect to his work on jelly fish.]

14. Galton F (1869) Hereditary Genius. Macmillan, London

15. Bateson G (1973) Steps to an Ecology of Mind. Paladin, St. Albans, p 127

16. Lipset D (1980) Gregory Bateson: The Legacy of a Scientist. Prentice-Hall, Englewood Cliffs, p 208 [This is also a splendid mini-biography of William Bateson.]

17. Bateson G (1963) The role of somatic change in evolution. Evolution 17:529–539

18. Wilson NG (1999) Archimedes: the palimpsest and the tradition. Byzantinische Zeitschrift 92:89–101

19. Crick F (1988) What Mad Pursuit. A Personal View of Scientific Discovery. Basic Books, New York

20. Forsdyke DR (2006) Heredity as transmission of information. (submitted for publication)

Chapter 4 – Chargaff's Second Parity Rule

1. Dickens C (1837) The Posthumous Papers of the Pickwick Club. Chapman-Hall, London

2. Moliere (1670) Le Bourgeous Gentilhomme. In: Oliver TE (ed) Ginn, New York (1914)

3. Bök C (2001) Eunoia. Coach House Books, Toronto

4. Perec G (1972) Les Revenentes. Julliard, Paris

5. Perec G (1969) La Disparition. Denoël, Paris

6. Ohno S (1991) The grammatical rule of DNA language: messages in palindromic verses. In: Osawa S, Honjo T (eds) Evolution of Life. Springer-Verlag, Berlin, pp 97–108

7. Rudner R, Karkas JD, Chargaff E (1968) Separation of *B. subtilis* DNA into complementary strands. III. Direct analysis. Proceedings of the National Academy of Sciences USA 60:921–922

8. Bell SJ, Forsdyke DR (1999) Accounting units in DNA. Journal of Theoretical Biology 197:51–61

9. Darwin C (1871) The Descent of Man and Sex in Relation to Sex. John Murray, London, p 316

10. Edwards AWF (1998) Natural selection and the sex ratio. American Naturalist 151:564–569

11. Forsdyke DR (2002) Symmetry observations in long nucleotide sequences. Bioinformatics 18:215–217

12. Prabhu VV (1993) Symmetry observations in long nucleotide sequences. Nucleic Acids Research 21:2797–2800

13. Forsdyke DR (1995) Relative roles of primary sequence and (G+C)% in determining the hierarchy of frequencies of complementary trinucleotide pairs in DNAs of different species. Journal of Molecular Evolution 41:573–581

14. Russell GJ, Subak-Sharpe JH (1977) Similarity of the general designs of protochordates and invertebrates. Nature 266:533–535

15. Bultrini E, Pizzi E, Guidice P Del, Frontali C (2003) Pentamer vocabularies characterizing introns and intron-like intergenic tracts from *Caenorhabditis elegans* and *Drosophila melanogaster*. Gene 304:183–192

16. Forsdyke DR, Bell SJ (2004) Purine-loading, stem-loops, and Chargaff's second parity rule: a discussion of the application of elementary principles to early chemical observations. Applied Bioinformatics 3:3–8

17. Baisnée P-F, Hampson S, Baldi P (2002) Why are complementary strands symmetric? Bioinformatics 18:1021–1033

18. Sueoka N (1995) Intrastrand parity rules of DNA base composition and usage biases of synonymous codons. Journal of Molecular Evolution 40:318–325

19. Nussinov R (1981) Eukaryotic dinucleotide preference rules and their implications for degenerate codon usage. Journal of Molecular Biology 149:125-131

Chapter 5 – Stems and Loops

1. Delbrück M (1949) Transactions of the Connecticut Academy of Arts and Sciences 38: 173–190

2. Salser W (1970) Discussion. Cold Spring Harbor Symposium in Quantitative Biology 35: 19

3. Ball LA (1972) Implications of secondary structure in messenger RNA. Journal of Theoretical Biology 36:313–320

4. Forsdyke DR (1998) An alternative way of thinking about stem-loops in DNA. Journal of Theoretical Biology 192:489–504

5. Seffens W, Digby D (1999) mRNAs have greater negative folding free energies than shuffled or codon choice randomized sequences. Nucleic Acids Research 27:1578–1584

6. Tinoco I, Uhlenbeck OC, Levine MD (1971) Estimating secondary structure in ribonucleic acids. Nature **230**:362–367

7. Allawi HT, SantaLucia J (1997) Thermodynamics and NMR of internal GT mismatches in DNA. Biochemistry 36:10581-10589

8. Zuker M (1990) Predicting optimal and suboptimal secondary structure for RNA. Methods in Enzymology 183:281–306

9. Bass BL (2002) RNA editing by adenosine deaminases that act on RNA. Annual Review of Biochemistry 71:817–846

Chapter 6 – Chargaff's Cluster Rule

1. Chargaff E (1963) Essays on Nucleic Acids. Elsevier, Amsterdam, p 148

2. Szybalski W, Kubinski H, Sheldrick P (1966) Pyrimidine clusters on the transcribing strands of DNA and their possible role in the initiation of RNA synthesis. Cold Spring Harbor Symposium in Quantitative Biology 31:123–127

3. Smithies O, Engels WR, Devereux JR, Slightom JL, Shen S (1981) Base substitutions, length differences and DNA strand asymmetries in the human Gλ and Aλ fetal globin gene region. Cell 26:345–353

4. Saul A, Battistutta D (1988) Codon usage in *Plasmodium falciparum*. Molecular Biochemistry and Parasitology 27:35–42

5. Bell SJ, Forsdyke DR (1999) Deviations from Chargaff's second parity rule correlate with direction of transcription. Journal of Theoretical Biology 197:63–76

6. Lao PJ, Forsdyke DR (2000) Thermophilic bacteria strictly obey Szybalski's transcription direction rule and politely purine-load RNAs with both adenine and guanine. Genome Research 10:1–20

7. Schattner P (2002) Searching for RNA genes using base-composition statistics. Nucleic Acids Research 30:2076–2082

8. Szybalski W, et al. (1969) Transcriptional controls in developing bacteriophages. Journal of Cellular Physiology 74, supplement 1:33–70

9. Frank AC, Lobry JR (1999) Asymmetric substitution patterns: a review of possible underlying mutational or selective mechanisms. Gene 238:65–77

10. Tillier ERM, Collins RA (2000) Replication orientation affects the rate and direction of bacterial gene evolution. Journal of Molecular Evolution 51:459–463

11. Brewer BJ (1988) When polymerases collide. Cell 53:679-686; French S (1992) Consequences of replication fork movement through transcription units in vivo. Science 258:1362–1365

12. Olavarrieta L, Hernández P, Krimer DB, Schvartzman JB (2002) DNA knotting caused by head-on collision of transcription and replication. Journal of Molecular Biology 322: 1–6

13. Chargaff E (1951) Structure and function of nucleic acids as cell constituents. Federation Proceedings 10:654–659

14. Elson D, Chargaff E (1955) Evidence of common regularities in the composition of pentose nucleic acids. Biochemica Biophysica Acta 17:367–376

15. Wang H-C, Hickey DA (2002) Evidence for strong selective constraints acting on the nucleotide composition of 16S ribosomal RNA genes. Nucleic Acids Research 30:2501–2507

16. Forsdyke DR, Bell SJ (2004) Purine-loading, stem-loops, and Chargaff's second parity rule: a discussion of the application of elementary principles to early chemical observations. Applied Bioinformatics 3:3–8

17. Eguchi Y, Itoh T, Tomizawa J (1991) Antisense RNA. Annual Reviews of Biochemistry 60:631–652

18. Brunel C, Marquet R, Romby P, Ehresmann C (2002) RNA loop-loop interactions as dynamic functional motifs. Biochimie 84:925–944

19. Cristillo AD, Heximer SP, Forsdyke DR (1996) A "stealth" approach to inhibition of lymphocyte activation by oligonucleotides complementary to the putative G0/G1 switch regulatory gene *G0S30/EGR1/ZFP6*. DNA and Cell Biology 15:561–570

20. Paz A, Mester D, Baca I, Nevo E, Korol A (2004) Adaptive role of increased frequency of polypurine tracts in mRNA sequences of thermophilic prokaryotes. Proceedings of the National Academy of Sciences USA 101:2951–2956

21. Spees JL, Olson SD, Whitney MJ, Prockop DJ (2006) Mitochondrial transfer between cells can rescue aerobic respiration. Proceedings of the National Academy of Sciences USA 103:1283–1288

Chapter 7 – Species Survival and Arrival

1. Bennett JH (1983) Natural Selection, Heredity and Eugenics. Including Selected Correspondence of R. A. Fisher with Leonard Darwin and Others. Clarendon Press, Oxford, p 122 [Mathematicians with an interest in biologi-

cal problems need a biologist to help them along. Fisher had Leonard Darwin. His American adversary, Sewall Wright, had Theodosius Dobzansky.]

2. Butler S (1862) Darwin and the origin of species. Reproduced from *The Press* of Christchurch. In: Streatfeild RA (ed) The First Year in Canterbury Settlement with Other Early Essays. Fifield, London, (1914) pp 149–164
3. Fisher RL (1930) The Genetical Theory of Natural Selection. Oxford University Press, Oxford
4. Darwin C (1856) Letter to J. D. Hooker. In: Darwin F (ed) Life and Letters of Charles Darwin. Volume 1. Appleton, New York (1887) p 445
5. Bateson W (1909) Heredity and variation in modern lights. In: Bateson B (ed) William Bateson FRS, Naturalist. His Essays and Addresses. Cambridge University Press, Cambridge (1928) pp 215–232
6. Bateson W (1894) Materials for the Study of Variation Treated with Especial Regard for Discontinuity in the Origin of Species. Macmillan, London, pp 85, 573
7. Bossi L, Roth JR (1980) The influence of codon context on genetic code translation. Nature 286:123–127
8. Simonson AB, Lake JA (2002) The transorientation hypothesis for codon recognition during protein synthesis. Nature 416:281–285
9. Eigen M, Schuster P (1978) The hypercycle. A principle of natural self-organization. Part C. The realistic hypercycle. Naturwissenschaften 65:341–369
10. Shepherd JCW (1981) Method to determine the reading frame of a protein from the purine/pyrimidine genome sequence and its possible evolutionary justification. Proceedings of the National Academy of Sciences USA 78:1596-1600
11. Akashi H (2001) Gene expression and molecular evolution. Current Opinion in Genetics and Development 11:660-666
12. Forsdyke DR (2002) Selective pressures that decrease synonymous mutations in *Plasmodium falciparum.* Trends in Parasitology 18:411–418
13. Forsdyke DR (2006) Positive Darwinian selection. Does the comparative method rule? (submitted for publication).
14. Fitch WM (1974) The large extent of putative secondary nucleic acid structure in random nucleotide sequences or amino acid-derived messenger-RNA. Journal of Molecular Evolution 3:279–291
15. Bernardi G, Bernardi G (1986) Compositional constraints and genome evolution. Journal of Molecular Evolution 24:1–11

16. Wolfe KH, Sharp PM (1993) Mammalian gene evolution: nucleotide sequence divergence between mouse and rat. Journal of Molecular Evolution 37:441–456

17. Novella IA, Zarate S, Metzgar D, Ebendick-Corpus, BE (2004) Positive selection of synonymous mutations in vesicular stomatitis virus. Journal of Molecular Biology 342: 1415–1421

18. Darwin C (1872) The Origin of Species by Means of Natural Selection. 6th Edition. John Murray, London, Introduction [By the time of the 6th edition, Darwin was more seriously thinking that some acquired characters might be inherited (Lamarckism). This is probably why he then questioned the sufficiency of natural selection as an explanation for evolutionary advance.]

19. Darwin C (1857) Letter to T. H. Huxley. In: Darwin F (ed) More Letters of Charles Darwin, Vol 1. Appleton, New York (1903) p 102

20. Hooker J (1860) On the origination and distribution of species. Introductory essay on the flora of Tasmania. American Journal of Science and Arts 29:1–25, 305–326

21. Mendel G (1865) Versuche uber Pflanzen Hybriden. Verhandlung des naturforschenden Vereines in Brunn 4:3–47 [The pea plant was a happy choice for Mendel. In this species the height character can be treated largely as a unigenic trait.]

22. Romanes GJ (1894) Letter to Schafer, 18th May. Wellcome Museum of the History of Medicine, London. [Romanes cited Mendel in an article on "Hybridism" in the *Encyclopaedia Britannica*, 1881.]

23. Darwin C (1866) Letter to A. R. Wallace. In: Marchant J (1916) Alfred Russel Wallace. Letters and Reminiscences. Harper, New York

24. Butler S (1923) *Life and Habit*, 3rd Edition. Jonathan Cape, London, pp. 135-160

25. Bateson W, Saunders ER (1902) Report 1. Reports to the Evolution Committee of the Royal Society. Harrison, London

26. Forsdyke DR (2001) The Origin of Species, Revisited. McGill-Queen's University Press, Montreal

27. Bateson W (1886) Letter to his sister Anna, dated 22nd November. The William Bateson Archive. University of Cambridge [His sister had sent him copies of *Nature* with Romanes' articles.]

28. Galton F (1872) On blood relationship. Proceedings of the Royal Society 20:394–402 [Bateson's concept of a "residue," may have derived from Galton's earlier postulate that "patent elements" responsible for the hereditary transmission of characters constituting "the person manifest to our senses" (i.e. phenotype), had a material base, which was separate from that of a much larger "residue" containing the "latent elements." These were respon-

sible for the hereditary transmission of ancestral characters that were not manifest in the current phenotype.]

29. Richmond ML, Dietrich MR (2002) Richard Goldschmidt and the crossing-over controversy. Genetics 161:477–482 [In his *Natural Inheritance* (1889) Galton used the metaphor of a necklace as "the main line of hereditary connection," and equated the elements responsible for personal characters (genes to the modern reader) to "pendants attached to its links."]

30. Goldschmidt R (1940) The Material Basis of Evolution. Yale University Press, New Haven, pp 205-6, 245-248 [The term "reaction system" was introduced to distinguish large genetic units between which recombination was restricted (i.e. each was an individual "reaction system"), from individual genes that exhibited standard Mendelian behaviour; Goodspeed TH, Clausen RE (1917) *American Naturalist* 51:31–46, 92–101.]

31. Avery OT, Macloed CM, McCarty M (1944) Studies on the chemical transformation of pneumococcal types. Journal of Experimental Medicine 79:137–158

Chapter 8 – Chargaff's GC Rule

1. Chargaff E (1951) Structure and function of nucleic acids as cell constituents. Federation Proceedings 10:654–659

2. Sueoka N (1961) Compositional correlations between deoxyribonucleic acid and protein. Cold Spring Harbor Symposium on Quantitative Biology 26:35–43

3. Forsdyke DR (2001) The Origin of Species, Revisited. McGill-Queen's University Press, Montreal

4. Dawkins R (1986) The Blind Watchmaker. Longman, Harlow, p 267

5. Gratia JP, Thiry M (2003) Spontaneous zygogenesis in *Escherichia coli*, a form of true sexuality in prokaryotes. Microbiology 149:2571–84

6. Bellgard M, Schibeci D, Trifonov E, Gojobori T (2001) Early detection of G + C differences in bacterial species inferred from the comparative analysis of the two completely sequenced *Helicobacter pylori* strains. Journal of Molecular Evolution 53: 465–468

7. Grantham R, Perrin P, Mouchiroud D (1986) Patterns in codon usage of different kinds of species. Oxford Surveys in Evolutionary Biology 3:48–81

8. Page AW, Orr-Weaver TL (1996) Stopping and starting the meiotic cell cycle. Current Opinion in Genetics and Development 7:23–31

9. Metz CW (1916) Chromosome studies on the Diptera. II. The paired association of chromosomes in the Diptera, and its significance. Journal of Experimental Zoology 21: 213–279

10. Muller HJ (1922) Variation due to change in the individual gene. American Naturalist 56: 32–50

11. Carlson EA (1981) Genes, Radiation and Society. The Life and Work of H. J. Muller, Cornell University Press, Ithaca, NY, p 390

12. Crick F (1971) General model for chromosomes of higher organisms. Nature 234:25–27

13. Sobell HM (1972) Molecular mechanism for genetic recombination. Proceedings of the National Academy of Sciences USA 69:2483–2487

14. Wagner RE, Radman M (1975) A mechanism for initiation of genetic recombination. Proceedings of the National Academy of Sciences USA 72:3619–3622

15. Doyle GG (1978) A general theory of chromosome pairing based on the palindromic DNA model of Sobell with modifications and amplification. Journal of Theoretical Biology 70:171–184

16. Gerton JL, Hawley RS (2005) Homologous chromosome interactions in meiosis: diversity amidst conservation. Nature Reviews Genetics 6:477–487

17. Weiner BM, Kleckner N (1994) Chromosome pairing via multiple interstitial interactions before and during meiosis in yeast. Cell 77:977–991

18. Wong BC, Chiu S-K, Chow SA (1998) The role of negative superhelicity and length of homology in the formation of paranemic joints promoted by RecA protein. Journal of Biological Chemistry 273:12120–12127

19. Forsdyke DR (1996) Different biological species "broadcast" their DNAs at different (G+C)% "wavelengths." Journal of Theoretical Biology 178:405–417

20. Dong F, Allawi HT, Anderson T, Neri BP, Lyamichev VI (2001) Secondary structure prediction and structure specific sequence analysis of single-stranded DNA. Nucleic Acids Research 29:3248–3257

21. Chen J-H, Le S-Y, Shapiro B, Currey KM, Maizel JV (1990) A computational procedure for assessing the significance of RNA secondary structure. CABIOS 6:7–18

22. Forsdyke DR (1998) An alternative way of thinking about stem-loops in DNA. A case study of the human *G0S2* gene. Journal of Theoretical Biology 192:489–504

23. Bronson EC, Anderson JN (1994) Nucleotide composition as a driving force in the evolution of retroviruses. Journal of Molecular Evolution 38:506-532

24. Levy DN, Aldrovandi GM, Kutsch O, Shaw GM (2004) Dynamics of HIV-1 recombination in its natural target cells. Proceedings of the National Academy of Sciences USA 101:4204-4209

25. Michel N, Allespach I, Venzke S, Fackler OT, Keppler OT (2005) The nef protein of human immunodeficiency virus established superinfection immunity by a dual strategy to downregulate cell-surface CCR5 and CD4. Current Biology 15:714–723 [The authors state: "Primate lentiviruses appear to have evolved time windows during which the permission or prevention of superinfection is regulated by gene expression. According to this model, after infection and prior to early HIV gene expression, superinfection can readily occur in order to permit recombination. As the most abundant early viral gene product, Nef defines the start point of a successful productive infection and functions as a master switch for the establishment of superinfection resistance by downregulating the entry receptor complex."]

26. Goldschmidt R (1940) The Material Basis of Evolution. Yale University Press, New Haven, p 220

27. Matassi G, Melis R, Macaya G, Bernardi G (1991) Compositional bimodality of the nuclear genome of tobacco. Nucleic Acids Research 19:5561-5567

28. Bernardi G (2001) Misunderstandings about isochores. Part 1. Gene 276:3–13

29. Cohen N, Dagan T, Stone L, Graur D (2005) GC composition of the human genome: in search of isochores. Molecular Biology and Evolution 22:1260–1272

30. Matsuo K, Clay O, Kunzler P, Georgiev O, Urbanek P, Schaffner W (1994) Short introns interrupting the Oct-2 POU domain may prevent recombination between POU family genes without interfering with potential POU domain 'shuffling' in evolution. Biological Chemistry Hoppe-Seyler 375:675–683

31. Newgard CB, Nakano K, Hwang PK, Fletterick RJ (1986) Sequence analysis of the cDNA encoding human liver glycogen phosphorylase reveals tissue-specific codon usage. Proceedings of the National Academy of Sciences USA 83:8132–8136

32. Moore R C, Purugganan MD (2003) The early stages of duplicate gene evolution. Proceedings of the National Academy of Sciences USA 100:15682–15687

33. Zhang Z, Kishino H (2004) Genomic background drives the divergence of duplicated *Amylase* genes at synonymous sites in *Drosophila*. Molecular Biology and Evolution 21:222–27

34. Montoya-Burgos JI, Boursot P, Galtier N (2003) Recombination explains isochores in mammalian genomes. Trends in Genetics 19:128–130

35. Skalka A, Burgi E, Hershey AD (1968) Segmental distribution of nucleotides in the DNA of bacteriophage lambda. Journal of Molecular Biology 34:1–16

36. Vizard DL, Ansevin AT (1976) High resolution thermal denaturation of DNA: thermalites of bacteriophage DNA. Biochemistry 15:741–750

37. Bibb MJ, Findlay PR, Johnson MW (1984) The relationship between base composition and codon usage in bacterial genes and its use for the simple and reliable identification of protein-coding sequences. Gene 30:157–166

38. Wada A, Suyama A (1985) Third letters in codons counterbalance the (G+C) content of their first and second letters. Federation of European Biochemical Societies Letters 188: 291–294

39. Kudla G, Helwak A, Lipinski L (2004) Gene conversion and GC-content evolution in mammalian Hsp70. Molecular Biology and Evolution 21:1438–1444

40. Williams GC (1966) Adaptation and Natural Selection. Princeton University Press, Princeton, pp 24–25

41. Dalgaard JZ, Garrett A (1993) Archaeal hyperthermophile genes. In: Kates M, Kushner DJ, Matheson AT (eds) The Biochemistry of Archaea (Archaebacteria). Elsevier, Amsterdam, pp 535–562

42. Forterre P, Elie C (1993) Chromosome structure, DNA topoisomerases, and DNA polymerases in archaebacteria (archaea). In: Kates M, Kushner DJ, Matheson AT (eds) The Biochemistry of Archaea (Archaebacteria). Elsevier, Amsterdam, pp 325–345

43. Galtier N, Lobry JR (1997) Relationships between genomic G+C content, RNA secondary structures, and optimal growth temperature in prokaryotes. Journal of Molecular Evolution 44:632–636

44. Lambros R, Mortimer JR, Forsdyke DR (2003) Optimum growth temperature and the base composition of open reading frames in prokaryotes. Extremophiles 7:443–450

45. Filipski J (1990) Evolution of DNA sequence. Contributions of mutational bias and selection to the origin of chromosomal compartments. Advances in Mutagenesis Research 2:1–54

46. Oshima T, Hamasaki N, Uzawa T, Friedman SM (1990) Biochemical functions of unusual polyamines found in the cells of extreme thermophiles. In: Goldembeg SH, Algranati ID (eds) The Biology and Chemistry of Polyamines. Oxford University Press, New York, pp 1–10

47. Friedman SM, Malik M, Drlica K (1995) DNA supercoiling in a thermotolerant mutant of *Escherichia coli*. Molecular and General Genetics 248:417–422

Chapter 9 – Conflict Resolution

1. Holliday R (1968) Genetic recombination in fungi. In: Peacock WJ, Brock RD (eds) Replication and Recombination of Genetic Material. Australian Academy of Science, Camberra, pp 157–174

2. Galton F (1876) A theory of heredity. Journal of the Anthropological Institute 5:329–348 [He did not think that the "germs" would be linearly arranged: "It is difficult to suppose the directions of the mutual influences of the germs to be limited to lines, like those that cause the blood corpuscles to become attached face to face, in long rouleaux, when coagulation begins."]

3. Schaap T (1971) Dual information in DNA and the evolution of the genetic code. Journal of Theoretical Biology 32:293–298

4. Grantham R (1972) Codon base randomness and composition drift in coliphage. Nature New Biology 237:265

5. Grantham R, Perrin P, Mouchiroud D (1986) Patterns in codon usage of different kinds of species. Oxford Surveys in Evolutionary Biology 3:48–81

6. Sharp PM, Stenico M, Peden JF, Lloyd AT (1993) Codon usage: mutational bias, translation selection, or both? Biochemical Society Transactions 21:835–841

7. Cox EC, Yanofsky C (1967) Altered base ratios in the DNA of an *Echerichia coli* mutator strain. Proceedings of the National Academy of Sciences USA 58:1895–1902

8. Muto A, Osawa S (1987) The guanine and cytosine content of genomic DNA and bacterial evolution. Proceedings of the National Academy of Sciences USA 84:166–169

9. Forsdyke DR (2004) Regions of relative GC% uniformity are recombinational isolators. Journal of Biological Systems 12:261–271

10. Lee J-C, Mortimer JR, Forsdyke DR (2004) Genomic conflict settled in favour of the species rather than of the gene at extreme GC% values. Applied Bioinformatics 3:219–228

11. Wada A, Suyama A, Hanai R (1991) Phenomenological theory of GC/AT pressure on DNA base composition. Journal of Molecular Evolution 32:374–378

12. D'Onofrio G, Bernardi G (1992) A universal compositional correlation among codon positions. Gene 110:81–88

13. Orr HA (2004) A passion for evolution. The New York Review of Books 51: no. 3, pp 27–29

14. Paz A, Kirzhner V, Nevo E, Korol A (2006) Coevolution of DNA-interacting proteins and genome "dialect." Molecular Biology and Evolution 23:56–64

15. Lao PJ, Forsdyke DR (2000) Thermophilic bacteria strictly obey Szybalski's transcription direction rule and politely purine-load RNAs with both adenine and guanine. Genome Research 10:228–236

16. Mortimer JR, Forsdyke DR (2003) Comparison of responses by bacteriophage and bacteria to pressures on the base composition of open reading frames. Applied Bioinformatics 2:47–62

17. Rayment JH, Forsdyke DR (2005) Amino acids as placeholders. Base composition pressures on protein length in malaria parasites and prokaryotes. Applied Bioinformatics 4:117–130

18. Lin F-H, Forsdyke DR (2006) Prokaryotes that grow optimally in acid have purine-poor codons in long open reading frames. (submitted for publication)

Chapter 10 – Exons and Introns

1. Hamming RW (1980) Coding and Information Theory. Prentice-Hall, Englewood Cliffs

2. Federoff NV (1979) On spacers. Cell 16:687–710

3. Scherrer K (2003) The discovery of 'giant' RNA and RNA processing. Trends in Biochemical Sciences 28:566–571

4. Gilbert W (1978) Why genes in pieces? Nature 271:501

5. Weber K, Kabsch W (1994) Intron positions in actin genes seem unrelated to the secondary structure of the protein. EMBO Journal 13:1280–1288

6. Stoltzfus A, Spencer DF, Zuker M, Logsdon JM, Doolittle WF (1994) Testing the exon theory of genes: evidence from protein structure. Science 265:202–207

7. Sakharkar M, Passetti F, Souza JE de, Long M, Souza SJ de (2002) ExInt: an Exon Intron Database. Nucleic Acids Research 30:191–194

8. Blake C (1983) Exons – present from the beginning. Nature 306:535–537

9. Naora H, Deacon NJ (1982) Relationship between the total size of exons and introns in protein-coding genes of higher eukaryotes. Proceedings of the National Academy of Sciences USA 79:6196–6200

10. Raible F, et al. (2005) Vertebrate-type intron-rich genes in the marine annelid *Platynereis dumerilii*. Science 310:1325–1326

11. Liu M, Grigoriev A (2004) Protein domains correlate strongly with exons in multiple eukaryotic genomes – evidence of exon shuffling? Trends in Genetics 20:399–403

12. Forsdyke DR (1981) Are introns in-series error-detecting codes? Journal of Theoretical Biology 93:861–866

13. Bernstein C, Bernstein H (1991) Aging, Sex and DNA Repair. Academic Press, San Diego

14. Williams GC (1966) Adaptation and Natural Selection. Princeton University Press, Princeton, pp 133–138

15. Forsdyke DR (1995). A stem-loop "kissing" model for the initiation of recombination and the origin of introns. Molecular Biology and Evolution 12:949–958

16. Forsdyke DR (1995) Conservation of stem-loop potential in introns of snake venom phospholipase A_2 genes. An application of FORS-D analysis. Molecular Biology and Evolution 12:1157–1165

17. Forsdyke DR. (1996) Stem-loop potential: a new way of evaluating positive Darwinian selection? Immunogenetics 43:182–189

18. Forsdyke DR (1995) Reciprocal relationship between stem-loop potential and substitution density in retroviral quasispecies under positive Darwinian selection. Journal of Molecular Evolution 41:1022–1037

19. Zhang C-Y, Wei J-F, He S-H (2005) The key role for local base order in the generation of multiple forms of China HIV-1 B/C intersubtype recombinants. BMC Evolutionary Biology 5:53

20. Alvarez-Valin F, Tort JF, Bernardi G (2000) Nonrandom spatial distribution of synonymous substitutions in the GP63 gene from Leishmania. Genetics 155:1683–1692

21. Bustamente CD, Townsend JP, Hartl DL (2000) Solvent accessibility and purifying selection within proteins of *Escherichia coli* and *Salmonella enterica*. Molecular Biology and Evolution 17:301–308

22. Heximer SP, Cristillo AD, Russell L, Forsdyke DR (1996) Sequence analysis and expression in cultured lymphocytes of the human *FOSB* gene (*G0S3*). DNA Cell Biology 12:1025–1038

23. Forsdyke DR (1991) Programmed activation of T-lymphocytes. A theoretical basis for short term treatment of AIDS with azidothymidine. Medical Hypothesis 34:24–27

24. Williams SA, Chen L-F, Kwon H, Fenard D, Bisgrove D, Verdin E, Greene WC (2005) Prostratin antagonizes HIV latency by activating NF-kappaB. Journal of Biological Chemistry 279:42008–42017 [HIV may have an Achilles heel, but first latent HIV must be "flushed" from the genome using "inductive therapy."]

25. Kurahashi H, Inagaki H, Yamada K, Ohye T, Taniguchi M, Emanuel BS, Toda T (2004) Cruciform DNA structure underlies the etiology for palindrome-mediated human chromosomal translocations. Journal of Biological Chemistry 279:35377–35383

26. Lang DM (2005) Imperfect DNA mirror repeats in *E. coli TnsA* and other protein-coding DNA. Biosystems 81:183–207

27. Barrette IH, McKenna S, Taylor DR, Forsdyke DR (2001) Introns resolve the conflict between base order-dependent stem-loop potential and the encoding of RNA or protein. Further evidence from overlapping genes. Gene 270:181–189

28. Grantham R (1974) Amino acid difference formula to help explain protein evolution. Science 185:862–864 [To construct a "PAM matrix", the observed frequency of interchanges between two amino acids is divided by the expected interchanges calculated by multiplying the respective frequencies of each amino acid in the data set. There being 20 amino acids, a 20 x 20 matrix is generated. Of the 400 values, 20 are on the diagonal and the remaining 380 are duplicates, so that 190 values form the final matrix. Two proteins whose amino acid differences generate a low total PAM score would be held to be closely related evolutionarily.]

Chapter 11 – Complexity

1. Bateson G (1964) The logical categories of learning and communication. In: Steps to an Ecology of Mind. Paladin, St. Albans (1973) pp 250–279

2. Sibbald PR (1989) Calculating higher order DNA sequence information measures. Journal of Theoretical Biology 136:475–483

3. Wan H, Wootton JC (2000) A global complexity measure for biological sequences. AT-rich and GC-rich genomes encode less complex proteins. Computers and Chemistry 24; 71–94

4. Cristillo AD, Mortimer JR, Barrette IH, Lillicrap TP, Forsdyke DR (2001) Double-stranded RNA as a not-self alarm signal: to evade, most viruses purine-load their RNAs, but some (HTLV-1, EBV) pyrimidine-load. Journal of Theoretical Biology 208:475–491

5. Forsdyke DR (2002) Selective pressures that decrease synonymous mutations in *Plasmodium falciparum*. Trends in Parasitology 18:411–418

6. Xue HY, Forsdyke DR (2003) Low complexity segments in *Plasmodium falciparum* proteins are primarily nucleic acid level adaptations. Molecular and Biochemical Parasitology 128:21–32

7. Pizzi E, Frontali C (2001) Low-complexity regions in *Plasmodium falciparum* proteins. Genome Research 11:218–229

8. Forsdyke DR (1996) Stem-loop potential: a new way of evaluating positive Darwinian selection? Immunogenetics 43:182–189

9. McMurray CT, Kortun LV (2003) Repair in haploid male cells occurs late in differentiation as chromatin is condensing. Chromosoma 111:505–508

10. Suhr ST, Senut M-C, Whitelegge JP, Faull KF, Cuizon DB. Gage FH. (2001) Identities of sequestered proteins in aggregates from cells with induced polyglutamine expression. The Journal of Cell Biology 153:283–294

11. Tian B, White RJ, Xia T, Welle S, Turner DH, Mathews MB, Thornton CA (2000) Expanded CUG repeat RNAs form hairpins that activate the double-stranded RNA-dependent protein kinase PKR. RNA 6:79–87

12. Peel AL, Rao RV, Cottrell BA, Hayden MR, Ellerby LM, Bredesen DE (2001) Double-stranded RNA-dependent protein kinase, PKR, binds preferentially to Huntington's disease (HD) transcripts and is activated in HD tissue. Human Molecular Genetics 10: 1531–1538

13. Ranum LPW, Day JW (2004) Pathogenic RNA repeats: an expanding role in genetic diseases. Trends in Genetics 20:506–512

14. Trifonov EN, Sussman JL (1980) The pitch of chromatin DNA is reflected in its nucleotide sequence. Proceedings of the National Academy of Sciences USA 77:3816–3820

15. Schieg P, Herzel H (2004) Periodicities of 10-11 bp as indicators of the supercoiled state of genomic DNA. Journal of Molecular Biology 343:891–901

16. Trifonov EN (1998) 3-, 10.5-, 200-, and 400-base periodicities in genome sequences. Physica A 249:511–516

17. Li W, Holste D (2004) An unusual 500,000 base long oscillation of guanine and cytosine content in human chromosome 21. Computational Biology and Chemistry 28:393–399

Chapter 12 – Self/Not-Self?

1. Shaw HW (1866) Josh Billings, His Sayings. Carleton, New York

2. Mira A (1998) Why is meiosis arrested? Journal of Theoretical Biology 194:275–287

3. Johnson J, Canning J, Kaneko T, Pru JK, Tilly JL (2004) Germ line stem cells and follicular renewal in the post-natal mammalian ovary. Nature 428:145–150

4. Granovetter M (1983) The strength of weak ties. A network theory revisited. Sociological Theory 1:201–233

5. Pancer Z, Amemiya CT, Ehrhardt GRA, Ceitlin J, Gartland GL, Cooper MD (2004) Somatic diversification of variable lymphocyte receptors in the agnathan sea lamprey. Nature 430:174–180

6. Zhang S-M, Adema CM, Kepler TB, Loker ES (2004) Diversification of Ig Superfamily Genes in an Invertebrate. Science 305:251–254

7. Brücke E (1861) Die Elementarorganismen. Sitzungsberichte der Akademie der Wissenschaften Wein, Mathematische-wissenschaftliche Classe 44:381–406

8. Forsdyke DR, Madill CA, Smith SD (2002) Immunity as a function of the unicellular state: implications of emerging genomic data. Trends in Immunology 23:575–579

9. Ohno S (1972) So much "junk" DNA in our genome. Brookhaven Symposium on Biology 23:366–370

10. Plant KE, et al. (2001) Intergenic transcription in the human ß-globin gene cluster. Molecular and Cellular Biology 21:6507–6514

11. Kapranov P, et al. (2002) Large-scale transcriptional activity in chromosomes 21 and 22. Science 296:916–919

12. Johnson JM, Edwards S, Shoemaker D, Schadt EE (2005) Dark matter in the genome: evidence of widespread transcription detected by microarray tiling experiments. Trends in Genetics 21:93–102

13. Darwin C (1871) Descent of Man, and Selection in Relation to Sex. Appleton, New York, pp 156–157

14. Cristillo AD, Mortimer JR, Barrette IH, Lillicrap TP, Forsdyke DR (2001) Double-stranded RNA as a not-self alarm signal: to evade, most viruses purine-load their RNAs, but some (HTLV-1, EBV) pyrimidine-load. Journal of Theoretical Biology 208:475–491

15. Wilkins C, Dishongh R, Moore SC, Whitt MA, Chow M, Machaca K (2005) RNA interference is an antiviral defence mechanism in *Caenorhabditis elegans*. Nature 436:1044–1047

16. Lauffer MA (1975) Entropy-driven Processes in Biology. Springer-Verlag, New York

17. Levanon EY, et al. (2004) Systemic identification of abundant A-to-I editing sites in the human transcriptome. Nature Biotechnology 22:1001–1005

18. Ota T, et al. (2004) Complete sequencing and characterization of 21243 full-length human cDNAs. Nature Genetics 36:40–45

19. Waddington CH (1952) Selection of the genetic basis for an acquired character. Nature 169:278

Chapter 13 – The Crowded Cytosol

1. Kipling R (1891) If. In: Rudyard Kipling's Verse. Inclusive Edition 1885–1918, Copp Clark, Toronto (1919) pp 645

2. Fulton AB (1982) How crowded is the cytoplasm? Cell 30:345–347

3. Wainwight M (2003) Early history of microbiology. Advances in Applied Microbiology 52:333–355

4. Forsdyke DR (1995) Entropy-driven protein self-aggregation as the basis for self/not-self discrimination in the crowded cytosol. Journal of Biological Systems 3:273–287

5. Lauffer MA (1975) Entropy-driven Processes in Biology. Springer-Verlag, New York [In addition to entropy, "volume exclusion" may play a role in some crowding phenomena; see Minton AP (2001) The influence of macromolecular crowding and macromolecular confinement on biochemical reactions in physiological media. *Journal of Biological Chemistry* 276:10577–10580.]

6. Forsdyke DR (2001) Adaptive value of polymorphism in intracellular self/not-self discrimination. Journal of Theoretical Biology 210:425–434

7. Moreau-Aubry A, Le Guiner S, Labarrière N, Gesnel M-C, Jotereau F, Breathnach R (2000) A processed pseudogene codes for a new antigen recognized by a CD8+ T cell clone on melanoma. Journal of Experimental Medicine 191:1617–1623

8. Goldschmidt R (1940) The Material Basis of Evolution. Yale University Press, New Haven, pp 266–271

9. Hickman HD, et al. (2003) Class 1 presentation of host peptides following human immunodeficiency virus infection. Journal of Immunology 171:22–26

10. Darnell RB (1996) Onconeural antigens and the paraneoplastic neurological disorders: at the intersection of cancer, immunity and the brain. Proceedings of the National Academy of Sciences USA 93:4529–4536

11. Pardoll D (2002) T cells take aim at cancer. Proceedings of the National Academy of Sciences USA 99:15840–15842

12. Lane C, Leitch J, Tan X, Hadjati J, Bramson JL, Wan Y (2004) Vaccination-induced autoimmune vitiligo is a consequence of secondary trauma to the skin. Cancer Research 64:1509–1514

13. Heaman EA (2003) St. Mary's. The History of a London Teaching Hospital. McGill-Queen's University Press, Montreal, p 322

14. Shull GH (1909) The "presence and absence" hypothesis. American Naturalist 43:410–419

15. Forsdyke DR (1994) The heat-shock response and the molecular basis of genetic dominance. Journal of Theoretical Biology 167:1–5

16. Sangster TA, Lindquist S, Queitsch C (2004) Under cover: causes, effects and implications of Hsp90-mediated genetic capacitance. BioEssays 26:348–362

17. Zinkernagel RM, Doherty PC (1974) Restriction of *in vitro* T cell-mediated cytotoxicity in lymphocytic choriomeningitis within a syngeneic or semiallogeneic system. Nature 248:701–702

18. Forsdyke DR (1975) Further implication of a theory of immunity. Journal of Theoretical Biology 52:187–198 [The "affinity/avidity" model for positive repertoire selection presented here is now generally accepted.]
19. Forsdyke DR (1991) Early evolution of MHC polymorphism. Journal of Theoretical Biology 150:451–456
20. Forsdyke DR (2005) "Altered-self" or "near-self" in the positive selection of lymphocyte repertoires. Immunology Letters 100:103–106

Chapter 14 – Rebooting the Genome

1. Naveira HF, Maside XR (1998) The genetics of hybrid male sterility in *Drosophila*. In: Howard DJ, Berlocher SH (eds) Endless Forms and Speciation. Oxford University Press, Oxford, pp 329–338
2. Delboeuf J (1877) Les mathématiques et le transformisme. Une loi mathématique applicable a la théorie du transformisme. La Revue Scientifique 29:669–679
3. Bernstein C, Bernstein H (1991) Aging, Sex and DNA Repair. Academic Press, San Diego, CA
4. Ridley M (2000) Mendel's Demon. Gene Justice and the Complexity of Life. Orion Books, London, pp 167–201
5. Medvinsky A, Smith A (2003) Fusion brings down barriers. Nature 422:823–825
6. Butler S (1914) The Humour of Homer and Other Essays. Kennerley, New York, pp 209–313
7. Noort V van, Worning P, Ussery DW, Rosche WA, Sinden RR (2003) Strand misalignments lead to quasipalindrome correction. Trends in Genetics 19:365–369
8. Lolle SJ, Victor JL, Young JM, Pruitt RE (2005) Genome-wide non-Mendelian inherence of extra-genomic information in Arabidopsis. Nature 434:505–509
9. Forsdyke DR (2001) The Origin of Species, Revisited. McGill-Queen's University Press, Montreal
10. Darwin C (1871) Descent of Man, and Selection in Relation to Sex. Appleton, New York, pp 245–311
11. Haldane JBS (1922) Sex ratio and unidirectional sterility in hybrid animals. Journal of Genetics 12:101–109
12. Coyne JA (1992) Genetics and speciation. Nature 355:511–515
13. Forsdyke DR (1995) Fine tuning of intracellular protein concentrations, a collective protein function involved in aneuploid lethality, sex-determination and speciation. Journal of Theoretical Biology 172:335–345

14. Chandley AC, Jones RC, Dott HM, Allen WR, Short RV (1974) Meiosis in interspecific equine hybrids. 1. The male mule (*Equus asinus* X *E. caballus*) and hinny (*E. caballus* X *E. asinus*). Cytogenetics and Cell Genetics 13:330–341

15. Vries H de (1889) Intracellular Pangenesis. Open Court, Chicago, (1910) pp 18–19

16. Darwin C (1851) A Monograph on the Subclass Cirripedia, vol. 1. The Ray Society, London, pp 281-293

17. Bateson W (1922) Evolutionary faith and modern doubts. Science 55:55–61

18. Goldschmidt R (1940) The Material Basis of Evolution, Yale University Press, New Haven, pp 233–236

19. Forsdyke DR (2000) Haldane's rule: hybrid sterility affects the heterogametic sex first because sexual differentiation is on the path to species differentiation. Journal of Theoretical Biology 204:443–452

20. Romanes GJ (1886) Physiological selection: an additional suggestion on the origin of species. Journal of the Linnean Society (Zoology) 19:337–411

21. Romanes GJ (1897) Darwin, and After Darwin: 3. Isolation and Physiological Selection. Longmans Green, London

22. Koller PC, Darlington CD (1934) The genetical and mechanical properties of the sex chromosomes. 1. *Rattus norvegicus*. Journal of Genetics 29:159–173

23. Montoya-Burgos JI, Boursot P, Galtier N (2003) Recombination explains isochores in mammalian genomes. Trends in Genetics 19:128–130

24. Willard HF (2003) Tales of the Y chromosome. Nature 423:810–813

25. Warburton PE, Giordano J, Cheung F, Gelfand Y, Benson G (2004) Inverted repeat structure of the human genome: the X chromosome contains a preponderance of large, highly homologous inverted repeats that contain testes genes. Genome Research 14:1861–1869

26. Ironside JE, Filatov DA (2005) Extreme population structure and high interspecific divergence of the *Silene* Y chromosome. Genetics 171:705–713

27. Bachtrog D (2003) Adaptation shapes patterns of genome evolution on sexual and asexual chromosomes in *Drosophila*. Nature Genetics 34:215–219

28. Carrel L, Cottle AA, Goglin KC, Willard HF (1999) A first-generation X-inactivation profile of the human X chromosome. Proceedings of the National Academy of Sciences USA 96:14440–14444

29. Forsdyke DR (1994) Relationship of X chromosome dosage compensation to intracellular self/not-self discrimination: a resolution of Muller's paradox? Journal of Theoretical Biology 167:7–12

Chapter 15 – The Fifth Letter

1. Bateson W (1924) Letter to G. H. Hardy. Bateson Archive, Cambridge University

2. Johnson TB, Coghill RD (1925) The discovery of 5-methyl-cytosine in tuberculinic acid, the nucleic acid of the tubercle bacillus. Journal of the American Chemical Society 47: 2838–2844

3. Bird A (2002) DNA methylation patterns and epigenetic memory. Genes and Development 16:6–21

4. Galagan JE, Selker EU (2004) RIP: the evolutionary cost of genome defence. Trends in Genetics 20:417–423

5. Takai D, Jones PA (2002) Comprehensive analysis of CpG islands in human chromosomes 21 and 22. Proceedings of the National Academy of Sciences USA 99: 3740–3745

6. Krieg AM (2002) CpG motifs in bacterial DNA and their immune effects. Annual Reviews of Immunology 20:709–760

7. Fraga MF, et al. (2005) Epigenetic differences arise during the lifetime of monozygotic twins. Proceedings of the National Academy of Sciences USA 102:10604–10609

8. Surani MA (2001) Reprogramming of genome function through epigenetic inheritance. Nature 414:122–127

9. Holmgren C, Kanduri C, Dell G, Ward A, Mukhopadhya R, Kanduri M, Lobanenkov V, Ohlsson R (2001) CpG methylation regulates the *Igf2/H19* insulator. Current Biology 11: 1128–1130

10. Bateson W, Pellew C (1915) On the genetics of "rogues" among culinary peas (*Pisum sativum*). Journal of Genetics 5:15–36

11. Stam M, Mittelsten Scheid O (2005) Paramutation: an encounter leaving a lasting impression. Trends in Plant Science 10:283–290

12. Pembrey ME, Bygren LO, Kaati, G, Edvinsson S, Northstone K, Sjostrom M, Golding J (2006) Sex-specific, male-line transgenerational responses in humans. European Journal of Human Genetics 14:159-166

Epilogue – To Perceive in Not To Select

1. Eliot G (1876) Daniel Deronda. William Blackman, London; Eliot G (1879) Impressions of Theophrastus Such. Blackwood, Edinburgh [These were Eliot's last major works. Eliot's partner was George Lewes, who with Ro-

manes and Michael Foster, played a major role in establishing the Physiological Society. Romanes, who attacked Butler venomously, was one of those privileged to attend Eliot's "court" at The Priory on Sunday afternoons. Butler probably served as a model for one of the friends of Theophrastus who, like Butler, antagonized Grampus (Darwin), and spoke of humans being superseded by machines (see anonymous review of Romanes' *Mental Evolution in Animals*, in *The Atheneum* (March 1st, 1884, pp 282–283). In an article in *Nineteenth Century* (1890) Romanes pointed to Theophrastus as "the earliest botanist whose writings have been preserved."]

2. Chargaff E (1978) Heraclitean Fire. Sketches from a Life before Nature. Warner Books, New York [After being prematurely "retired," Chargaff obtained some laboratory space at the Roosevelt Hospital, New York, until 1992.]
3. Forsdyke DR (2000) Tomorrow's Cures Today? Harwood Academic, Amsterdam
4. Wilszek F, Devine B (1987) Longing for the Harmonies. Norton, New York, pp 111, 209
5. Kant I (1781) The Critique of Pure Reason. Guyer P, Wood AW (eds) Cambridge University Press, Cambridge (1998)
6. Haldane JS (1891) Letter to Louisa Trotter. 3 December. In: Romano T (2002) Making Medicine Scientific. John Burdon Sanderson and the Culture of Victorian Science. John Hopkins University Press, Baltimore (2002), pp 128
7. Darwin C (1868) Letter to Alfred Wallace. 27 February. In: Darwin F, Seward AC (eds) More Letters of Charles Darwin. John Murray, London (1903), pp 301
8. Olby RC (1966) Origins of Mendelism. Schocken Books, New York [They may have been unaware of Mendel, but in 1876 in an article in *Nature* E. R. Lankester drew the Victorians' attention to the work of Ewald Hering.]
9. Dawkins R (1983) Universal Darwinism. In: Bendall DS (ed) Evolution from Molecules to Man. Cambridge University Press, Cambridge, pp 403–425
10. Forsdyke DR (2001) The Origin of Species, Revisited. McGill-Queen's University Press, Montreal
11. Gould SJ (1982) The uses of heresy. Forward to reprint of: Goldschmidt R (1940) The Material Basis of Evolution. Yale University Press, New Haven, pp xiii–xlii
12. Gould SJ (1980) Is a new and general theory of evolution emerging? Paleobiology 6: 119–130

13. Gould SJ (2002) The Structure of Evolutionary Thought. Harvard University Press, Cambridge, MA, pp 1002–1003

14. Forsdyke DR (2004) Grant Allen, George Romanes, Stephen Jay Gould and the evolution establishments of their times. Historic Kingston 52:94–98

15. Smith, JM (1995) Genes, memes and minds. The New York Review of Books 42: no. 19, pp 17–19 [The terms "ultra-Darwinian" and "neo-Darwinian" were used by Romanes to disparage Wallace's and Weismann's inflexible advocacy of the power of natural selection.]

16. Tooby J, Cosmides L (1997) [A letter to the editor of *The New York Review of Books* that was not accepted for publication: see http://cogweb.ucla.edu/Debate/]

17. Adams MB (1990) La génétique des populations était-elle une génétique évolutive? In: Fischer J-L, Schneider WH (eds) Histoire de la Génétique, pp 153-171. ARPEM, Paris [See also: Adams MB. Little evolution, big evolution. Rethinking the history of population genetics. (Personal communication, 2003)]

18. Provine WB (1992) Progress in evolution and the meaning of life. In: Waters CK, Helden A van (eds) Julian Huxley, Biologist and Statesman of Science. Rice University Press, Houston, pp 165–180

19. Bateson P (2002) William Bateson: a biologist ahead of his time. Journal of Genetics 81: 49–58

20. Butler S (1914) The Humour of Homer and Other Essays. Kennerley, New York, pp 245–313

21. Galison P (2003) Einstein's Clocks, Poincaré's Maps: Empires of Time. Norton, New York

22. Butler S (1985) The Notebooks of Samuel Butler. Jones HF (ed) Hogarth Press, London, pp 360–378

23. Barber B (1961) Resistance by scientists to scientific discovery. Science 134:596–602

24. Sacks O (2001) Uncle Tungsten. Knopf, New York, pp 104–105

25. Hook EB (2002) Prematurity in Scientific Discovery. On Resistance and Neglect. University of California Press, Berkeley

26. Dawkins R (2003) A Devil's Chaplain: Reflections on Hope, Lies, Science and Love. Houghton Mifflin, Boston, p 48

27. Fisher RA (1932) Letter to T. H. Morgan. In: Bennett JH (ed) Natural Selection, Heredity and Eugenics. Including Selected Correspondence of R. A. Fisher with Leonard Darwin and Others. Clarendon Press, Oxford (1983), p 239

28. Somerville MA (2002) A postmodern moral tale: the ethics of research relationships. Nature Reviews Drug Discovery 1:316–320 [Somerville's point was also independently made in the *Globe and Mail* of Toronto by myself (May 5, 2001) and John Polanyi (July 7, 2005), and in a letter to *Science* (2005) by forty Canadian scientists. Commenting on the latter Polanyi remarked: "What is excellent ... is a revelation. It is precisely because it surprises us that it is resistant to being planned. To find 40 scientists willing to challenge authority is also a surprise. Canadian science is coming of age."]

29. Huxley AL (1931) Letter to R. A. Fisher. In: Bennett JH (ed) Natural Selection, Heredity and Eugenics. Including Selected Correspondence of R. A. Fisher with Leonard Darwin and Others. Clarendon Press, Oxford (1983), p 220

30. Forsdyke DR (1966) Letter to Editor. Survival 8:36

31. Forsdyke DR (1969) Book review. Survival 11: 69–70

32. Roll-Hansen N (2005) The Lysenko effect: undermining the autonomy of science. Endeavour 29:143–147

33. Nirenberg M (2004) Historical review: deciphering the genetic code – a personal account. Trends in Biochemical Sciences 29:46–54

34. Meadows AJ (1972) Science and Controversy. A Biography of Sir Norman Lockyer. MIT Press, Cambridge, p 209–237

35. Punnett RC (1950) Early days of genetics. Heredity 4:1–10

36. Bennett JH (1983) Notes. In: Natural Selection, Heredity and Eugenics. Including Selected Correspondence of R. A. Fisher with Leonard Darwin and Others. Clarendon Press, Oxford, p 118

37. Forsdyke DR (2003) William Bateson, Richard Goldschmidt, and non-genic modes of speciation. Journal of Biological Systems 11:341–350

38. Forsdyke DR (2004) Chromosomal speciation: a reply. Journal of Theoretical Biology 230:189–196

39. Orr HA (2000) In: Crow JF, Dove WF (eds) Perspectives in Genetics. Anecdotal, Historical and Critical Commentaries 1987-1998. University of Wisconsin Press, Wisconsin, pp 555–559

40. Cove D (2002) Book review. Genetics Research 79:265

41. Voltaire (1770) Letter to F. L. H. Leriche. 6th February. In: The Complete Works of Voltaire, Vol 120. The Voltaire Foundation, Banbury, 1975, p 18 [The full translation reads: "The number of wise men will always be small. It is true that it is increasing, but it is nothing compared with the number of fools and, although they say it is regrettable, God is always for the big battalions. It is necessary that honest people quietly stick together. There is no way their little force can attack the host of the closed-minded who occupy the high ground."]

42. Eliot G (1874) Middlemarch. A Study of Provincial Life. Haight GS (ed). Houghton Mifflin, Boston (1956) p 613

Appendix 3 – No Line?

1. Rushdie S (2003) Step Across the Line. Random House, Toronto
2. Gould SJ (1999) Rocks of Ages. Science and Religion in the Fullness of Life. Ballantine, New York
3. Ruse M (2001) Can a Darwinian be a Christian? The Relationship between Science and Religion. Cambridge University Press, Cambridge
4. Jones HF (1919) Samuel Butler. A Memoir. Macmillan, London
5. Bateson W (1889) Letter to his future wife. 13th January. The William Bateson Archive, University of Cambridge
6. Machiavelli N (1950) The Prince and the Discourses. Random House, New York
7. Huxley TH (1948) Selections from the Essays. Huxley. Castell A (ed) AHM Publishing, Northbrook
8. Forsdyke DR (2001) The Origin of Species, Revisited. McGill-Queen's University Press, Montreal
9. Wilczek F, Devine B (1987) Longing for the Harmonies. Norton, New York
10. Crick F (1988) What Mad Pursuit. Basic Books, New York
11. Dawkins R (1986) The Blind Watchmaker. Longman, Harlow, p ix
12. Chardin PT de (1959) The Phenomenon of Man. Collins, London
13. Butler S (1985) The Notebooks of Samuel Butler. Jones HF (ed) Hogarth Press, London, pp 59, 299
14. Huxley TH (1896) Darwiniana Essays. Macmillan, London, pp 447–475
15. Shakespeare W (1599) Julius Caesar. In: Rowse AL (ed) The Annotated Shakespeare. Orbis, New York (1988)

Index